Microcontroller and Smart Home Networks

RIVER PUBLISHERS SERIES IN COMMUNICATIONS

Series Editors:

ABBAS JAMALIPOUR
The University of Sydney
Australia

MARINA RUGGIERI
University of Rome Tor Vergata
Italy

JUNSHAN ZHANG
Arizona State University
USA

Indexing: All books published in this series are submitted to the Web of Science Book Citation Index (BkCI), to SCOPUS, to CrossRef and to Google Scholar for evaluation and indexing.

The "River Publishers Series in Communications" is a series of comprehensive academic and professional books which focus on communication and network systems. Topics range from the theory and use of systems involving all terminals, computers, and information processors to wired and wireless networks and network layouts, protocols, architectures, and implementations. Also covered are developments stemming from new market demands in systems, products, and technologies such as personal communications services, multimedia systems, enterprise networks, and optical communications.

The series includes research monographs, edited volumes, handbooks and textbooks, providing professionals, researchers, educators, and advanced students in the field with an invaluable insight into the latest research and developments.

For a list of other books in this series, visit www.riverpublishers.com

Microcontroller and Smart Home Networks

Dawoud Shenouda Dawoud

International University of East Africa
Uganda

Peter Dawoud

Microsoft
USA

River Publishers

Routledge
Taylor & Francis Group

LONDON AND NEW YORK

Published 2020 by River Publishers
River Publishers
Alsbjergvej 10, 9260 Gistrup, Denmark
www.riverpublishers.com

Distributed exclusively by Routledge
4 Park Square, Milton Park, Abingdon, Oxon OX14 4RN
605 Third Avenue, New York, NY 10158

First published in paperback 2024

Microcontroller and Smart: Home Networks / by Dawoud Shenouda Dawoud, Peter Dawoud.

Routledge is an imprint of the Taylor & Francis Group, an informa business

Publisher's Note
The publisher has gone to great lengths to ensure the quality of this reprint but points out that some imperfections in the original copies may be apparent.

While every effort is made to provide dependable information, the publisher, authors, and editors cannot be held responsible for any errors or omissions.

ISBN: 978-87-7022-156-6 (hbk)
ISBN: 978-87-7004-333-5 (pbk)
ISBN: 978-1-003-33883-3 (ebk)

DOI: 10.1201/9781003338833

To Nadia, Dalia, Dina, Bahya, and Adam

D.S.D.

To Dajana

P.D

Contents

IV Bluetooth   413

9 Bluetooth   415

Preface

During the last few decades, industries faced the problem of how to connect devices to speak together with minimum wiring. Philips Semiconductors (now NXP Semiconductors) faced this problem when needing to connect many integrated circuits (ICs) together, and Bosch, in the automotive industry, faced similar problems when connecting tens of microcontrollers in each car. Recently, with smart homes, the problem started to be part of every home: you may want to build your smart home with accessories from different manufacturers and you want these devices to speak together, that is, to communicate with each other. This problem is compounded when you consider central control through an app or a voice assistant.

Philips found the solution in late 1970s and introduced the Inter-integrated Circuits (I2C) network protocol. I2C was designed as a method to provide an interface between microprocessors and peripheral devices (like keyboard, mouse, memory, and IO/serial/parallel ports) without wiring full address, data, and control busses between devices.

Bosch solved the problem in 1986 by introducing Controller Area Network (CAN) protocol as a multi-master, message broadcast robust vehicle bus standard that allows microcontrollers and devices to communicate with each other in applications without a host computer. I2C and CAN are wired communication standards.

Concerning home automation, a number of common home automation protocols that allow different devices to speak and communicate together appeared during the last few decades. This is started by "Universal Powerline Bus (UPB)" and reached to Z-Wave protocol and now Thread. Some of the most popular protocols in use today include the following:

- Ethernet
- Wi-Fi
- WiMax
- Z-Wave
- ZigBee
- Bluetooth

  – Thread
  – Insteon
  – X10
  – Universal Powerline Bus (UPB).

Each technology has its own benefits and limitations.

Wi-Fi, WiMax and Insteon were part of our book "Serial Communication Protocols and Standards". This book introduces to the reader comprehensive knowledge on the following protocols: I2C (IIC), I3C, CAN, ZigBee, Bluetooth, and Z-Wave. To complete the knowledge of the reader, we give a short summary for the remaining protocols: Ethernet, Thread, Insteon, X10, and UPB.

The book is divided into sections; each consists of a number of chapters. Each section covers one technology.

Section I: Inter-integrated Circuits (IIC)

This section covers IIC technology in three chapters:

  Chapter 1: Inter-integrated Circuits (IIC)
  Chapter 2: I2C Design and Operation
  Chapter 3: Improved Inter-integrated Circuits (I3C)

Section 2: Controller Area Network (CAN)

This section covers Controller Area Network protocol in two chapters:

  Chapter 4: Controller Area Network (CAN)
  Chapter 5: CAN Data Link Layer

Section 3: ZigBee

This section covers ZigBee protocol in three chapters:

  Chapter 6: IEEE 802.15.4 Standard and ZigBee
  Chapter 7: ZigBee Stack Layers
  Chapter 8: ZigBee Pro and ZigBee Security

Section 4: Bluetooth

This section covers Bluetooth technology in one chapter:

Chapter 9: Bluetooth

Section 5: Z-Wave

This section covers Z-Wave technology in one chapter:

Chapter 10: Z-Wave Technology

Section 6: Comparisons

The last chapter of the book gives a comprehensive comparison between the different smart home protocols:

Chapter 11: Smart Home Protocols: Comparison

Acknowledgment

The authors of this book would like to express their special thanks of gratitude to Microchip for giving them the permission and the opportunity to use materials that came from Microchip scientific notes.

List of Figures

List of Tables

List of Abbreviations

ACK	Acknowledge
AES	Advanced Encryption Standard
AODV	On-demand Distance Vector
AODVjr	On-demand Distance Vector Junior
APL	Application Layer
APS	Application Support Sublayer
ARP	Address Resolution Protocol
BOD	Brown-Out Detection
CAN	Controller Area Network
CCC	Common Command Code
CD	Collision Detection
CD	Coordinator Device
CoAP	Constrained Application Protocol
CR	Collision Resolution
CRC	Cyclic Redundancy Check
CSMA/CD	Carrier Sense Multiple Access with Collision Detection
CSMA/CD +AMP	CSMA/CD and Arbitration on Message Priority
CTS	Conformance Test Suite
DDA	Dynamic Address Assignment
DDC	Display Data Channel
DDoS	Distributed Denial-of-Service (attack)
DLC	Data Length Code
DoS	Denial–of-Service
DSR	Dynamic Source Routing
DTLS	Datagram Transport Layer Security
ED	End Device
EMI	Electromagnetic Interface
EOF	End of Frame
ESD	Electrostatic Discharge
ESI	Electrostatic Interface

FFD	Fully Functional Device
Fm	Fast Mode
Fm+	Fast-mode Plus
GAP	Generic Access Profile
GFSK	Gaussian Frequency Shift Keying
HAN	Home-Area Networks
HCI	Host Controller Interface
HDR	High Data Rate (a set of I3C bus modes)
HDR-DDR	HDR Double Data Rate (an I3C bus mode)
HDR-TSP	HDR Ternary Symbol for Pure Bus (an I3C bus mode)
HLP	Higher Layer Protocol
Hs	High-speed Mode
HSP	Headset Profile
IBHR	In-band Hardware Reset (an expected I3C bus feature)
IBI	In-band Interrupt (an I3C bus feature)
IIC	Inter-integrated Circuit
IID	Interface Identifier
IIIC	Improved Inter-integrated Circuit
IFS	Inter Frame Space
IPMI	Intelligent Platform Management Interface
IPT	Information Processing Time
ISM	Industrial, Scientific, and Medical (Radio frequency)
KEK	Key Encryption Key
L2CAP	Logical Link Control and Adaptation Protocol
LLL	Logical Link Control
LMP	Link Manager Protocol
LR-WPAN	Low Rate Wireless Personal Area Network
LQI	Link Quality Indication
MAC	Medium Access Control (Media access control)
MAC	Message Authentication Code
MANET	Mobile Ad Hoc Network
MIC	Message Integrity Code
MQTT	Message Queuing Telemetry Transport
NIB	Network Layer Information Base
NLME	Network Layer Management Entity
NRZ	Non-return to Zero
NWK	Network Layer
OBEX	Object Exchange Protocol
ODR	Output Data Rate

PAN	Personal Area Network
PHY	Physical Layer
PLC	Power-Line Communications
PMBus	Power Management Bus
POR	Power-on Reset
PPDU	PHY Protocol Data Unit
PPP	Point-to-point Protocol
PSDU	PHY Service Data Unit
RFCOMM	Radio Frequency Communication
RFD	Reduced Functional Device
RTR	Remote Transmission Request
SCL	Serial Clock Line
SDA	Serial Data (an I3C bus line)
SDDB	Service Discovery Database
SDR	Single Data Rate
SDL	Serial Data Line
SDAP	Service Discovery Application Protocol
SDP	Service Discovery Protocol
SDR	Single Data Rate (an I3C bus mode)
SFD	Start-of-frame Delimiter
SJW	Synchronous Jump Width
SKKE	Symmetric-Key Key Establishment protocol
SMBus	System Management Bus
SOF	Start of Frame
SPI	Serial Peripheral Interface (an interface specification)
SPP	Serial Port Profile
SRR	Substitute Remote Request
STP	Shielded Twisted Pair
SUC	Static Update Controller
TDC	Transmitter Delay Compensation
TTC	Time-triggered Communication
TTCAN	Time-triggered CAN
TWI	Two-wire Interface
UFm	Ultra-fast Mode
UPB	Universal Powerline Bus
UTP	Un-shielded Twisted Pair
WPAN	Wireless Personal Area Network
ZDO	ZigBee Device Objects
ZiCL	ZigBee Cluster Label
ZTC	ZigBee Trust Center

Section I

Inter-integrated Circuits (IIC)

This section deals with IIC. It consists of three chapters:

Chapter 1: Inter-integrated Circuits (IIC)
Chapter 2: IIC Design and Operation
Chapter 3: Improved Inter-integrated Circuits (I3C)

Some of the topics covered in this section are:

- I2C protocol
- The I2C bus hardware structure
- Electrical consideration
- The physical bus–I2C bus interface
- Masters and slaves
- I2C data validity
- Voltage levels and resistor values
- Addressing–addressing standards
- I2C bus transaction
- Clock stretching
- Applicability of I2C bus features
- I2C modes: bus speeds
- I2C as a multi-master bus: bus arbitration
- Clock synchronization and handshaking
- I2C interface: connecting I2C bus to a PC
- Design of I2C bus
- Calculation of the pull-up resistor
- Maximum clock frequency of I2C bus
- Use of repeaters

- Series protection resistors
- Managing the delays in large systems
- Use of multiplexor/switches to solve address conflict
- Common problems in I2C bus systems
- Obscure problems in systems
- Typical problems on the I2C bus
- Derivative technologies
- Improved inter-integrated circuits (I3C)
- I3C bus configuration and device roles
- I3C features
- Dynamic addressing
- I3C bus activity states
- Error detection and recovery methods in I3C
- Resolving communication conflicts on the I3C bus
- MIPI I3C specifications and software development

1

Inter-integrated Circuits (IIC/I2C)

1.1 Introduction

The most popular form of serial bus protocol is the "Inter-integrated Circuit" (IIC). Sometimes, the bus is called I2C bus (pronounced eye-to-see) or I^2C (pronounced eye-squared see). I2C was developed as a communication protocol to interact between different ICs on a motherboard, a simple internal bus system. This standard was originally developed by Philips Semiconductors (now NXP Semiconductors) in the late 1970s as a method to provide an interface between microprocessors and peripheral devices (like keyboard, mouse, memory, and IO/serial/parallel ports) without wiring full address, data, and control busses between devices; more specifically, it was designed to provide an easy way to connect a CPU to peripheral chips in a TV set. The problem faced by the designers at that time until now is that the peripheral devices in embedded systems are often connected to the microcomputer unit (normally a microcontroller) as memory mapped I/O devices, using the microcontroller's parallel address and data bus. This results in lots of wiring on the printed circuit boards (PCBs) to route the address and data lines. The number of wires increases if we consider the need to connect also a number of address decoders and glue logic to connect everything. In mass production items such as TV sets, videocassette recorders (VCRs), and audio equipment, this is not acceptable. In such applications, every component that can be saved means increased profitability for the manufacturer and more affordable products for the end customer. Another problem directly related to the use of a lot of wires is the possible interference: lots of control lines imply that the system is more susceptible to disturbances by electromagnetic interference (EMI) and electrostatic discharge (ESD). The need for a more effective and economical way of connecting different devices (normally ICs) resulted in introducing the 2-wire communication bus that we call it now as

I2C bus protocol. This protocol enables peripheral ICs in electronic systems to communicate with each other using simple communication hardware. The ICs can be on the same board or linked via cables. The length of the cable is limited by the total bus capacitance and the noise generated on the bus.

Today, I2C has become a *de facto* world standard that is now implemented in over 1000 different ICs and is licensed to more than 50 companies including some of the leading chip manufacturers like Xicor, ST Microelectronics, Infineon Technologies, Intel, Texas Instruments, Maxim, Atmel, Analog Devices, and others.

Additionally, the versatile I2C bus is used in various control architectures such as System Management Bus (SMBus), Power Management Bus (PMBus), Intelligent Platform Management Interface (IPMI), Display Data Channel (DDC), and Advanced Telecom Computing Architecture (ATCA). Figure 1.1 shows an example of I2C bus applications.

Concerning SMBus, it is defined by Intel in 1995 and is a subset of I2C, defining a stricter usage. One purpose of SMBus is to promote robustness and interoperability. Accordingly, modern I2C systems incorporate some policies and rules from SMBus, sometimes supporting both I2C and SMBus, requiring only minimal reconfiguration either by commanding or by output pin use.

The **Power Management Bus** (PMBus) is a variant of the SMBus which is targeted at digital management of power supplies. Like SMBus, it is a relatively slow-speed two-wire communication protocol based on I2C. Unlike either of those standards, it defines a substantial number of domain-specific commands rather than just saying how to communicate using commands defined by the reader.

Since October 10, 2006, no licensing fees are required to implement the I2C protocol. However, fees are required to obtain IšC slave addresses allocated by NXP.

1.1.1 I2C Revisions

The history of I2C specification releases:

- In 1982, the original 100 kHz I2C system was created as a simple internal bus system for building control electronics with various Philips chips.
- In 1992, Version 1 added 400 kHz *Fast mode (Fm)* and a 10-bit addressing mode to increase capacity to 1008 nodes. This was the first standardized version.

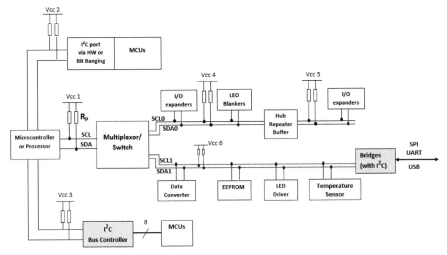

Figure 1.1 Example of I2C bus applications.

- In 1998, Version 2 added 3.4 MHz *High-speed mode (Hs)* with power-saving requirements for electric voltage and current.
- In 2000, Version 2.1 clarified version 2, without significant functional changes.
- In 2007, Version 3 added 1 MHz *Fast mode plus (Fm+)* (using 20 mA drivers), and a device ID mechanism.
- In 2012, Version 4 added 5 MHz *Ultra-fast mode (UFm)* for new USDA (data) and USCL (clock) lines using push–pull logic without pull-up resistors, and added an assigned manufacturer ID table. It is only a unidirectional bus.
- In 2012, Version 5 corrected mistakes.
- In 2014, Version 6 corrected two graphs. This is the most recent standard.

This chapter helps the reader to understand this very popular protocol. The knowledge given in the chapter will help device and system designers to understand how the I2C bus works and to help them to implement a working application. Various operating modes are described. It contains a comprehensive introduction to the I2C bus data transfer, handshaking, and bus arbitration schemes. Detailed sections cover the timing and electrical specifications for the I2C bus in each of its operating modes.

The main reference of this chapter is the NXP Semiconductor (former Philips Semiconductors) note UM10204 I2C bus specification and user manual. Other references are used and given in the "Reference" section of the chapter.

1.2 I2C Bus Terminology

Transmitter: The device which sends the data or message to the receiver.

Receiver: The device which receives the data or message from the bus.

Master: The device which generates clock signals, initiates a transfer, that is, start condition, and terminates a transfer, that is, stop condition. It may also act as a transmitter/receiver.

Slave: The device addressed by a master. Also it may act as a transmitter/receiver.

Multi-master: More than one master can attempt to control the bus at the same time without demeaning the data/message.

Arbitration: It is a procedure to ensure that if more than one master simultaneously try to control the bus, only one is allowed to do so and the data are not corrupted.

Synchronization: It is a procedure to synchronize the clock signals of two or more devices.

Serial data line (SDA): The **serial data line (SDA)** is the line on which all the data transfer among the devices takes place.

Serial clock line (SCL): The **serial clock line (SCL)** is the serial clock. I2C is a synchronous protocol, and hence, SCL is used to synchronize all the devices and the data transfer together. We'll learn how it works a little later in this post.

1.3 I2C Protocol

The I2C protocol is a very simple protocol. The protocol describes the format of a message, while communication between ICs connected to the bus. Also it gives the timing diagram of an entire transaction–from start to stop.

1.3.1 Transaction Format–Message Format

An I2C transaction consists of one or more messages. Each message (Figure 1.2) begins with a start symbol, and the transaction ends with a stop

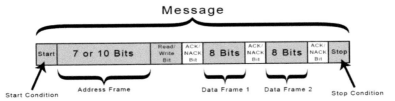

Figure 1.2 The I2C protocol format.

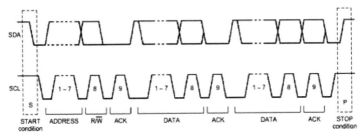

Figure 1.3 Data transfer timing diagram.

symbol. Start symbols after the first, which begin a message but not a transaction, are referred to as repeated start symbols.

Each message is a read or a write. A transaction consisting of a single message is called either a read or a write transaction. A transaction consisting of multiple messages is called a combined transaction. The most common form of the latter is a write message providing intra-device address information, followed by a read message.

Many I2C devices do not distinguish between a combined transaction and the same messages sent as separate transactions, but not all. The device ID protocol requires a single transaction; slaves are forbidden from responding if they observe a stop symbol. Configuration, calibration, or self-test modes which cause the slave to respond unusually are also often automatically terminated at the end of a transaction.

1.3.2 Timing Diagram

Figure 1.3 shows the real shape of the signals while changing state from high to low and from low to high. Ideally, whenever a signal changes its state (from high to low or vice versa), it is supposed to do so immediately. But in real scenario, it is almost impossible for it to happen without a time lag. This time lag is shown by the skew lines. The protocol is as follows:

- The transaction starts with a START sequence. During the start condition, the serial data line (SDA) switches from a high-voltage level to a

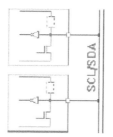

Figure 1.4   I2C hardware.

low-voltage level before the serial clock line (SCL) switches from high to low (see Figure 1.4 for SDA and SCL and also Section 1.6).
- After the START sequence, the *Master* has to send the address of the *Slave* it wants to talk to, that is, the 7-bit ADDRESS (MSB first) followed by R/W' bit determining whether it is required to read from the device or write into it. If the master wants to send data to the slave, the read/write bit is a low-voltage level. If the master is requesting data from the slave, the bit is a high-voltage level
- The *Slave* responds by acknowledging the address using ACK/NACK bit. In case if there is no ACK, the *Master* will act as mentioned before.
- Once the *Slave* acknowledges the address, it means that it is now ready to send/receive data to/from the *Master*. Thus begins the data transfer. The DATA are always of 8 bits (MSB first), and the receiver has to send the ACK signal after each byte of data received.
- When the transaction is over, the *Master* ends it by generating a STOP sequence. Alternatively, *Master* could also begin with a repeated START.

The above describes the standard format. Possible modification can take place. Such modifications will be discussed in Section 1.15.

1.4 The I2C Bus Hardware Structure

The I2C bus physically consists of two active wires: a data line wire called serial data line (SDA) and a clock wire called serial clock line (SCL), and a ground connection, as shown in Figure 1.4. Sometimes, I2C is referred to as **Two-wire Interface** or the **TWI,** since it consists of two active wires. Both SDA and SCL are bidirectional. This means that in a particular device, these lines can be driven by the IC itself or from an external device. To achieve this functionality, these signals use open-collector or open-drain outputs (depending on the technology). The bus interface is built around an input buffer and an open drain or open collector transistor (Figure 1.4). When the bus is IDLE, the bus lines are in the logic HIGH state. This is achieved by using an external pull-up resistor, which is a small current source. To put a signal on the bus, the chip drives its output transistor, thus pulling the bus to a LOW level. The two wires are acting accordingly as wired-AND. This concept achieves a "built-in" bus mastering technique. If the bus is "occupied" by a chip that is sending a 0, then all other chips lose their right to access the bus. This will be used for "bus arbitration" (see Section 1.19 for Bus Arbitration).

1.4.1 Electrical Consideration

Some electrical considerations must be taken by the designer of an I2C system. The two main considerations are as follows:

- The electrical considerations when connecting different devices to I2C bus
- The electrical considerations that limits the length of the bus

1.4.1.1 Types of devices that can be connected to I2C bus

The design of the I2C bus places no restrictions on the type of devices that may be connected to the bus. Figure 1.5 represents the hardware needed to connect two devices of the same logic family (e.g., both are using TTL or CMOS) using I2C bus. The two devices use the same supply voltage. The SCL is used to strobe data transmitted on SDL from or to the master that currently has control over the bus. Because the devices are wired-ANDed, this means that the bus is free if SDA and SCL are high and also that the device output is ANDed with the signal on the bus. This, as mentioned before, represents the basics of achieving the bus arbitration when there is more than one master. It is possible, in general to configure a system with a mixture of devices from different logic families, that is, a mixture of various TTL and CMOS logic families. In this case, each device has its own different supply voltage. To accommodate bidirectional communication with devices operating on different voltages, for example, 5 V and 3.5 V, it is necessary to incorporate into the SCL and SDA lines a form of buffering circuit as shown in Figure 1.6. The two devices, operating at different voltage levels, are now separated from one another.

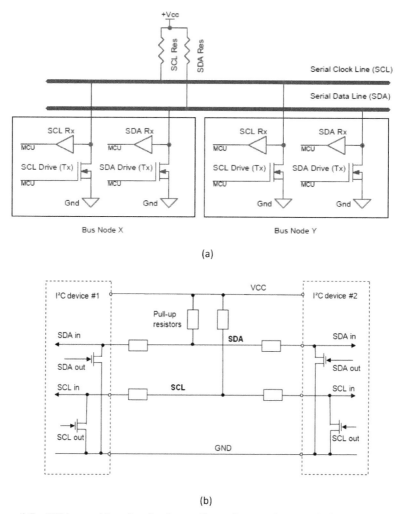

(a)

(b)

Figure 1.5   I2C bus and interface hardware: Case of connecting two devices with the same supply voltage (www.freescale.com).

1.4.1.2 Electrical considerations that limits I2C bus length

There is no maximum length specified in the I2C specification. The limit of the length is a function of several factors including capacitance (usual maximum limit is 400 pF), the minimum value for Rp, propagation delays along the cables, the type of cable used and the integrity of the logic signals in the presence of noise. There are bus rise and fall times that must be met with limits set by the static and dynamic sink capability of standard I2C devices

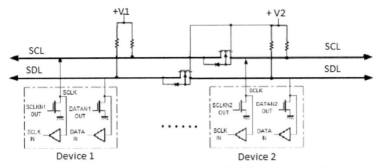

Figure 1.6 Connecting two devices with different supply voltage.

(3 mA sink current). Here we briefly discuss the main factors that affect the length of I2C bus which are as follows:

- Bus capacitance limitation
- Clock rise time limitation
- Noise pickup limitation

i. Limitation due to bus capacitance

The open-collector technique in use with I2C has a drawback that limits the length of the bus. Long lines present a capacitive load for the output drivers. Since the pull-up is passive, the RC constant of the bus will affect the wave shape of the signals propagating on the bus, which in turn limits the speed that the bus can handle; the higher the RC constant of the bus, the slower the speed of the bus. This results because of the effect of the RC constant on the slew rate of the edge of the signals. At a certain point, and because of the slew rate, the ICs will not be able to distinguish clearly between logic 1 and logic 0. Increasing the length of the line has another effect: it increases the possibility of getting reflections at high speed. The reflected waves can corrupt the data on the bus to the extent that it becomes unreadable.

It is important to note here that the bus capacitance includes not only the capacitance between each signal line (clock and data) to return but also the input capacitance of each device attached to the bus. This input capacitance can usually be found in a device's data sheet. For common ICs made with a CMOS process, the input capacitance is typically on the order of 10 pF. I2C specifications allows up to a maximum of 400 pF for the bus capacitance, the matter that put limitation on the maximum number to be connected to the bus.

To avoid such problems, a number of strict electrical specifications are to be followed when using I2C bus. According to the specification, the length of the bus is limited by the total bus capacitance: the total capacitance of the bus remains under 400 pF.

ii. Clock rise time limitation

The RC constant of the bus will affect the wave shape of the signals propagating on the bus, which in turn limits the speed that the bus can handle; the higher the RC constant of the bus, the slower the speed of the bus. This results because of the effect of the RC constant on the slew rate of the edge of the signals. At a certain point, and because of the slew rate, the ICs will not be able to distinguish clearly between logic 1 and logic 0. Increasing the length of the line has another effect: it increases the possibility of getting reflections at high speed. The reflected waves can corrupt the data on the bus to the extent that they become unreadable.

For standard mode, the maximum rise time of the clock and data signals is 1000 ns. Note that the rise time is specified as the time the signal takes to rise from 30% to 70% of the supply voltage.

There are some ways to improve the rise time by using, for example, a switched pull-up circuit. In Section 2.4 of Chapter 2, some discussions concerning improving the rise time is given.

iii. Noise pickup limitation

Another issue is that a lot of devices generally means a long bus (more than about 20–25 cm), and a long bus often has problems with noise pickup. A common problem on noisy I2C buses is that a spike on the data line when the clock is high can cause the devices on the bus to see an improper START or STOP condition. This confuses the devices on the bus and can cause the bus to hang.

The original I2C specification allows a data transfer rates of up to 100 kbps and 7-bit addressing. Seven-bit addressing allows a total of 128 devices to communicate over a shared 1 bus. With increased data transfer rate requirements, the I2C specification has been recently enhanced (Version 1.0–1992 and Version 2.0–1998) to include fast mode, 3.4 Mbps, with 10-bit addressing. At present, I2C range includes more than 150 CMOS and bipolar I2C bus compatible types for performing communication functions between intelligent control devices (e.g., microcontrollers), general-purpose circuits (e.g., LCD drivers, thermal sensors, remote I/O ports, RAM, flash, and EPROM memories), and application-oriented circuits (e.g., digital tuning and signal processing circuits for radio and video systems). Other common devices capable of interfacing to an I2C bus include real-time clocks and watchdog timers.

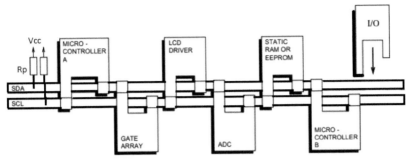

Figure 1.7 Basic configuration of I2C.

1.5 The Physical Bus–I2C Bus Interface

Figure 1.7 shows a typical I2C bus for an embedded system, where many different peripherals, acting as multiple slave devices, are sharing the bus. The microcontroller represents the I2C master, and controls the IO expanders, various sensors, EEPROM, ADCs/DACs, and much more. All of which are controlled with only 2 pins from the master. All the devices connected to the bus are classified as either being *Master* or *Slave* (just like SPI). The number of ICs that can be connected to the same bus is limited only by a maximum bus capacitance. More capacitance may be allowed under some conditions.

The possibility that many different peripherals may share one bus is one of the largest benefits that the I2C bus can give when compared to other interfaces. I2C bus is bidirectional data transfer with data rate depends on the mode of operation (standard mode, Fast mode, Fast-mode Plus).

Each device connected to the bus is software addressable by a unique address and simple master/slave relationships exist at all times; masters can operate as **master-transmitters** or as **master-receivers**.

Figure 1.8 highlights the master–slave and receiver–transmitter relationships found on the I2C bus. Note that these relationships are not permanent, but only depend on the direction of data transfer at that time. The transfer of data would proceed as follows:

1. Suppose microcontroller A wants to send information to microcontroller B:
   - Microcontroller A (master) addresses microcontroller B (slave).
   - Microcontroller A (**master-transmitter**) sends data to microcontroller B (slave-receiver)–microcontroller A terminates the transfer.

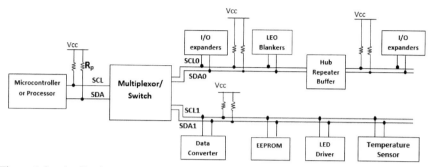

Figure 1.8   Application example showing I2C communication between the different ICs on a system and with pull-up resistors on I2C bus.

2. If microcontroller A wants to receive information from microcontroller B:

   – Microcontroller A (master) addresses microcontroller B (slave).
   – Microcontroller A (**master-receiver**) receives data from microcontroller B (slave-transmitter).
   – Microcontroller A terminates the transfer. Even in this case, the master (microcontroller A) generates the timing and terminates the transfer.

The possibility of connecting more than one microcontroller to the I2C bus means that more than one master could try to initiate a data transfer at the same time. To avoid the chaos that might ensue from such an event, I2C includes collision detection and arbitration to prevent data corruption if two or more masters simultaneously initiate data transfer.

Arbitration procedure relies on the wired-AND connection of all I2C interfaces to the I2C bus. If two or more masters try to put information onto the bus, the first to produce a "one" when the other produces a "zero" loses the arbitration. The clock signals during arbitration are a synchronized combination of the clocks generated by the masters using the wired-AND connection to the SCL line (for more detailed information concerning arbitration, see Section 1.19). Generation of clock signals on the I2C bus is always the responsibility of master devices; each master generates its own clock signals when transferring data on the bus. Bus clock signals from a master can only be altered when they are stretched by a slow slave device holding down the clock line or by another master device when arbitration occurs.

1.6 SDA and SCL Signals

Typically, both SDA and SCL are each connected to a 3.3 or 5 V power line through a single "pull-up" resistor, one on each line. This is necessary, because I2C devices have open-drain (or open-collector) lines to perform wired-AND function: they can force the voltage on the line to 0 V or "low," but can't raise it to 3.3 V or "high." The pull-up resistor is used to ensure no component-harming high currents are able to flow when two devices try and signal simultaneously. High and low are the electrical representations of the 1s and 0s that are the fundamental components of digital information. Adding these two resistors–and the bus needs only two, no matter how many devices are connected to it–ensures the voltage rises back to 3.3 V without a short circuit.

In most cases, the designer will be expected to add these pull-up resistors, but some devices, typically those that operate at 3.3 V, include them in order to be compatible with devices supplying 5 V. Remembering that only one pair of pull-up resistors are needed per bus, it may be necessary to remove pull-up resistors attached to other devices on the bus.

The value of the pull-up resistor has direct effect on the rise time of the signals. Some companies like Texas Instrument introduced TCA980x Family to replace the resistance and to improve the rise time (see Chapter 2, Sections 2.2 and 2.3 for detail).

Data on the I2C bus can be transferred at rates of up to 100 kbps in the Standard mode, up to 400 kbps in the Fast mode, up to 1 Mbps in Fast mode Plus, or up to 3.4 Mbps in the High-speed mode. The bus capacitance limits the number of interfaces connected to the bus. For a single master application, the master's SCL output can be a push–pull driver design if there are no devices on the bus which would stretch the clock.

1.7 Masters and Slaves

The I2C bus, as in case of ISP, separates devices into "masters" and "slaves." A master is the device which initiates a data transfer on the bus and generates the clock signals to permit that transfer. All the other devices synchronize their timings to the master and are thus considered slaves. The master and its slaves can all transmit and receive data, but only the master can tell the slave when to transmit data back.

Both lines are connected to a positive supply via a ***pull-up resistor*** and remain HIGH when the bus is not busy. Each device is recognized by a unique address–whether it is a microcomputer, LCD driver, memory or keyboard interface–and can operate as either a transmitter or receiver, depending on the function of the device. A device generating a message or data is a transmitter, and a device receiving the message or data is a receiver. Obviously, a passive function like an LCD driver could only be a receiver, while a microcontroller or a memory can both transmit and receive data. When a data transfer takes place on the bus, a device can either be a master or a slave. The device which initiates the transfer, and generates the clock signals for this transfer, is the master. At that time any device addressed is considered a slave. It is important to note that a master could either be a transmitter or a receiver; a master microcontroller may send data to a RAM acting as a transmitter and then interrogate the RAM for its contents acting as a receiver in both cases performing as the master initiating the transfer. In the same manner, a slave could be both a receiver and a transmitter. The I2C is a multi-master bus. It is possible to have, in one system, more than one device capable of initiating transfers and controlling the bus. A microcontroller may act as a master for one transfer, and then be the slave for another transfer, initiated by another processor on the network. The master/slave relationships on the bus are not permanent and may change on each transfer.

Just like SPI, it is possible to have multiple *Masters* connected to the same I2C bus. As more than one master may be connected to the bus, it is possible that two devices will try to initiate a transfer at the same time. Obviously, in order to eliminate bus collisions and communications chaos, an arbitration procedure is necessary. The I2C design has an inherent arbitration and clock synchronization procedure relying on the wired-AND connection of the devices on the bus. In a typical multi-master system, a microcontroller program should allow it to gracefully switch between master and slave modes and preserve data integrity upon loss of arbitration. Figure 1.8 is an application example of communication between different ICs.

1.7.1 Buffering and Multiplexing

When there are many I2C devices in a system, there can be a need to include bus buffers or multiplexers (see Figure 1.8) to split large bus segments into smaller ones. This can be necessary to keep the capacitance of a bus segment below the allowable value or to allow multiple devices with the same address to be separated by a multiplexer. Many types of multiplexers and buffers exist and all must take into account the fact that I2C lines are specified to be bidirectional. Multiplexers can be implemented with analog switches, which can tie one segment to another. Analog switches maintain the bidirectional nature of the lines but do not isolate the capacitance of one segment from another or provide buffering capability.

Buffers can be used to isolate capacitance on one segment from another and/or allow I2C to be sent over longer cables or traces. Buffers for bidirectional lines such as I2C must use one of several schemes for preventing latch-up. I2C is open-drain, so buffers must drive a low on one side when they see a low on the other. One method for preventing latch-up is for a buffer to have carefully selected input and output levels such that the output level of its driver is higher than its input threshold, preventing it from triggering itself. For example, a buffer may have an input threshold of 0.4 V for detecting a low, but an output low level of 0.5 V. This method requires that all other devices on the bus have thresholds which are compatible and often means that multiple buffers implementing this scheme cannot be put in series with one another.

Alternatively, other types of buffers exist that implement current amplifiers or keep track of the state (i.e., which side drove the bus low) to prevent latch-up. The state method typically means that an unintended pulse is created during a hand-off when one side is driving the bus low, then the other drives it low, then the first side releases (this is common during an I2C acknowledgment). Buffering and multiplexing will be considered in Chapter 2.

1.8 I2C Data Validity

For the data to be valid on the SDA line, it must not change while the SCL is high. The data on the SDA line should change only and only when the SCL line goes low: Data on the SDA line must remain stable during the high phase of the clock period (Figure 1.9). If this standard is not followed, the data transfer becomes flawed, in which case it becomes a start/stop sequence.

1.9 Voltage Levels and Resistor Values

I2C supports a wide range of voltage levels, and hence, the designer can provide +5 V or +3.3 V as Vcc easily. The levels of logical "0" (LOW) and "1" (HIGH) are also not fixed and depend on the associated V_{cc}.

This gives the designer a wide range of choices for the values of the pull-up resistors (R_p). Anything within the range of 1 k to 47 k should be fine; however, values lower than 10 k are usually preferred.

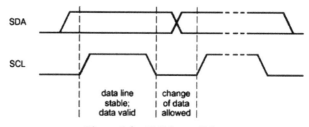

Figure 1.9 I2C data validity.

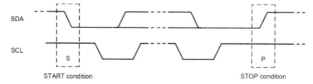

Figure 1.10 Start and stop conditions.

1.10 Start/Stop Sequence

All transactions on the bus begin with a START condition (S) and end with a STOP condition (P). In order for the *Master* to start talking to the *Slave(s)*, it must notify the *Slave(s)* about it. This is done using a special start sequence. While discussing I2C data validity, it was stated that the SDA should not change while the SCL is high. This rule doesn't hold for the start/stop sequence, which makes the start/stop sequence a special sequence.

When the SCL is high and SDA goes from high to low (as shown in the Figure 1.10), it marks the beginning of the transaction of *Master* with the *Slave(s)*.

And when the SDA goes from low to high while the SCL is still high (as shown in Figure 1.10), it marks the end of the transaction of that *Master* with the *Slave(s)*.

Detection of START and STOP conditions by devices connected to the bus is easy if they incorporate the necessary interfacing hardware. However, microcontrollers with no such interface have to sample the SDA line at least twice per clock period to sense the transition. START and STOP conditions are always generated by the master.

In-between the start and stop sequences, the bus is busy and no other *Master(s)* (if any) should try to initiate a transfer. The bus is considered to be free again a certain time after the STOP condition. This bus free situation is specified in I2C specifications. The bus stays busy if a repeated START

(Sr) is generated instead of a STOP condition. In this respect, the START (S) and repeated START (Sr) conditions are functionally identical. For the remainder of this document, therefore, the S symbol is used as a generic term to represent both the START and repeated START conditions, unless Sr is particularly relevant.

1.11 Repeated START Condition

A repeated START condition is similar to a START condition and is used in place of a back-to-back STOP then START condition. It looks identical to a START condition, but differs from a START condition because it happens before a STOP condition (when the bus is not idle). This is useful for when the master wishes to start a new communication, but does not wish to let the bus go idle with the STOP condition, which has the chance of the master losing control of the bus to another master (in multi-master environments).

1.12 Addressing Structure

1.12.1 7-bit Addressing

In order for one I2C device to communicate with another on a one-to-one basis, both devices need to be uniquely identifiable. This identity is the device's I2C address. The address comes directly after the START condition. I2C addresses are usually 7-bit numbers (Figure 1.11), so a bus can comprise up to 127 devices in all. As will be discussed, 10-bit address is also used.

When the *Master* calls out to the *Slave(s)*, it still needs to send out 8 bits of data. In this case, the *Master* appends an extra Read/Write (R/W') bit to the 7 bits of address (note that W' means Write complemented). The extra bit is used to indicate whether the signal is being sent by the master to the slave–a "write"–or in the other direction–a "read." The R/W' bit is added to the LSB of the address byte. The 7-bit address is placed in bits one through

Figure 1.11 7-bit address.

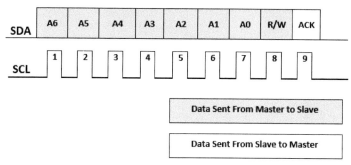

Figure 1.12 I2C device addressing.

Note: The legend shown at the bottom is only for SDA. SCL is always generated by the Master (whether transmitter or receiver).

Field:	S	I2C address field							R/W'	A	I2C message sequences...	P
Type	Start	Byte 1								ACK	Byte X etc...	Stop
Bit position in byte X		7	6	5	4	3	2	1	0		Rest of the read or write	
7-bit address pos		7	6	5	4	3	2	1			message goes here	
Note		MSB					LSB		1 = Read			
									0=Write			

Figure 1.13 7-bit addressing structure.

seven of the address byte, and the data sent look as in Figure 1.12. The 7-bit addressing structure is shown in Figure 1.13.

1.12.2 Acknowledge Scheme

I2C transfers 8 bits (1 byte) of data at a time. After the transfer of each byte is complete, the receiver must acknowledge it. To acknowledge, the receiver sends an ACK bit back to the transmitter. This goes as follows:

- The transmitter (could be either *Master* or *Slave*) transmits 1 byte of data (MSB first) to the receiver during 8 clock pulses of SCL, after which it releases the SDA line, that is, the SDA line becomes HIGH for the ACK clock pulse.
- The receiver (could be either *Master* or *Slave*, it depends) is obliged to generate an acknowledgment after each byte sent by the transmitter by pulling the SDA line LOW for the ACK clock pulse (9th clock pulse) of SCL.

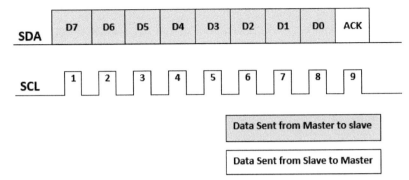

Figure 1.14   I2C acknowledgment scheme (Assumption: Master is the transmitter).

Note: The legend shown at the bottom is only for SDA. SCL is always generated by the Master (whether transmitter or receiver).

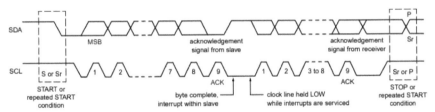

Figure 1.15   Data transfer on the I2C bus.

• So overall, there are 9 SCL clock pulses required to transmit a byte of data. This is shown in Figure 1.14 with the assumption that *Master* is the transmitter.

The number of bytes that can be transmitted per transfer is unrestricted. As mentioned before, each byte must be followed by an Acknowledge bit (see Figure 1.15).

Consider now the case of receiver fails to acknowledge receiving the data sent to it. There are two cases to that situation.

Case 1: Slave is at the receiver's end.

Under this case, it is possible to identify two possible cases:

• **CASE 1A:** The *Slave-receiver* does not acknowledge the *Slave* address (see "Address" latter). In that case, it simply leaves the SDA line HIGH. Now the *Master-transmitter* either generates a Stop sequence or attempts a repeated Start sequence.

- **CASE 1B:** The *Slave-receiver* acknowledges the *Slave* address, but after some time it is unable to receive any data and leaves the SDA line HIGH during the ACK pulse. Even in this case, the *Master-transmitter* does the same–either generates a Stop sequence or attempts a repeated Start sequence.

Case 2: Master is at the receiver's end.

In this case, the *Master* is the one generating ACK, as well as responsible for generating Start/Stop sequence. Now how does that work out, especially when the transaction ends?

In this case, in order to signal the *Slave-transmitter* the end of data, the *Master-receiver* does NOT generate any ACK on the last byte clocked out of the *Slave-transmitter*. In this case, the *Slave-transmitter* must let go of the SDA line to allow *Master* to generate a Stop or a repeated Start sequence.

1.12.3 I2C Addresses Standard: 10-bit Addressing

In order to prevent address clashes, due to the limited range of the 7-bit addresses, a new 10-bit address scheme has been introduced. This enhancement can be mixed with 7-bit addressing and increases the available address range about ten times. The 10-bit address uses two bytes (Figure 1.16). After the start condition, a leading "11110" introduces the 10-bit addressing scheme. The last two address bits of the first byte concatenated with the eight bits of the second byte of the whole 10-bit address. Devices which only use 7-bit addressing simply ignore messages with the leading "11110." The following picture shows the first two bytes of a transfer with a 10-bit address.

A master transmitter addresses the slave with two address bytes as described above with the R/W' bit = "0" followed by data bytes from the master. The master receiver transfer is only possible with a Combined Transfer due to the fact that the second address byte can only be transmitted

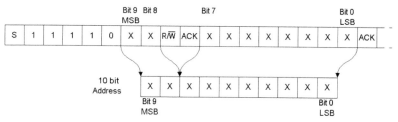

Figure 1.16   10-bit addressing.

Field:	S	10-bit mode indicator					Upper addr		R/W'	A	Lower address field								I2C message sequences	P
Type	Start	Byte 1								ACK	Byte 2								Byte X etc.	Stop
Bit position in byte X		7	6	5	4	3	2	1	0		7	6	5	4	3	2	1	0		
Bit value		1	1	1	1	0	X	X	X		X	X	X	X	X	X	X	X		
10-bit address pos							10	9			8	7	6	5	4	3	2	1	Rest of the read or write message goes here	
Note		Indicates mode		10-bit			MSB		1 = Read 0 = Write									LSB		

Figure 1.17 10-bit addressing structure.

if the R/W' bit of the first address byte is "0." Hence, the start of a master receiver transfer will be the same as a master transmitter transfer followed by a repeated start condition and the first byte of address byte with R/W' bit = "1" (switching to slave transmitter mode). Please refer to the following master receiver sequence:

- Start condition
- First address byte, R/W' bit = "0," ACK from the slave
- Second address byte, ACK from the slave
- Repeated start condition (no stop condition!)
- First address byte again, R/W' bit = "1," ACK from the slave, slave switches to transmit mode
- Slave transmits data bytes, ACK from master
- After the last data byte, the master sends a NACK
- Stop condition

Figure 1.17 shows the 10-bit addressing structure.

Practically, the 10-bit addressing has the following impact on the I2C protocol (refer to Figure 1.16):

- Two address words are used for device addressing instead of one.
- The first address word MSBs are conventionally coded as "11110" so any device on the bus is aware that the master sends a 10-bit device address.

1.12.3.1 I2C bus transactions in case of 10-bit address

All combinations of read/write formats previously described for 7-bit addressing (Section 1.13) are possible with 10-bit addressing. Two are

Figure 1.18 A master-transmitter addresses a slave-receiver with a 10-bit address.

Figure 1.19 A master-receiver addresses a slave-transmitter with a 10-bit address.

detailed here:

- Master-transmitter transmits to slave-receiver with a 10-bit slave address.

The transfer direction is not changed (see Figure 1.18). When a 10-bit address follows a START condition, each slave compares the first seven bits of the first byte of the slave address (1111 0XX) with its own address and tests if the eighth bit (R/W direction bit) is 0. It is possible that more than one device finds a match and generate an acknowledge (A1). All slaves that found a match compare the eight bits of the second byte of the slave address (XXXX XXXX) with their own addresses, but only one slave finds a match and generates an acknowledge (A2). The matching slave remains addressed by the master until it receives a STOP condition (P) or a repeated START condition (Sr) followed by a different slave address.

- Master-receiver reads slave-transmitter with a 10-bit slave address.

The transfer direction is changed after the second R/W bit (Figure 1.19). Up to and including acknowledge bit A2, the procedure is the same as that described for a master-transmitter addressing a slave-receiver. After the repeated START condition (Sr), a matching slave remembers that it was addressed before. This slave then checks if the first seven bits of the first byte of the slave address following Sr are the same as they were after the START condition (S) and tests if the eighth (R/W) bit is 1. If there is a match, the slave considers that it has been addressed as a transmitter and generates acknowledge A3. The slave-transmitter remains addressed until it receives a STOP condition (P) or until it receives another repeated START condition (Sr) followed by a different slave address. After a repeated START condition (Sr), all the other slave devices will also compare the first seven bits of the first byte of the slave address (1111 0XX) with their own addresses and test the eighth (R/W) bit. However, none of them will be addressed because R/W = 1 (for 10-bit devices), or the 1111 0XX slave address (for 7-bit devices) does not match.

Slave devices with 10-bit addressing react to a "general call" in the same way as slave devices with 7-bit addressing. Hardware masters can transmit their 10-bit address after a "general call." In this case, the "general call" address byte is followed by two successive bytes containing the 10-bit address of the master-transmitter. The format is as shown in Figure 1.21 where the first DATA byte contains the eight least-significant bits of the master address.

The START byte 0000 0001 (01h) can precede the 10-bit addressing in the same way as for 7-bit addressing.

1.12.4 I2C Addresses Standard: Special Addresses and Exceptions in 7-bit Address Space

As mentioned before, the first byte of an I2C transfer contains the slave address and the data direction. The address is 7 bits long, followed by the direction bit (W/R bit). A 7-bit wide address allows, theoretically 128 I2C addresses–however, some addresses are reserved for special purposes. Only 112 addresses are available with the 7-bit address scheme.

As a matter of fact, I2C gives a loose standard for the address. It uses the most significant four bits to identify the type of the device and the next three bits are used to specify one of eight devices of this type (or further specify the device type). Also, some devices require certain patterns for the last three bits, while others (such as large serial EEPROMs) use these bits to specify an address inside the device. This shows the importance of mapping out the devices to be put on the bus and all their addresses. In the I2C address map, there are also what is called "*reserved addresses.*"

1.12.4.1 Reserved and none-reserved addresses in 7-bit address space
1.12.4.1.1 Reserved addresses in 7-bit address space
In case of 7-bit, two groups of eight addresses (0000 XXX and 1111 XXX) are reserved for the purposes shown in Table 1.1.

Assignment of addresses within a local system is up to the system architect who must take into account the devices being used on the bus and any future interaction with other conventional I2C buses. For example, a device with seven user-assignable address pins allows all 128 addresses to be assigned. If it is known that the reserved address is never going to be used for its intended purpose, a reserved address can be used for a slave address.

Table 1.1   Reserved addresses (X = don't care; 1 = high; 0 = low)

Slave Address	R/W'	Description
0000 000	0	General call address [1]
0000 000	1	START byte [2]
0000 001	X	CBUS address [3]
0000 010	X	Reserved for different bus format [4]
0000 011	X	Reserved for future purposes
0000 1XX	X	HS mode master code
1111 1XX	1	Device ID
1111 0XX	X	10-bit slave address

Notes:

[1] The general call address is used for several functions including software reset (see next).

[2] No device is allowed to acknowledge at the reception of the START byte.

[3] The CBUS address has been reserved to enable the inter-mixing of CBUS compatible and I2C bus compatible devices in the same system. I2C bus compatible devices are not allowed to respond on reception of this address.

[4] The address reserved for a different bus format is included to enable I2C and other protocols to be mixed. Only I2C bus compatible devices that can work with such formats and protocols are allowed to respond to this address.

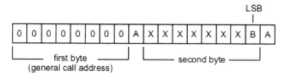

Figure 1.20   General call address format.

i. General call address

The general call address is for addressing every device connected to the I2C bus at the same time. However, if a device does not need any of the data supplied within the general call structure, it can ignore this address by not issuing an acknowledgment. If a device does require data from a general call address, it acknowledges this address and behaves as a slave-receiver. The master does not actually know how many devices are acknowledged if one or more devices respond. The second and following bytes are acknowledged by every slave-receiver capable of handling these data. A slave who cannot process one of these bytes must ignore it by not acknowledging. Again, if one or more slaves acknowledge, the not acknowledged will not be seen by the master. The meaning of the general call address is always specified in the second byte (see Figure 1.20).

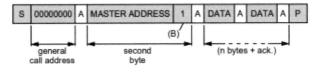

Figure 1.21 Data transfer from a hardware master-transmitter.

There are two cases to consider:

- When the least significant bit B is a "zero."
- When the least significant bit B is a "one."

When bit B is a "zero," the second byte has the following definition:

- 0000 0110 (06h): Reset and write programmable part of slave address by hardware. On receiving this 2-byte sequence, all devices designed to respond to the general call address reset and take in the programmable part of their address. Precautions must be taken to ensure that a device is not pulling down the SDA or SCL line after applying the supply voltage, since these low levels would block the bus.
- 0000 0100 (04h): Write programmable part of slave address by hardware. Behaves as above, but the device does not reset.
- 0000 0000 (00h): This code is not allowed to be used as the second byte.

Sequences of programming procedure are published in the appropriate device data sheets. The remaining codes have not been fixed and devices must ignore them.

When bit B is a "one," the 2-byte sequence is a "hardware general call." This means that the sequence is transmitted by a hardware master device, such as a keyboard scanner, which can be programmed to transmit a desired slave address. Since a hardware master does not know in advance to which device the message has to be transferred, it can only generate this hardware general call and its own address–identifying itself to the system (see Figure 1.21).

The seven bits remaining in the second byte contain the address of the hardware master. This address is recognized by an intelligent device (e.g., a microcontroller) connected to the bus which then accepts the information from the hardware master. If the hardware master can also act as a slave, the slave address is identical to the master address.

In some systems, an alternative could be that the hardware master transmitter is set in the slave-receiver mode after the system reset. In this way, a system configuring master can tell the hardware master-transmitter (which is now in slave-receiver mode) to which address data must be sent (see

a. Configuring master sends dump address to hardware master

b. Hardware master dumps data to selected slave

Figure 1.22   Data transfer by a hardware-transmitter capable of dumping data directly to slave devices.

Figure 1.22). After this programming procedure, the hardware master remains in the master-transmitter mode.

ii. Software reset

Following a general call address (0000 0000), sending 0000 0110 (06h) as the second byte causes a software reset. This feature is optional and not all devices respond to this command. On receiving this 2-byte sequence, all devices designed to respond to the general call address reset and take in the programmable part of their address. Precautions must be taken to ensure that a device is not pulling down the SDA or SCL line after applying the supply voltage, since these low levels would block the bus.

iii. START byte

Microcontrollers can be connected to the I2C bus in two ways. A microcontroller with an on-chip hardware I2C bus interface can be programmed to be only interrupted by requests from the bus. When the device does not have such an interface, it must constantly monitor the bus via software. Obviously, the more times the microcontroller monitors, or polls the bus, the less time it can spend carrying out its intended function. There is therefore a speed difference between fast hardware devices and a relatively slow microcontroller which relies on software polling. In this case, data transfer can be preceded by a start procedure which is much longer than normal (see Figure 1.23). The start procedure consists of:

• A START condition (S)
• A START byte (0000 0001)

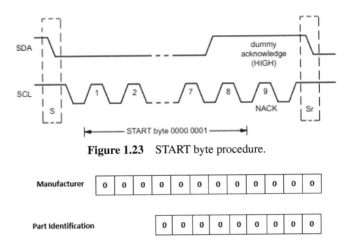

Figure 1.23 START byte procedure.

Figure 1.24 Device ID field.

- An acknowledge clock pulse (ACK)
- A repeated START condition (Sr).

After the START condition S has been transmitted by a master which requires bus access, the START byte (0000 0001) is transmitted. Another microcontroller can therefore sample the SDA line at a low sampling rate until one of the seven zeros in the START byte is detected. After the detection of this LOW level on the SDA line, the microcontroller can switch to a higher sampling rate to find the repeated START condition Sr which is then used for synchronization. A hardware receiver resets upon receipt of the repeated START condition Sr and therefore ignores the START byte. An acknowledge-related clock pulse is generated after the START byte. This is present only to conform with the byte handling format used on the bus. No device is allowed to acknowledge the START byte.

iv. Device ID

The Device ID field (see Figure 1.24) is an optional 3-byte read-only (24 bits) word giving the following information:

- Twelve bits with the manufacturer name, unique per manufacturer (e.g., NXP)
- Nine bits with the part identification, assigned by manufacturer (e.g., PCA9698)

• Three bits with the die revision, assigned by manufacturer (e.g., RevX)

The Device ID is read-only, hard-wired in the device and can be accessed as follows:

1. START condition
2. The master sends the Reserved Device ID I2C bus address followed by the R/W bit set to "0" (write): "1111 1000."
3. The master sends the I2C bus slave address of the slave device it must identify. The LSB is a "Don't care" value. Only one device must acknowledge this byte (the one that has the I2C bus slave address).
4. The master sends a Re-START condition. Remark: A STOP condition followed by a START condition resets the slave state machine and the Device ID Read cannot be performed. Also, a STOP condition or a Re-START condition followed by an access to another slave device resets the slave state machine and the Device ID Read cannot be performed.
5. The master sends the Reserved Device ID I2C bus address followed by the R/W bit set to "1" (read): "1111 1001."
6. The Device ID Read can be done, starting with the 12 manufacturer bits (first byte + four MSBs of the second byte), followed by the nine part identification bits (four LSBs of the second byte + five MSBs of the third byte), and then the three die revision bits (three LSBs of the third byte).
7. The master ends the reading sequence by NACKing the last byte, thus resetting the slave device state machine and allowing the master to send the STOP condition. Remark: The reading of the Device ID can be stopped anytime by sending a NACK.

If the master continues to ACK the bytes after the third byte, the slave rolls back to the first byte and keeps sending the Device ID sequence until a NACK has been detected.

1.12.4.2 Non-reserved addresses in 7-bit address space
Table 1.3 shows the non-reserved addresses in the 7-bit address space.

Although MSB 1111 is reserved for Device ID and 10-bit slave addressing, it is also used by VESA DDC display-dependent devices such as pointing devices.

1.13 I2C Bus Transaction

After discussing the I2C bus interface, it is important to explain how the data transfer actually takes place through that interface. I2C supports

Table 1.2   Assigned manufacturer IDs

				Manufacturer Bits							
11	10	9	8	7	6	5	4	3	2	1	Company
0	0	0	0	0	0	0	0	0	0	0	NXP Semiconductors
0	0	0	0	0	0	0	0	0	0	1	NXP Semiconductors (reserved)
0	0	0	0	0	0	0	0	0	1	0	NXP Semiconductors (reserved)
0	0	0	0	0	0	0	0	0	1	1	NXP Semiconductors (reserved)
0	0	0	0	0	0	0	0	1	0	0	Ramtron International
0	0	0	0	0	0	0	0	1	0	1	Analog Devices
0	0	0	0	0	0	0	0	1	1	0	STMicroelectronics
0	0	0	0	0	0	0	0	1	1	1	ON Semiconductor
0	0	0	0	0	0	0	1	0	0	0	Sprintek Corporation
0	0	0	0	0	0	0	1	0	0	1	ESPROS Photonics AG
0	0	0	0	0	0	0	1	0	1	0	Fujitsu Semiconductor
0	0	0	0	0	0	0	1	0	1	1	Flir
0	0	0	0	0	0	0	1	1	0	0	O_2 Micro
0	0	0	0	0	0	0	1	1	0	1	Atmel

Table 1.3   Non-reserved addresses in the 7-bit address space

MSB (4-bit)	Typical Usage
0001	Digital receivers, SMBus
0010	TV video line decoders, IPMB
0011	AV codecs
0100	Video encoders, GPIO expanders
0101	ACCESS bus, PMBus
0110	VESA DDC, PMBus
0111	Display controller
1000	TV signal processing, audio processing, SMBus
1001	AV switching, ADCs and DACs, IPMB, SMBus
1010	Storage memory, real-time clock
1011	AV processors
1100	PLLs and tuners, modulators and demodulators, SMBus
1101	AV processors and decoders, audio power amplifiers, SMBus
1110	AV color space converters

unidirectional as well as bidirectional data transfer. There are three possible cases of data transfer, normally called bus events:

- Unidirectional Data Transfer

    ○ *I2C Bus Events 1: Master-transmitter* to *Slave-receiver* (Case 1)
    ○ *I2C Bus Events 2: Slave-transmitter* to *Master-receiver* (Case 2)

- Bidirectional Data Transfer

    ○ *I2C Bus Events 3: Master* to *Slave* and *Slave* to *Master* (Case 3)

1.13.1 I2C Bus Events 1: Master (Transmitter) to Slave (Receiver) Data Transfer

This is the event when the master transmitting a byte to a slave device. In other words, it is the case when the master writes to the slave. The transaction in this case will be as in Figure 1.25.

To write to a slave, the write cycle is as follows:

- The master device initiates the transfer by a start condition. This acts as an "Attention" signal to all of the connected devices. All ICs on the bus will listen to the bus for incoming data.
- Once the start condition has sent, the master sends a byte to the slave. This first byte is used to identify the slave on the bus and to select the mode of operation. As shown in Figure 1.25, the 7 bits forming the address of the slave starting with the most significant bit (MSB) is sent at first followed by the eighth bit that defines the direction of the data. The eighth bit is labeled R/W in the figure with 1 for "read" and 0 for "write". In this case, the eighth bit is set to zero because the Master is writing to the slave. Here, the bit value is placed on the SDA line by the master device, while the SCL line is low and maintained stable until after a clock pulse on SCL.

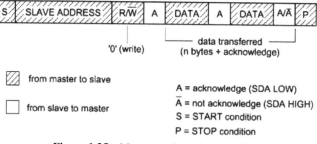

Figure 1.25   Master-to-slave data transfer.

- Having received the address, all the ICs will compare it with their own address. If there is no match, the ICs will ignore the rest of this transaction and wait for the next, that is, wait until the bus is released by the stop condition. On the other hand, if the address matches, the receiving device will respond by producing an acknowledge signal. The receiving device produces this signal by holding the SDA line low during the first ACK clock cycle.
- Once the *Slave* acknowledges the address, *Master* can now send data to the *Slave* byte-by-byte. The *Slave has to* send the ACK bit after every byte it receives. The master device transmits a byte of data starting with the most significant bit down to the least significant bit. The receiving device, in this case one of the slaves, acknowledges the reception of data by holding the SDA line low during the second ACK clock cycle. This means that for every 8 bits transferred, the device receiving the data sends back an acknowledge bit, so there are actually 9 SCL clock pulses to transfer each 8 bit byte of data. If the receiving device sends back a low ACK bit, then it has received the data and is ready to accept another byte.
- This goes on till *Slave* can no longer receive data and does NOT send the ACK bit.
- This indicates that the slave cannot accept any further data and the master should terminate the transfer by sending a stop sequence.

We see that the data transfer never changes its direction. Data always flow from *Master* to *Slave*, which makes the setup quite easy.

An example of this case would be like performing page write operations on an EEPROM chip.

1.13.2 I2C Bus Events 2: Slave (Transmitter) to Master (Receiver) Data Transfer

This is the Bus Event when the master reads a byte from a slave device. The transaction in this case is shown in Figure 1.26.

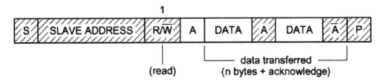

Figure 1.26 Slave-to-master data transfer.

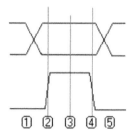

Figure 1.27   Read event.

To read from a slave device, the read cycle will be as follows:

- If performing a read operation, the master initiates the transfer by a start condition, sends the address of the device that is being read, sends a one (logic high on SDA line) requesting a read and waits to receive an acknowledgment.
- The *Slave* acknowledges the address, thus ready to send data now.
- The protocol format is the same as in transmitting a byte to a slave, except that now the master is not allowed to touch the SDA line. Prior to sending the 8 clock pulses needed to clock in a byte on the SCL line, the master releases the SDA line. The slave will now take control of this line. The line will then go high if it wants to transmit a "1," or if the slave wants to send a "0," the line will remain low.
- All the master has to do (see Figure 1.27) is to generate a rising edge on the SCL line (2), read the level on SDA (3) and generate a falling edge on the SCL line (4). The slave will not change the data during the time that SCL is high. (Otherwise a Start or Stop condition might inadvertently be generated.) During (1) and (5), the slave may change the state of the SDA line.
- In total, this sequence has to be performed 8 times to complete the data byte.

The meaning of all bytes being read depends on the slave. There is no such thing as a "universal status register." The user needs to consult the data sheet of the slave being addressed to know the meaning of each bit in any byte transmitted.

- *Slave* keeps on sending data to the *Master*, and the *Master* keeps on sending ACK to the *Slave* after each byte until it can no longer accept any more data.

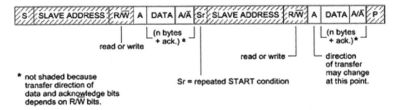

Figure 1.28   Bidirectional data transfer.

- When the *Master* feels like ending the transaction, it *does not* send the ACK, thus ending with the STOP sequence.

An example of this case could be an Analog to Digital Converter (ADC) sending data to the microcontroller continuously. The microcontroller accepts data as long as it wants to, after which it stops/finishes execution.

1.13.3 I2C Bus Events 3: Bidirectional Read and Write in Same Data Transfer

The transaction is as in Figure 1.28.

- The *Master* sends out the START sequence, followed by the 7-bit *Slave* address and the R/W' bit.
- The *Slave* acknowledges the address.
- Depending upon the value of the R/W' bit, read/write operations are performed (like the above two cases).
- Whatever the case it may be, it always ends with the receiver *not* sending the ACK.
- Until now, in the previous two cases, we have seen that the *Master* would close the connection. But in this case, the *Master* attempts a repeated START.
- And the entire process repeats again, until the *Master* decides to STOP.

As we can see, a change of direction of data transfer might happen depending upon the R/W' bits in the entire transaction.

An example of this case could be performing sequential read from an EEPROM chip. It is bidirectional because the CPU first *writes* the address from where it would like to start reading, followed by *reading* from the device. It is like, unless you tell the device from where you would like to start reading, how would it start sending you the data?

1.14 Clock Stretching

The above discussion concerning timing is valid as the *slave* is able to return to the *Master* the date needed. In some cases, the *Slave* is not ready yet to return the data to the *Master*. With devices like ADC or EEPROM, this is not an issue, but it is an issue with devices as microcontroller.

What if the *Slave* is a microcontroller, and the *Master* requests for a data which is not there in its cache? This would require the microcontroller to perform context switching, force it to search for it in the RAM, store it back in cache and then send it to the *Master*. This could (and definitely would) take a much longer time than the clock pulses of the SCL, and everything would just go wrong!

To solve such situation, I2C introduces what is called *"Clock Stretching."* An addressed slave device may hold the clock line (SCL) low after receiving (or sending) a byte, indicating that it is not yet ready to process more data. The master communicating with the slave will try to raise the clock to transfer the next bit but will verify that the clock line will remain low. The master will then have to wait for the slave to release the line (SCL goes high) so that the clock signal can be transmitted. Normally, the master waits an additional minimum time of 4 μs for standard 100 kbps I2C before pulling the clock low again. If a master wants to slow down the rate of data transfer, it just delays the next clock edge.

Although the master may also hold the SCL line low for as long as it desires (this is not allowed in newest Rev. 6 of the protocol–Section 3.1.1), the term "clock stretching" is normally used only when slaves do it. Although in theory any clock pulse may be stretched, generally it is the intervals before or after the acknowledgment bit which are used. For example, if the slave is a microcontroller, its I2C interface could stretch the clock after each byte, until the software decides whether to send a positive acknowledgment or a NACK.

Clock stretching is the only instance in the entire I2C protocol where the *Slave* drives the clock line SCL. In many processors and microcontrollers, the low-level hardware does this for us, so that we don't have to worry about it while writing the code.

Many slaves do not need to clock stretch and thus treat SCL as strictly an input with no circuitry to drive it. Some masters, such as those found inside custom ASICs may not support clock stretching; often, these devices will be labeled as a "two-wire interface" and not I2C.

To ensure a minimal bus throughput, SMBus places limits on how far clocks may be stretched. Hosts and slaves adhering to those limits cannot

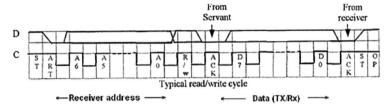

Figure 1.29 Read/write cycle.

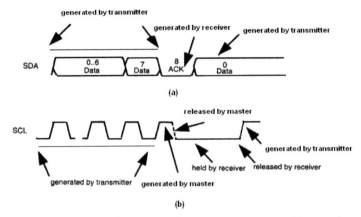

Figure 1.30 Acknowledge extension (a) without and (b) with extension.

block access to the bus for more than a short time, which is not a guarantee made by pure I2C systems.

1.15 Possible Modifications on the Timing Diagram

Figure 1.29 depicts a timing diagram of a typical read/write cycle. Sometimes, this timing diagram may change. Some of the possible changes are (see Figure 1.30):

- *Acknowledge Extension:* Under normal circumstances, following the ACK bit time, the master will release the SCL line so that transmission may continue with the next byte. If, however, the receiver (which is usually a "slave device" or "master") is temporarily unable to proceed, it will hold the SCL line LOW, thereby extending the ACK interval. In other words, it is allowed for the acknowledge bit to float high. When the receiver is ready to proceed again, it will release the SCL line and transmission continues. The timing diagram in Figure 1.30 illustrates the ACK interval extension (see "Clock Stretching").

- *Multi-register Slaves:* While explaining the write/read cycle, we assumed that the slave is simple and has only one register. In such a case, just sending the address of the slave is enough. In many cases, the slave has many registers, and each has its own address within the slave device. If the master wants to write at specific register within the slave, it must send this address. In such cases, having addressed the slave device, the master must now send out the internal location or register number inside the slave that it wishes to write to or read from. This number is obviously dependent on what the slave actually is and how many internal registers it has. In such a case, the master can continue to send data bytes to the slave, and these will normally be placed in the following registers because the slave will automatically increment the internal register address after each byte. When the master has finished writing all data to the slave, it sends a stop sequence which completes the transaction.

- *Repeated Start:* During an I2C transfer, there is often the need to first send a command and then read back an answer right away. This has to be done without the risk of another (multi-master) device interrupting this atomic operation. The I2C protocol defines a so-called repeated start condition.

  After having sent the address byte (address and read/write bit), the master may send any number of bytes followed by a stop condition. If, however, it wishes to establish a connection with a different slave, rather than issue the Stop, the master will issue another Start, using the address of the new device and then sends more data. This is defined recursively allowing any number of start conditions to be sent. The purpose of this is, in general, to allow combined write/read operations to one or more devices without releasing the bus and thus with the guarantee that the operation is not interrupted.

- Also we have to mention here that, in some devices, a start bit has to be resent to reset the receiving device for the next command, for example, in a serial EEPROM read, the first command sends the address to read from, the second reads the data at that address. Multiple address is shown in Figure 1.31.

1.16 Bus Clear

In the unlikely event where the clock (SCL) is stuck LOW, the preferential procedure is to reset the bus using the HW reset signal if I2C devices in

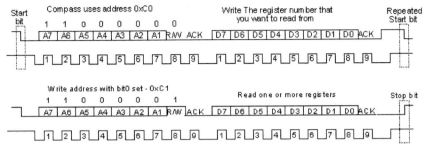

Figure 1.31 Multiple address.

use have HW reset inputs. If the I2C devices do not have HW reset inputs, cycle power of the devices activate the mandatory internal Power-On Reset (POR) circuit. In such cases, data line (SDA) is stuck LOW, the master should send nine clock pulses. The device that held the bus LOW should release it sometime within those nine clocks. If not, then use the HW reset or cycle power to clear the bus.

1.17 Applicability of I2C Bus Features

The above sections discussed the main features of the I2C bus protocol. Some of the features are mandatory and some are optional based on if the bus is single master or multi-master. Table 1.4 summarizes the use of mandatory and optional portions of the I2C bus specification and which system configurations use them.

1.18 I2C Modes: Bus Speeds

The first I2C introduced by Philips is considered as the standard mode. It refers to the initial transfer speed mode of the I2C specification which allows up to 100 kbps. There are applications where the I2C transfer speed is a limiting factor. To allow for higher transmission rates while retaining a certain amount of compatibility, Philips in 1998 has introduced the high-speed (HS) I2C standard and continued to introduce new standards. Currently, there are five operating speed categories (modes):

1. Standard mode (Low-speed mode) with a bit rate up to 100 kbps
2. Fast mode (FM) with a bit rate up to 400 kbps
3. Fast-mode plus (FM+) with a bit rate up to 1 Mbps

Table 1.4   Applicability of I2C bus protocol features

Feature	Configuration		
	Single Master	Multi-master	Slave*
START condition	M	M	M
STOP condition	M	M	M
Acknowledge	M	M	M
Synchronization	n/a	M	n/a
Arbitration	n/a	M	n/a
Clock stretching	O **	O **	O
7-bit slave address	M	M	M
10-bit slave address	O	O	O
General call address	O	O	O
Software reset	O	O	O
START byte	n/a	O***	n/a
Device ID	n/a	n/a	O

M = mandatory; O = optional; n/a = not applicable.

Notes: [*] Also refers to a master acting as a slave. [**] Clock stretching is a feature of some slaves. If no slaves in a system can stretch the clock (hold SCL Low), the master need not be designed to handle this procedure. [***] "Bit banging" (software emulation) multi-master systems should consider a START byte.

4. High-speed mode (HS-mode) with a bit rate up to 3.4 Mbps
5. Ultra-fast mode (UFm) with a bit rate up to 5 Mbps

The first four are bidirectional and compatible: any device may be operated at a lower bus speed. Ultra-fast mode devices are not compatible with the first four modes since the bus is unidirectional.

1.18.1 Low-speed Mode or Standard Mode

The very first specification of I2C dates back to the year 1982. It only covered Standard mode: up to 100 kbps and 7-bit addressing.

1.18.2 Enhanced I2C (Fast Mode)

In the fast mode, the physical bus parameters are not altered. The protocol, bus levels, capacitive load, etc., remain unchanged. However, the data rate has been increased to 400 kbps and a constraint has been set on the level of noise that can be present in the system. To accomplish this task, a number of changes have been made to the I2C bus timing.

The fast-mode I2C bus specification has the following additional features compared with the standard mode:

- The maximum bit rate is increased to 400 kbps.
- Timing of the serial data (SDA) and serial clock (SCL) signals has been adapted. There is no need for compatibility with other bus systems such as CBUS because they cannot operate at the increased bit rate.
- The inputs of fast-mode devices incorporate spike suppression and a Schmitt trigger at the SDA and SCL inputs.
- The output buffers of fast-mode devices incorporate slope control of the falling edges of the SDA and SCL signals.
- If the power supply to a fast-mode device is switched off, the SDA and SCL I/O pins must be floating so that they do not obstruct the bus lines.

1.18.3 Fast-mode plus (FM+)

FM+ devices introduced by Philips Semiconductors (now: NXP) in April 2006 can transfer information at rates up to 1 Mbps. The FM+ devices are fully downward compatible with Fast- or Standard-mode devices for bidirectional communication in a mixed-speed bus system. The same serial bus protocol and data format is maintained as with the Fast- or Standard-mode system.

The I2C fast mode was originally specified to run at 400 kHz. Ignoring the 400 kilobit limit and running the same bus with the same logic at higher speeds faces a couple of factors which limit the maximum speed on the bus. First, there's the capacitance which introduces a rise time for the signals. As a "counter measure" the current on the bus can be increased by lowering the resistor values for the termination pull-ups.

Second, the bus has some timing tolerances (tolerant of the 1 μs rise time), and if the speed is increased, these tolerances need to be more restrictive.

In applications where only Fast-mode Plus parts are present, the high drive strength and tolerance for slow rise and fall times allow the use of larger bus capacitance as long as set-up, minimum LOW time and minimum HIGH time for Fast-mode Plus are all satisfied, and the fall time and rise time do not exceed the 300 ns tf and 1 μs tr specifications of Standard mode. Bus speed can be traded against load capacitance to increase the maximum capacitance by about a factor of ten.

Unlike the high speed, there is no additional logic to implement while using FM+. Therefore, fast-mode plus devices are downward compatible with standard and fast-mode devices. In fact, many I2C slaves on the market like RAMs and EEPROMs will already tolerate higher bus frequencies.

The specification for FM+ is now part of the current IšC-bus specification and user manual.

1.18.4 High-speed Mode (HS mode)

High-speed mode (HS mode) devices offer a quantum leap in I2C bus transfer speeds. HS-mode devices can transfer information at bit rates of up to 3.4 Mbps, yet they remain fully downward compatible with Fast- or Standard mode (F/S-mode) devices for bidirectional communication in a mixed-speed bus system. With the exception that arbitration and clock synchronization is not performed during the HS-mode transfer, the same serial bus protocol and data format is maintained as with the F/S-mode system. Depending on the application, new devices may have a Fast- or Hs-mode I2C bus interface, although HS-mode devices are preferred as they can be designed-in to a greater number of applications.

1.18.4.1 Electrical characteristics of HS mode

To achieve a bit transfer of up to 3.4 Mbps, HS devices have the following electrical characteristics:

- The high-speed variant of the I2C bus allows communication up to 3.4 Mbps.
- Both master and slave devices must be high-speed-enabled in order to benefit from this increase.
- High-speed IC devices are downward compatible allowing for mixed bus systems.
- In order to shorten signal rise time, HS-mode master devices have an open-drain output buffer for the SDAH signal and a combination of an open-drain pull-down and current-source pull-up circuit on the SCL output. It also has an open-drain output buffer for SDA signal. Only the current source of one master is enabled at any one time and only during HS mode.
- HS IC masters can actually source current to the bus which is referred to as boosting.
- This current source is enabled only during HS operation and just for one master.
- HS-mode master devices generate a serial clock signal with a HIGH to LOW ratio of 1 to 2. This relieves the timing requirements for set-up and hold times.

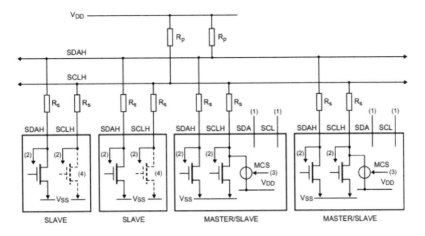

(1)   SDA and SCL are not used here but may be used for other functions.
(2)   To input filter.
(3)   Only the active master can enable its current-source pull-up circuit.
(4)   Dotted transistors are optional open-drain outputs which can stretch the serial clock signal SCLH.

Figure 1.32   I2C bus configuration with HS devices.

- HS-mode master devices can have a built-in bridge to separate lower speed devices from the bus during HS transfer. The main purpose of such bridge is to reduce the capacitive load on the bus and to avoid conflicts caused by low-speed devices.
- No arbitration or clock synchronization is performed during HS-mode transfer in multi-master systems, which speeds-up bit handling capabilities. The arbitration procedure always finishes after a preceding master code transmission in F/S-mode.

Figure 1.32 shows the physical I2C bus configuration in a system with only HS-mode devices. The pins, SDA and SCL, on the master devices are only used in mixed-speed bus systems and are not connected in an Hs-mode only system. In such cases, these pins can be used for other functions.

In Figure 1.32, an optional series resistor R_s is used. The optional series resistors R_s protect the I/O stages of the I2C bus devices from high-voltage spikes on the bus lines and minimize ringing and interference. Pull-up resistors Rp maintain the SDAH and SCLH lines at a HIGH level when the bus is free and ensure that the signals are pulled up from a LOW to a HIGH level within the required rise time. For higher capacitive bus-line loads (>100 pF), the resistor Rp can be replaced by external current source pull-ups to meet the rise time requirements. Unless proceeded by an acknowledge bit, the rise time

Figure 1.33   Details of the first part of high speed transfer format.

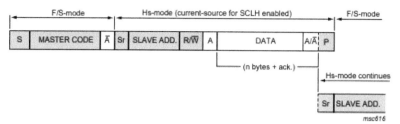

Figure 1.34   Data transfer format in HS-mode.

of the SCLH clock pulses in HS-mode transfers is shortened by the internal current-source pull-up circuit MCS of the active master.

1.18.4.2 Transmission format of high-speed mode

Serial data transfer format in HS-mode meets the Standard-mode I2Cbus specification. HS-mode can only commence after the following conditions (all of which are in F/S-mode):

1. START condition (S)
2. 8-bit master code (0000 1XXX)
3. Not-acknowledged bit (A)

Figure 1.33 shows the beginning of a high-speed transfer. This master code has two main functions:

- It allows arbitration and synchronization between competing masters at F/S-mode speeds, resulting in one winning master.
- It indicates the beginning of an HS-mode transfer.

Figures 1.34 and 1.35 show the transmission format in more detail. The format is as follows:

- Start condition is sent.
- After the start condition a so-called master code is transmitted "00001XXX" (see Table 1.1), followed by a mandatory not-acknowledge bit. The master code is sent in Fast- or Standard-mode (this is with at most 400 kbps).
- The three lowest bits are used to identify different I2C masters on the same bus–each one has its unique identifier. During transmission of the

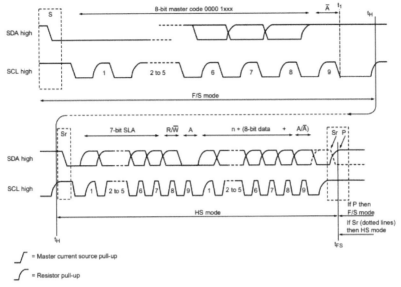

Figure 1.35 A complete HS-mode transfer.

master code, arbitration takes place (see arbitration latter), so that only the winning master can perform the following high-speed transfer. The winning master is the "active master."

- The master codes are selectable by the designer and allow up to eight high-speed masters to be connected in one system (master code "00001000" should be reserved for test and diagnostic purposes).
- Active master switches to high-speed communication.
- Arbitration and clock synchronization only take place during master code transmission, not during HS transfer.
- After the acknowledge phase following the master code, the high-speed transfer begins with a repeated start condition, followed by the slave address and the succeeding data, just like in Fast or Standard mode, but with a higher bit rate.
- After each address, the slave must respond by *ack* or *nack* signal.
- The current source circuit is disabled after each repeated start condition and after each *ack* or *nack* to give slaves a chance to stretch the clock.
- The high-speed mode remains active until a stop condition is transmitted, on which the connected high-speed devices switch back to slow transmission rates like Fast or Standard mode.
- All devices return to fast mode operation after a stop condition is sent.

Note: Clock Stretching

Clock stretching during this mode has a special rule: It is only allowed after the ACK bit (and before the 1st bit of the next byte). Stretching between 2 and 9 bits is illegal because the edges of these bits are boosted with an additional current source.

1.18.5 Ultra-fast Mode (UFm)

The I2C Ultra-fast mode (*UFm*) is the fastest I2C mode available. It offers 5 MHz. This mode is pretty different to other I2C modes:

- UFm works **unidirectionally**, and data are sent in one direction only.
- Ultra-fast mode lines are driven in push–pull technology.
- Slaves do not send an acknowledge bit.
- There is a single master only.
- No arbitration for multi-master environment is available.

Nevertheless, multiple devices can be addressed sharing an I2C UFm bus; 112 nodes are available.

Currently, there are just a couple of devices supporting this mode. NXPs PCU 9669 is a 3-channel bus controller with two 5 MHz UFm and one fast-mode plus (FM+) interface. Variations are available.

Since the ultra-fast mode bus basically focuses on LED devices, NXP comes with an LED controller PCU9655.

To have more information about UFm mode, NXP provides the application note "AN11250 *Practical guide for the use of ultra-fast mode I2C bus.*"

1.19 I2C as a Multi-master Bus: Bus Arbitration

So far we have seen the operation of the I2C bus from the master's point of view and using only one master on the bus. The I2C bus was originally developed as a multi-master bus. This means that more than one device initiating transfers can be active in the system.

When using only one master on the bus, there is no real risk of corrupted data, except if a slave device is malfunctioning or if there is a fault condition involving the SDA/SCL bus lines. This situation changes when using more than one master, for example, use of the 2 microcontrollers units CPU1 and CPU2 in Figure 1.36.

When CPU 1 issues a start condition and sends an address, all slaves will listen (including CPU 2 which at that time is considered a slave as well). If

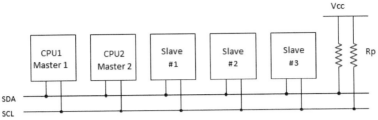

Figure 1.36 I2C bus interface–multiple master.

the address does not match the address of CPU 2, this device has to hold back any activity until the bus becomes idle again after a stop condition.

As long as the two CPUs monitor what is going on the bus (start and stop) and as long as they are aware that a transaction is going on because the last issued command was not a STOP, there is no problem. The problems arise if the two CPUs decide to initiate the "start condition" at the same time or if one of the CPUs miss the start condition and still think that the bus is idle. As a fact this is not the only problem that may arise when a system has multiple masters. A second problem that can arise is the case when having multiple clocks in the system. The first problem is resolved by *"arbitration"* and the second by *"synchronization."* The two problems and their solution are discussed next.

1.19.1 Arbitration

For proper functioning in case of multi-master, each device needs to be able to cooperate with the fact that another device is currently talking and the bus is therefore busy. This can be translated into (Figure 1.37):

- *Being able to follow arbitration logic*. If two devices start to communicate at the same time, the one writing more zeros to the bus (or the slower device) wins the arbitration and the other device immediately discontinues any operation on the bus.
- *Bus busy detection*. Each device must detect an ongoing bus communication and must not interrupt it. This is achieved by recognizing traffic and waiting for a stop condition to appear before starting to talk on the bus.

The physical I2C bus setup is designed not only to help the devices to monitor what is going on the bus but more importantly, it is designed to prevent any risk of data corruption that may arise from data collision. The bus monitoring and collision avoidance are discussed next.

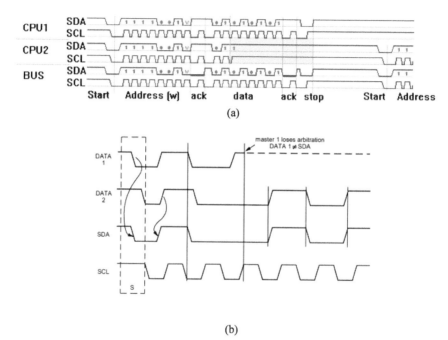

Figure 1.37   Bus arbitration.

There is an undefined condition if the arbitration procedure is still in progress at the moment when one master sends a repeated START or a STOP condition while the other master is still sending data. In other words, the following combinations result in an undefined condition:

- Master 1 sends a repeated START condition and master 2 sends a data bit.
- Master 1 sends a STOP condition and master 2 sends a data bit.
- Master 1 sends a repeated START condition and master 2 sends a STOP condition.

1.19.2 Bus Monitoring

The I2C bus structure is a wired AND. This means that if one device pulls a line low, it stays low and accordingly any device can test if the bus is idle or occupied. When a master (the sender) changes the state of a line to HIGH, it MUST always check that the line really has gone to HIGH. If it stays low, then this is an indication that the bus is occupied and some other device is pulling the line low.

Therefore, the general rule of thumb in I2C bus is: If a master (a sender) cannot get a certain line to go high, it loses arbitration and needs to back off and wait until a stop condition is seen before making another attempt to start transmitting.

1.19.3 Possibility of Collision

Since the previous rule says that a master loses arbitration when it cannot get either SCL or SDA to go high when needed, the problem of data corruption (or data collision) does not exist. It is the device that is sending the "0" that rules the bus. One master cannot disturb the transmission of other master because if it cannot detect one of the lines to go high, it backs off, and if it is the other master that cannot do so, it will behave the same.

This kind of back-off condition will only occur if the two levels transmitted by the two masters are not the same. As an example, let us consider Figure 1.30, where two CPUs start transmitting at the same time.

The two CPUs are accessing a slave in write mode at address 1111001. The slave acknowledges this. So far, both masters are under the impression that they "own" the bus. Now CPU1 wants to transmit 01010101 to the slave, while CPU2 wants to transmit 01100110 to the slave. The moment the data bits do not match anymore (because what the CPU sends is different than what is present on the bus), one of them loses arbitration and backs off. Obviously, this is the CPU which did not get its data on the bus. For as long as there has been no STOP present on the bus, it won't touch the bus and leave the SDA and SCL lines alone (shaded zone). The moment a STOP was detected, CPU2 can attempt to transmit again.

From the example above, we can conclude that it is the master that is pulling the line LOW in an arbitration situation that always wins the arbitration. The master which wanted the line to be HIGH when it is being pulled low by the other master loses the bus. We call this a loss of arbitration or a back-off condition. When a CPU loses arbitration, it has to wait for a STOP condition to appear on the bus. Then, it knows that the previous transmission has been completed.

1.19.4 Clock Synchronization and Handshaking

1.19.4.1 Clock synchronization

All masters generate their own clock on the SCL line to transfer messages on the I2C bus. Data are only valid during the HIGH period of the clock. A

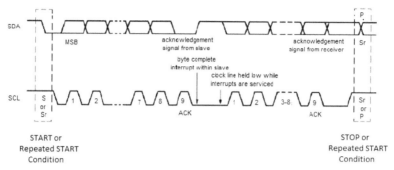

Figure 1.38   Clock synchronization.

defined clock is therefore needed for the bit-by-bit arbitration procedure to take place.

Clock synchronization is performed using the wired-AND connection of I2C interfaces to the SCL line. This means that a HIGH-to-LOW transition on the SCL line will cause the devices concerned to start counting off their LOW period, and once a device clock has gone LOW, it will hold the SCL line in that state until the clock HIGH state is reached. However, the LOW-to-HIGH transition of this clock may not change the state of the SCL line if another clock is still within its LOW period. The SCL line will therefore be held LOW by the device with the longest LOW period. Devices with shorter LOW periods enter a HIGH wait-state during this time, see Figure 1.38.

When all devices concerned have counted off their LOW period, the clock line will be released and go HIGH. There will then be no difference between the device clocks and the state of the SCL line, and all the devices will start counting their HIGH periods. The first device to complete its HIGH period will again pull the SCL line LOW. In this way, a synchronized SCL clock is generated with its LOW period determined by the device with the longest clock LOW period, and its HIGH period determined by the one with the shortest clock HIGH period.

1.19.4.2 Handshaking: Using the clock synchronizing mechanism as a handshake

The I2C protocol also includes a synchronization mechanism, which can be used as a handshake mechanism between slow and fast devices or between masters in a multi-master session.

When a slow slave (slow in terms of internal execution) is attached to the bus, then problems may occur. Let us consider a serial EEPROM. The actual writing process inside the EEPROM might take some time. Now if you send multiple bytes to such a device, the risk exists that you send new data to it before it has completed the write cycle. This would corrupt the data or cause data loss.

The slave must have some means to tell the master that it is busy. It could of course simply not respond to the acknowledge cycle. This would cause the master to send a stop condition and retry. (That is how it is done in hardware in EEPROMs.) Other cases might not be so simple. Think about an A/D converter.

It might take some time for the conversion to complete. If the master would just go on, it would be reading the result of the previous conversion instead of the newly acquired data.

Now the synchronization mechanism can come in handy. This mechanism works on the SCL line only. The slave that wants the master to wait simply pulls the SCL low as long as needed. This is like adding "wait states" to the I2C bus cycle. The master is then not able to produce the acknowledge clock pulse because it cannot get the SCL line to go high. Of course, the master software must check this condition and act appropriately. In this case, the master simply waits until it can get the SCL line to go HIGH and then just goes on with whatever it was doing.

There are a number of minor drawbacks involved when implementing this.

If the SCL gets stuck due to an electrical failure of a circuit, the master can go into deadlock. Of course, this can be handled by timeout counters. Plus, if the bus gets stuck like this, the communication is not working anyway.

Another drawback is speed. The bus is locked at that moment. If you have rather long delays (long conversion time in our example above), then this penalizes the total bus speed a lot. Other masters cannot use the bus at that time either.

This technique does not interfere with the previously introduced arbitration mechanism because the low SCL line will lead to back-off situations in other devices which possibly would want to "claim" the bus. So, there is no real drawback to this technique except the loss of speed / bandwidth and some software overhead in the masters.

It is possible to use this mechanism between masters in a multi-master environment. This can prevent other master from taking over the bus. In a two-master system, this is not useful. But as soon as the system has three or more masters, this is very handy. A third master cannot interrupt a transfer between master 1 and 2 in this way. For some mission-critical situations, this can be a very nice feature.

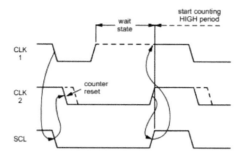

Figure 1.39   Clock synchronization during the arbitration procedure.

It is possible to make this technique more rigid by pulling not only the SCL line low but also the SDA line. Then, any master other than the two masters talking to each other will immediately back off. Before continuing, the master must first let SDA go back high, and then SCL, representing a stop condition.

Any master which attempted to communicate in the meantime would have detected a back-off situation and would be waiting for a STOP to appear

1.20 I2C Interface: Connecting I2C Bus to a PC

The term "I2C interface" usually refers to means for connecting an I2C bus to a PC. However, there are certainly other I2C interfaces which provide connectivity for this universally used bus to non-PC devices.

There are a number of ways to attach the I2C lines to a PC. There is even a zero-cost solution available which takes advantage of the PC parallel port pins. A bit-banging driver is then used to toggle the signals to implement a simple I2C single master device. Provided that the parallel port of the used PC is tolerant enough with respect to various electrical characteristics, such approach may well work for hobbyists. Since it is quite unreliable, a direct (hardwareless) parallel port interface should never be applied in commercial applications.

Philips provided schematics for an adapter called *Single master,* which essentially works like a hardwareless solution but compensates for some parallel port-related issues by using a driver chip between the PC port pins and the I2C SDA and SCL lines. Such interface can only be operated as single master, and the maximum speed highly depends on the drivers and operating system of the PC. Since each bit is toggled individually, no timing can be guaranteed under Windows or Linux with such connection. Another disadvantage is the potential danger of damaging the PC with electronic discharge or faulty connections of the external lines. This is particularly annoying if the parallel port functionality is integrated into a multifunctional non-replaceable chip as is the case with all modern PCs and notebooks.

A much more reliable, powerful and safe alternative are interfaces which provide their own I2C logic such as Connii MM 2.0 or Tracii XL 2.0. These devices are connected via USB; in addition to being able to operate in multi-master mode, they provide a number of important and helpful additional features. Since they communicate with the PC software on a more efficient level than bit banging, the transfer speed is considerably higher and the load on the PC is much lower.

It is important to discuss the considerations that the user has to look for when choosing a professional I2C interface. Some of the considerations are as follows:

- An I2C interface should not only consist of wiring to the parallel port of a PC. This approach can cause damage to your PC or notebook and is not very reliable.
- If the user is intending to use the interface with another I2C master simultaneously, he/she should opt for a multi-master interface in order to avoid arbitration conflicts on the bus.
- The interface that the user is going to choose should be capable of running at the same speed as the other I2C devices in his/her system.
- If the application has slaves which are implemented using a microcontroller or if the user know that a slave device requires clock stretching, he/she need to make sure that the selected interface supports it.
- Choose an interface which has a user interface and programming API which you can easily use and which is supported.
- If possible, an USB interface is preferable over a plug-in card for the user's PC. This reduces the cable length you need extra on the I2C bus.
- Choose a vendor who offers reliable support and long-term market commitment. As operating systems tend to change from time to time, you will need upgrades in order to be able to use such new versions.

Summary

- IIC/I2C is an 8-bit bidirectional synchronous serial communication protocol requiring only two wires for operation.

- The I2C bus consists of two open-drain lines: SDA (data) and SCL (clock).
- Several devices, being either *Master* or *Slave*, can be connected to the bus. The *Master* device must initiate the transfer and drive the clock line (SCL).
- I2C supports the standard speed of 100 kbps, up to a maximum speed of 3.4 Mbps.
- *Master* must generate unique Start and Stop conditions in order to mark the beginning and end of a transaction.
- The receiver must send the ACK bit after every byte that it receives, failing which the *Master* may either Stop the transaction or attempt a repeated Start.
- Every device connected to the I2C bus has either 7-bit or 10-bit address. An additional R/W' bit is added to the address by the *Master* to determine whether it wants to read or write from/to the device.
- Data transfer can be unidirectional (*Master* to *Slave* OR vice versa) or bidirectional.
- *Slave* can hold the clock line low until it is ready with the result to be sent to the *Master*, called Clock Stretching.

2

Design of I2C Bus and Operation

2.1 Design of I2C Bus

The two main parameters that needed to be calculated while designing I2C are the Pull-up resistor and the clock frequency. In some cases, a series protection resistor is needed to limit voltage spikes. In this section, we discuss at first the open-drain lines and then the calculation of the pull-up resistor, clock frequency limits, and the serial protection resistor are given.

2.1.1 Open-drain Lines

2.1.1.1 Open-drain for Bidirectional Communication

As mentioned before, SDA and SCL are *open-drain* (actually open drain/*open collector*) lines pulled up with resistors. What does that mean? It means that the devices connected to the I2C bus are capable of pulling any of these two lines low, but they cannot drive them high. If any of the devices would ever want to drive the lines high, they would simply need to *let go* of that line, and it would be driven high by the pull-up resistors (R_p). In other words, in the event of the bus being released by the master or a slave, the pull-up resistor (RP) on the line is responsible for pulling the bus voltage up to the power rail. Since no device may force a high on a line, this means that the bus will never run into a communication issue where one device may try to transmit a high, and another transmits a low, causing a short (power rail to ground). I2C requires that if a master in a multi-master environment transmits a high, but sees that the line is low (another device is pulling it down), to halt communications because another device is using the bus. Push–pull interfaces do not allow for this type of freedom, which is a benefit of I2C.

Figure 2.1 shows a simplified view of the internal structure of the slave or master device on the SDA/SCL lines, consisting of a buffer to read input data

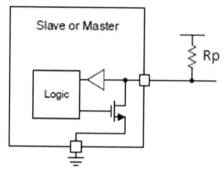

Figure 2.1   Basic internal structure of SDA/SCL lines.

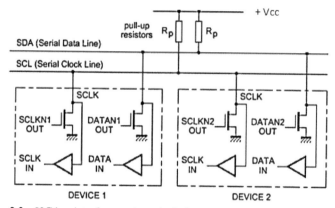

Figure 2.2   I2C bus interface–a closer look (Image source infoindustrielle.free.fr).

and a pull-down FET to transmit data. A device is only able to pull the bus line low (provide short to ground) or release the bus line (high impedance to ground) and allow the pull-up resistor to raise the voltage. This is an important concept to realize when dealing with I2C devices, since no device may hold the bus high. This property is what allows bidirectional communication to take place.

Note: Actually, using open drain/open collector with an input buffer on the same line allows a single data line to be used for bidirectional data flow.

Figure 2.2 gives a closer look of the internal structure. Figure 2.2 shows the NMOS transistors inside the devices. In order for the device to pull any of the two lines low, it needs to provide a high voltage to the gate of the transistor (that's how an NMOS transistor operates). If the gate voltage is

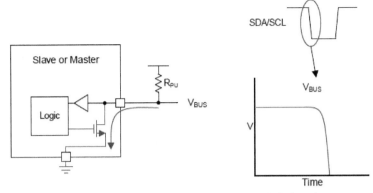

Figure 2.3 Pulling the bus low with an open-drain interface.

low, the NMOS transistor is not activated and the corresponding line is driven high.

a. Open-drain pulling low

As described in the previous section, the Open-drain setup may only pull a bus low, or "release" it and let a resistor pull it high. Figure 2.3 shows the flow of current to pull the bus low. The logic wanting to transmit a low will activate the pull-down FET, which will provide a short to ground, pulling the line low.

b. Open-drain releasing bus

When the slave or master wishes to transmit a logic high, it may only release the bus by turning off the pull-down FET. This leaves the bus floating, and the pull-up resistor will pull the voltage up to the voltage rail, which will be interpreted as high. Figure 2.4 shows the flow of current through the pull-up resistor, which pulls the bus high.

2.1.2 Calculation of the Pull-up Resistor

As mentioned before, at the physical layer, both SCL and SCA lines are in open drain, hence the pull-up resistors. Pull-up resistors need to be connected from the I2C lines to the supply voltage to enable communication.

   With regard to the capacitive load on the bus, there is a max resistance value associated with the bus capacitance, while the minimum value is determined from the maximum allowed sink current (per I2C slave devices on the bus), offset voltage, and supply voltage. Once the thresholds are determined,

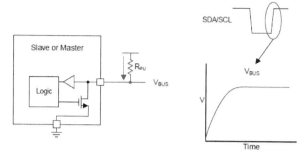

Figure 2.4 Releasing the bus with an open-drain interface.

the system designer must determine if a stronger pull up is required to account for the capacitance.

With a stronger pull up (lower resistance value), more current will be used to make rise times faster for SCL and SDA and ensure that I2C rise time specifications are met. At the same time, strong pull-up (small resistor) prevents the I2C pin on an IC from being able to drive low.

The three considerations to be taken while determining the pull-up resistor values (Rp) are discussed here:

- Supply voltage (Vcc)
- Total bus capacitance (C_b or C_{bus})
- Total high-level input current (I_{IH})

2.1.2.1 Supply voltage (Vcc)

The I2C specification defines a voltage below V_{IL}, or 30% of the supply voltage, as a logical low and, likewise, above V_{IH}, or 70% of the supply voltage, as a logical high, as shown in Figure 2.5. A voltage between these two levels leads to an undefined logic level. In reality, the pin will read either logical high or low in this range, but it may vary between devices, with temperatures, voltages, noise sources, and other environmental factors influencing the logic levels.

The supply voltage limits the minimum Rp value for which the bus can be pulled low. A strong pull up will prevent a device from being able to bring the line sufficiently low, to ensure a logical low is detected. This is caused by the potential divider formed between the pull-up resistor and the on-resistance of the transistor to ground, as shown in Figure 2.6.

The on resistance of the transistor is not typically specified. Instead, a maximum sink current (I_{OL}) is given for which the voltage drop across the transistor is below the output logical low-voltage level (V_{OL}).

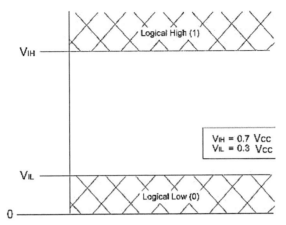

Figure 2.5 Specified voltage levels for logical high and low.

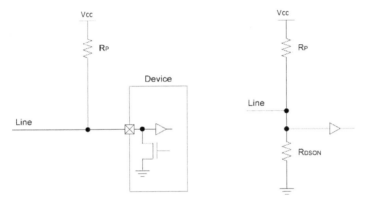

Figure 2.6 Open-collector topology and equivalent circuit. (R_{DSON} is the drain to source resistance when the transistor is ON.)

The V_{OL} level that can be read as a valid logical low by the input buffers of an IC determines the minimum pull-up resistance [R_p(min)]. R_p(min) is a function of V_{cc}, V_{OL}(max), and I_{OL}. Applying Ohm's law gives equation (2.1).

$$R_P(\min.) = \frac{(V_{cc} - V_{OL}(\max.))}{I_{OL}} \qquad (2.1)$$

2.1.2.2 Total bus capacitance (CBUS or Cb) and Rp(max)

On the SCL and SDA lines, the capacitance includes all pins, connections, PCB traces and wire. Combined, this is referred to as the bus capacitance

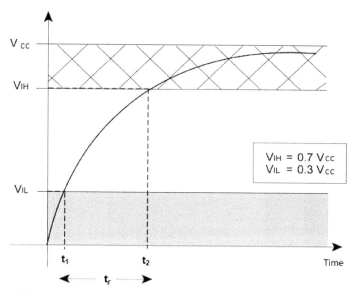

Figure 2.7　Charge time for transition between logical low to high.

and, for long traces and cabling, this can be significant. The open-collector topology requires the external resistor to pull the line high when released. The pull-up resistor, coupled with the bus capacitance, has an RC time constant, which limits the rise time. This becomes significant with increasing clock frequencies, as less time is available for the line to rise. If the selected resistor value is too high, the line may not rise to a logical high before it is next pulled low. This is an important consideration for designs that feature many devices on a single bus, which often have higher bus capacitance.

Bus capacitance can be calculated from PCB trace lengths and published pin capacitance, or measured using capacitance probes or smart tweezers. If a precise calculation or measurement of the bus capacitance is not possible, an overestimated worst-case reading should provide a safe maximum-resistance value.

The maximum pull-up resistance is limited by the bus capacitance (Cb) due to I2C standard rise time specifications. If the pull-up resistor value is too high, the I2C line may not rise to a logical high before it is pulled low. The response of an RC circuit to a voltage step of amplitude VCC, starting at time t = 0, is characterized by time constant RC. The voltage waveform can be written as:

$$V(t) = V_{cc} \, x \big(1 - e^{\frac{-t}{RC}}\big) \qquad (2.2)$$

For $V_{IH} = 0.7 \times V_{cc}$:

$$V_{IL} = 0.7 \; x \; V_{cc} = V_{cc}x(1 - e^{\frac{-t_1}{R_{P} \cdot C_b}}) \tag{2.3}$$

For $V_{IL} = 0.3 \; V_{cc}$:

$$V_{IL} = 0.3 \; x \; V_{cc} = V_{cc}x(1 - e^{\frac{-t_2}{R_{P} \cdot C_b}}) \tag{2.4}$$

The rise time for the I2C bus can be written as:

$$t_r = t_2 - t_1 = 0.8473 \times R_P \times C_b \tag{2.5}$$

The maximum pull-up resistance is a function of the maximum rise time (tr):

$$R_P(\text{max}) = \frac{t_r}{(0.8473 \times C_b)} \tag{2.6}$$

where the parameters t_r, C_b, and V_{OL} as given from I2C specifications are listed in Table 2.1.

From Table 2.1, the maximal rise times for 100 kHz, 400 kHz, and 1 MHz I2C are 1 μs, 300 ns, and 120 ns, respectively.

Equation (2.6) can be re arranged to take the form:

$$\text{Rp(max)} = t_r/0.8473 \times C_b \tag{2.7}$$

The R_p (min) is plotted as a function of V_{cc} in Figure 2.8. The RP (max) is plotted as a function of Cb in Figure 2.9 for standard-mode and fast-mode I2C.

Table 2.1 Parametric from I2C specifications

	Parameter	Standard Mode (Max)	Fast Mode (Max)	Fast-mode Plus (Max)	Unit
t_r	Rise time of both SDA and SCL signals	1000	300	120	ns
C_o	Capacitive load for each bus line	400	400	550	pF
V_{OL}	Low-level output voltage (at 3 mA current sink, $V_{CC} > 2$ V)	0.4	0.4	0.4	V
	Low-level output voltage (at 2 mA current sink, $V_{CC} \leq 2$ V)	–	$0.2 \times V_{cc}$	$0.2 \times V_{cc}$	V

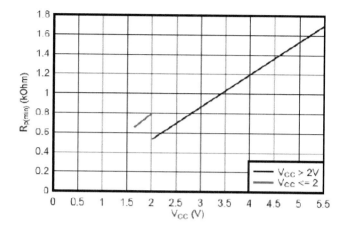

$$V_{OL} = 0.2 \times V_{cc}, \quad I_{OL} = 2 \ mA \ when \ V_{cc} \leq 2V$$
$$V_{OL} = 0.4 \ V, \quad I_{OL} = 3mA \ when \ V_{cc} > 2 \ V$$

Figure 2.8   Minimum pull-up resistance [RP (min)] versus pull-up reference voltage (VCC).

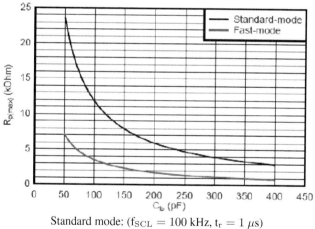

Standard mode: (f_{SCL} = 100 kHz, t_r = 1 μs)
Fast mode: (f_{SCL} = 400 kHz, t_r = 300 ns)

Figure 2.9   Maximum pull-up resistance [RP (max)] versus bus capacitance (Cb).

Example 2.1

For fast-mode I2C communication with the following parameters:

$$C_b = 200 \ pF, \quad C_{cc} = 3.3 \ V$$

Calculate the pull-up resistor value.

Solution:

Taking the values from Table 2.1:

$$R_P(\text{max}) = \frac{t_r}{(0.8473 \times C_b)} = \frac{(300 \times 10^{-9})}{(0.8473 \times 200 \times 10^{-12})} = 1.77 \ k\Omega$$

$$R_p(\text{min}) = \frac{V_{CC} - V_{OL}(\text{max.})}{I_{OL}} = \frac{(3.3 - 0.4)}{(3 \times 10^{-3})} = 966.667 \ \Omega$$

Therefore, we can select any available resistor value between 966.667 Ω and 1.77 kΩ. The value of the pull-up resistor can be selected based on the trade-off for the power consumption and speed.

2.1.2.3 Total high-level input current (I_{IH}): Input leakage

Even when no device is pulling down the line and it is a logical high, current continues to flow through the pull-up resistors. This current is caused by the leakage of the digital inputs of the devices on the bus, from low-quality PCB materials and possibly from soldering residues. Some of these cannot be foreseen, but, assuming quality materials and good manufacturing practices, the input pin leakage is dominant.

From Figure 2.5, the line needs to be above V_{IH} to be regarded as logical high, when there are no devices pulling the bus low. The leakage current limits the maximum value of Rp, such that the voltage drop across it does not prevent the line from being pulled above V_{IH}.

It is also prudent to allow some guard margin on the V_{IH} specification, to prevent noise spikes from bringing the voltage below the V_{IH} level. For robust operation in a high-noise environment, the I2C specification recommends 0.2 V_{cc} as a suitable margin above V_{IH}.

Equation (2.8) gives the additional margin over logical high input level:

$$V_{HMAR} = 0.2 \times V_{cc} \quad V_{HMAR} = 0.2 \times 5 \ V \quad V_{HMAR} = 1 \ V \quad (2.8)$$

The total HIGH-LEVEL input current, I_{IH}, is shown as a function of $R_p(\text{max})$ in Figure 2.10.

The leakage of digital inputs is normally given in the datasheet of devices, and, for Microchip's I2C EEPROM devices, the maximum input leakage current (I_{LI-EE}) is 1 μA. The minimum components for a system are a microcontroller I2C master and an I2C slave device. For this example, assuming a microcontroller with 1 μA input leakage (I_{LI-MCU}) and four I2C EEPROM devices, and allowing 100% margin, I_{IH} is 10 μA.

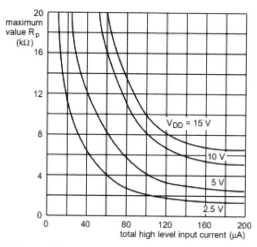

Figure 2.10   Total high-level input current as a function of the maximum value of Rp with supply voltage as a parameter.

Equation (2.9) gives the calculations of the leakage current due to pin leakages for defined bus:

$$I_{IH} = (1 + M\ argin) \times (I_{LIMCU} + (4 \times I_{LIEE}))$$
$$I_{IH} = (1 + 1) \times (1 + (4 \times 1)))\mu A = 10\ \mu A \qquad (2.9)$$

Applying Ohm's law, we can determine the maximum value for Rp that will meet these specifications. Equation 10 gives Maximum pull-up resistance value to ensure logical high:

$$R_p \leq \frac{(V_{dd} - (V_{IH} + V_{HMAR})}{I_{IH}} R_p \leq \frac{5.0 - (3.5 + 1.0)}{10 \times 10^{-6}} R_p \leq 50k\Omega$$
$$(2.10)$$

2.1.2.4 Bus speed versus power consumption

The pull-up resistors must be reduced in size, when increasing the bus speed or when there is significant bus capacitance. The lower-value resistors cause increased current draw, as each logical low on the bus creates a path to ground, negatively impacting power consumption.

The bus speed can become a trade-off between completing tasks quickly and returning a system to a low-power idle state, versus the additional current draw created by the higher bus speed requirements. For applications with very low power budgets, SPI may be a better-suited bus protocol, since it uses driven lines, instead of open collectors.

The next section discusses the maximum allowed clock frequency, maximum speed, of I2C bus.

2.1.3 Maximum Clock Frequency of I2C Bus

2.1.3.1 Using repeaters: Effect of increasing the number of devices–use of repeaters

The I2C standard limits the maximum allowed capacitance on the bus to 400 pF for I2C fast mode (FM) and 550 pF for I2C fast-mode plus (FM+). With the ever-growing system complexity, more and more integrated circuits (ICs) are added to the I2C bus, and complying with the I2C specifications capacitance limit has become a concern. Each IC added results in a capacitance increase up to approximately 15 pF on the I2C bus. I2C repeaters are circuits which provide a solution to the previously described problem by isolating the capacitance between two I2C bus, hence, allowing greater capacitance on an I2C bus for a given timing budget.

The maximum clock frequency (f_{SCL} (max)) is specified to be up to 400 kHz for I2C FM and up to 1000 kHz for FM+ spec. With the increasing number of devices, application requirements also tend to dictate faster operating frequencies to improve overall system response time. Since I2C repeaters typically buffer both the clock (SCL) and the data (SDA) lines, an I2C system utilizing I2C repeaters must properly account for the propagation delays through the repeater when determining the optimal operating frequency.

In this section, we show the calculations that can be used to determine the maximum clock frequency on an I2C bus-based on repeater propagation delays. The trade-offs that need to be considered between system design parameters and timing budget requirements are also discussed

2.1.3.2 I2C bus without a repeater

The timing diagram for an I2C bus with a master generating the SCL clock signal and a slave responding with data on SDA is shown in Figure 2.11 (refer to the I2C spec for a detailed description of the timing parameters).

The maximum clock frequency for this case can be calculated as:

$$f_{SCL}(\max) = \frac{1}{t_{LOW} + t_{HIGH} + t_r + T_F} \qquad (2.11)$$

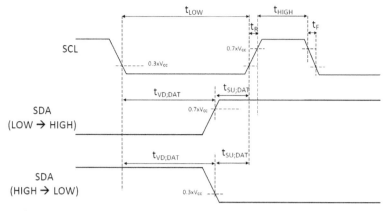

Figure 2.11   Timing diagram for an I2C system with the master generating the SCL clock signal and a slave responding with data on SDA.

Table 2.2 f_{SCL} (max) calculation for I2C systems based on FM or FM+ masters and slaves

Parameter	Based on I2C FM Spec Limits	Based on I2C FM+ Spec Limits
Low period of SCL clock (t_{LOW} (min))	1300 ns	500 ns
HIGH period of SCL clock (t_{HIGH}(max))	600 ns	260 ns
Rise time of both SDA and SCL signals (t_R(max))	300 ns	120 ns
Fall time of both SDA and SCL signals (t_f(max))	300 ns	120 ns
Maximum clock frequency (f_{SCL} (max))	400 kHz	1000 kHz

where:

t_{LOW} = low period of the clock
t_{HIGH} = high period of the clock
t_R = rise time of the clock
t_F = fall time of the clock

The timing parameters in Equation (2.11) and the f_{SCL} (max) calculation result based on these timing parameters for a system based on I2C FM and FM+ spec is shown in Table 2.2. As expected, the f_{SCL} (max) for FM spec is 400 kHz, and for FM+ spec, it is 1000 kHz, as also mentioned in the I2C spec.

From the timing diagram in Figure 2.11, t_{LOW} (min) must be greater than $t_{VD;DAT}$ (max) plus $t_{SU;DAT}$ to satisfy the data valid and setup time requirement:

$$t_{LOW}(\min) > t_{VD;DAT} + t_{SU;DAT} \tag{2.12}$$

Table 2.3 Relationship between t_{LOW} (min), $t_{VD;DAT}$, and $t_{SU;DAT}$

Parameter	Based on FM Spec Limits	Based on I2C FM+ FM+ Spec Limits
Low period of SCL clock (t_{LOW} (min))	1300 ns	500 ns
Data valid time ($t_{VD;DAT}$(max))	900 ns	450 ns
Data valid time ($t_{SU;DAT}$(min))	100 ns	50 ns
Timing margin ($= t_{LOW}$ (min) $- t_{VD;DAT}$ (max) $- t_{SU;DAT}$ (min))	300 ns	0 ns

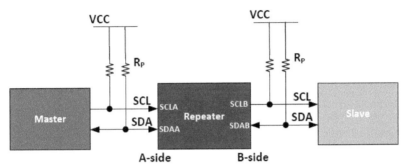

Figure 2.12 Generic I2C bus system showing a bus master, repeater, and a slave.

Based on numbers in Table 2.3, the FM spec has enough timing margin to satisfy Equation (2.12); however, there is no timing margin in FM+ spec to satisfy Equation (2.12). This can cause challenges when a repeater that adds propagation delays is used on the I2C FM+ bus.

2.1.3.3 I2C bus with a repeater

Consider an I2C bus system where a repeater is used to isolate the capacitance between the master and slave side similar to Figure 2.12.

The worst-case timing delay is seen for the case when the slave is sending the data to the master as in this case the data valid time ($t_{VD;DAT}$) and setup time ($t_{SU;DAT}$) have to be met with the repeater propagation delayed SCL (Figure 2.13). The master generates the I2C clock signal (SCL) on the A-side of the repeater.

This SCL is passed through the repeater and appears at the B-side with the repeater propagation delay added ($t_{PHL;AB}$). The data on the B-side (SDA) must be valid within the data valid time ($t_{VD;DAT}$) as per the I2C spec (note that the $t_{VD;DAT}$ is referenced to the delayed SCL on B-side). After the data become valid, they pass through the repeater from B-side to A-side and a propagation delay ($t_{PHL;BA}$) is added. For simplicity, we have assumed

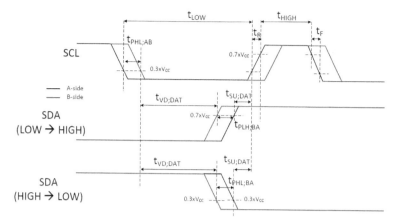

Figure 2.13 Timing diagram for an I2C system with the master generating the SCL clock signal, an I2C repeater providing capacitance buffering (and adding propagation delay), and a slave responding with data on SDA.

that the rise/fall time on the A-side and B-side of the repeater is the same; however, in practice they can differ. The data on the A-side must be available at setup time ($t_{SU;DAT}$) before the rising edge of SCL. For the LOW to HIGH transition on SDA, adding all of the mentioned time corresponds to the LOW period of the SCL as:

$$t_{LOW-LH} = t_{PHL;AB} + t_{VD;DAT} + t_{PLH;BA} + t_{SU;DAT} \qquad (2.13)$$

For the high-to-low transition on SDA, adding the previously mentioned timings corresponds to the LOW period of the SCL as:

$$t_{LOW-LH} = t_{PHL;AB} + t_{VD;DAT} + t_{PHL;BA} + t_{SU;DAT} \qquad (2.14)$$

The maximum clock (SCL) frequency that can be used on an I2C bus with a repeater can be calculated as:

$$f_{SCL}(\max) = \frac{1}{t_{LOW} + t_{HIGH} + t_R + t_F} \qquad (2.15)$$

Assuming the high-to-low transition on SDA results in greater timing delay compared to the low-to-high transition:

$$
\begin{aligned}
f_{SCL}(\max) &= \frac{1}{t_{LOW-HL} + t_{HIGH} + t_R + t_F} \\
&= \frac{1}{(t_{PHL;AB} + t_{VD;DAT} + t_{PHL;BA} + t_{SU;DAT})} \\
&\qquad + t_{HIGH} + t_R + t_F
\end{aligned}
\qquad (2.16)
$$

Table 2.4 Maximum clock frequency (f_{SCL} (max)) calculation based on I2C FM specifications and for FM specifications with TCA9617B repeater characteristics

Parameter	Based on I2C FM Spec with TCA9617B Propagation Delays	Based on I2C FM Spec with TCA9617B Propagation Delays and Rise/Fall Times
Data valid time ($t_{VD;DAT}$(max)), from I2C FM spec	900 ns	900 ns
Data setup time ($t_{SU;DAT}$(min)), from I2C FM spec	100 ns	100 ns
Repeater propagation delay from B to A defined as $0.3 \times V_{cc}$ on A and B side ($t_{PHL;BA}$), from repeater datasheet (TCA9617B)	140 ns	140 ns
Repeater propagation delay from A to B defined as $0.7 \times V_{cc}$ on A and B side ($t_{PHL;AB}$), from repeater datasheet (TCA9617B)	144 ns	144 ns
Fall time of SCL (t_F(max)), from I2C FM spec or repeater datasheet (TCA9617B)	300 ns	13.8 ns
Rise time of SCL (t_R(max)), from I2C FM spec or repeater datasheet (TCA9617B)	300 ns	88 ns
Maximum clock frequency (f_{SCL}(max))	402.6 kHz	503.6 kHz

Table 2.4 shows the values for the timing parameters shown in Equation (2.16) and the calculation result for the f_{SCL} (max) based on I2C FM spec and for I2C FM spec with TCA9617B repeater rise/fall time specs. The calculation shows that it is possible to meet the 400 kHz I2C FM f_{SCL} (max) spec in both cases (f_{SCL} (max) is higher than 400 kHz in both cases).

Table 2.4 shows the values for the timing parameters shown in Equation (2.6) and the calculation result for the f_{SCL} (max) based on I2C FM+ spec and for I2C FM+ spec with TCA9617B repeater rise/fall time specs.

These calculation results show that with the TCA9617B repeater, it may not be possible to meet the 1000 kHz I2C FM+ f_{SCL} (max) spec under all loading conditions (f_{SCL} (max) is lower than 1000 kHz in both cases). For smaller loading conditions than those specified in TCA9617B datasheet, the repeater propagation delays and rise/fall times are smaller; hence, higher f_{SCL} (max) than those mentioned in Table 2.5 can be achieved. However, because

Table 2.5 Maximum clock frequency (f_{SCL} (max)) calculation based on I2C FM+ specifications and for FM+ specifications with TCA9617B repeater characteristics

Parameter	Based on I2C FM+ Spec with TCA9617B Propagation Delays	Based on I2C FM+ Spec with TCA9617B Propagation Delays and Rise/Fall Times
Data valid time ($t_{VD;DAT}$(max)), from I2C FM+ spec	450 ns	450 ns
Data setup time ($t_{SU;DAT}$(min)), from I2C FM+ spec	50 ns	50 ns
High period of SCL (t_{HIGH}(min)), from I2C FM+ spec	260 ns	260 ns
Repeater propagation delay from B to A defined as $0.3 \times V_{cc}$ on A and B side ($t_{PHL;BA}$), from repeater datasheet (TCA9617B)	140 ns	140 ns
Repeater propagation delay from A to B defined as $0.7 \times V_{cc}$ on A and B side ($t_{PHL;AB}$), from repeater datasheet (TCA9617B)	144 ns	144 ns
Fall time of SCL (t_F(max)), from I2C FM+ spec or repeater datasheet (TCA9617B)	120 ns	13.8 ns
Rise time of SCL (t_R(max)), from I2C FM+ spec or repeater datasheet (TCA9617B)	120 ns	88 ns
Maximum clock frequency (f_{SCL}(max))	778.8 kHz	872.8 kHz

of no timing margin on the I2C FM+ spec, it still is not possible to meet the 1000 kHz f_{SCL} (max) spec.

If an experiment is carried out in the lab to measure the maximum frequency on SCL when using a repeater, it may be possible to clock the signals at a faster rate than what the calculations in Table 2.5 show. However, the system designer should take these calculations into consideration as all

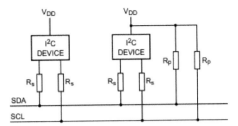

Figure 2.14   Series resistor (R_s) for protection against high-voltage spikes.

the slaves on the I2C bus may not be able to support that high frequency over all possible temperature / voltage corners conditions, hence, leading to yield fallout. The system designer has the flexibility to reduce the loading conditions and achieve the highest clock frequency possible.

2.1.4 Series Protection Resistors

As shown in Figure 2.14, series resistors (Rs) of, for example, 300 Ω can be used for protection against high-voltage spikes on the SDA and SCL lines (resulting from the flash-over of a TV picture tube, for example). If series resistors are used, designers must add the additional resistance into their calculations for Rp and allowable bus capacitance. The required noise margin of 0.1 VDD for the LOW level, limits the maximum value of Rs. Rs(max) as a function of Rp is shown in Figure 2.15. Note that series resistors affect the output fall time.

2.1.4.1 Effect of the serial resistance on static low level of I2C line
(Pull-up Resistance versus Serial Resistance)

Usually, if the I2C bus is used solely on one board, serial resistors Rs are very low or not existing at all. As mentioned before, in some circuits, for example, if the I2C bus is externally accessible by a connector, it is desirable to add serial resistors in the SDA and SCL line in order to protect the I2C devices against overcurrent. Figure 2.16 shows a simplified equivalent circuit diagram for an I2C connection between two devices (master or slave) containing all relevant factors for I2C.

The serial protection resistance Rs, together with the pull-up resistance Rp, affects the static low level of the I2C lines. Consider Figure 2.16 with the serial resistances $Rs_1 = 250$ Ω and $Rs_2 = 0$ Ω, and a pull-up resistance Rp = 1 kΩ. Figure 2.17 shows the transfer as measured at the clamps of I2C

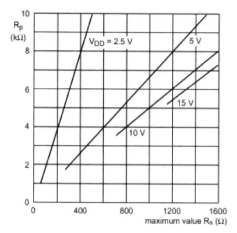

Figure 2.15   Maximum value of Rs as a function of the value of Rp with supply voltage as a parameter. The value of the series resistance can be defined practically based on the value of R_p used.

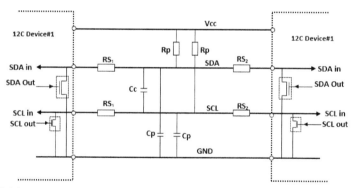

Figure 2.16   Simplified equivalent circuit diagram for an I2C connection between two devices (master or slave) containing all relevant factors for I2C.

device #1 (here: the master), and Figure 2.18 shows the transfer as measured at the clamps of I2C device #2 (here: the slave). Note the different low-level voltages; the low level measured at the slave is about Rs / (Rs + Rp) * Vcc = ~1/5 * Vcc, while the low level measured at the master is nearly GND.

A high low level on SCL and SDA can lead to problems under certain circumstances as some devices may not properly recognize this as a valid low level. If Rs cannot be reduced for some reason, it helps to raise the termination resistance Rp instead. Figure 2.19 shows the transfer at the slave clamps with an increased termination resistance of Rp = 10 kΩ.

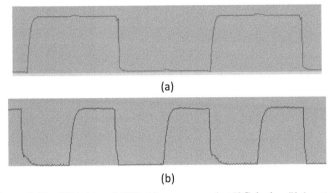

Figure 2.17 SDA (a) and SCL (b) as measured at I2C device #1 (master).

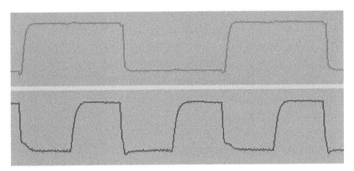

Figure 2.18 SDA (a) and SCL (b) as measured at I2C device #2 (slave).

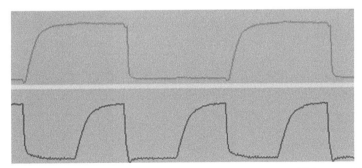

Figure 2.19 SDA and SCL measured at I2C device #2 (slave) with increased Rp.

2.1.4.2 Serial resistance and debugging

Despite their obvious downsides mentioned above, serial resistances can be useful sometimes when analyzing I2C bus traffic to track down I2C-related problems. As stated before, serial resistances Rs lead to different voltage

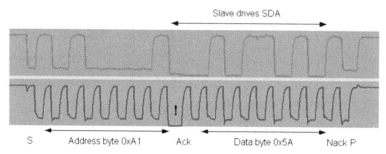

Figure 2.20 A complete master receiver transfer between an I2C master and I2C slave device with Rs = 200 Ω, measured at the slave device.

levels during the low phase, depending on which device currently pulls SDA or SCL low. Thus, it is possible to recognize which I2C device is currently active by analyzing the low level on the I2C lines. Figure 2.20 shows a complete master receiver transfer between an I2C master and a slave device. Due to serial resistors, the low level of the slave device is lower than one of the master device. This applies for both lines, SDA and SCL. Thus, it is possible to easily see that the slave device stretches the SCL line shortly during the acknowledge clock cycle (marked with "!" in Figure 2.20).

2.1.4.3 Termination versus capacitance

Cp and Rp effectively limit the maximum data rate which can be transferred over SDA and SCL. A high Cp can be compensated with a low Rp and vice versa.

Figure 2.21a–c shows the same part of an I2C transfer. Compared to Figure 2.21 part label (a), part label (b) shows the signals with a modified Rp of 2 kΩ and part label (c) with a lower Cp of 150 pF.

Note: Long wires increase Cp dramatically. I2C connections should always be as short as possible and connected by a suitable wiring pattern (c.f. I2C specification, Section 1.7.3).

Note that the I2C standard limits Cp to the maximum value of 400 pF. However, with an appropriate termination resistance, it is often possible (although not recommended) to operate I2C buses with higher capacitance.

2.1.5 Cross Channel Capacitance: Crosstalk between SDA and SCL

A simplified circuit diagram for an I2C connection between two devices (master or slave) is given before (Figure 2.16). Beside the pull-up resistance

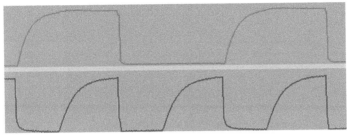

(a) SDA (above) and SCL (below) with Rp = 10 kΩ and Cp = 300 pF. The SCL clock runs with 100 kHz (nominal).

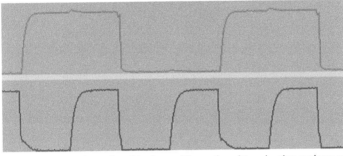

(b) The same transfer as above, but this time with a reduced termination resistance (Rp = 2 kΩ, Cp = 300 pF).

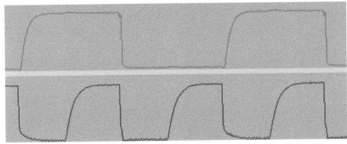

(c) The same transfer as above, but this time with a reduced wire capacitance (Rp = 10 kΩ and Cp = 150 pF).

Figure 2.21 Effect of C_p and R_p (Images from: Tracii XL 2.0).

R_p and the series resistance R_s, it shows a cross channel capacitance C_c. The image shown in Figure 2.22 shows a typical I2C transmission. A closer look at the signals discloses small spikes in the signals. These are consequences arising out of cross channel capacitances C_c. In general, crosstalk should not be a problem as long as the resulting spikes are not too large. C_c itself can be minimized by using as short interconnections as possible. Increasing the

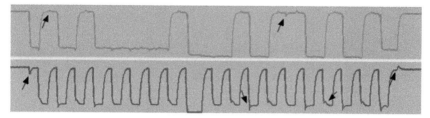

Figure 2.22 Crosstalk between SDA and SCL.

serial resistors Rs and termination Rp can help to reduce the effects of the cross channel capacitances C_c.

2.2 Operating above the Maximum Allowable Bus Capacitance

Bus capacitance limit is specified to limit rise time reductions and allow operating at the rated frequency. Adding more I2C and SMBus devices on the I2C bus may exceed the 400 pF limitation. Once the capacitance is exceeded, the rise times on the I2C bus may violate the timing requirements.

While most designs can easily stay within the limit of 400 pF, some applications may exceed it. There are several strategies available to system designers to cope with excess bus capacitance.

- Reduced f_{SCL} (see Section 2.1.3): The bus may be operated at a lower speed (lower f_{SCL}).
- Higher drive outputs (see Section 2.4.2): Devices with higher drive current such as those rated for fast-mode plus can be used (PCA96xx).
- Bus buffers (see Section 2.4.1): There are a number of bus buffer devices available that can divide the bus into segments so that each segment has a capacitance below the allowable limit, such as the PCA9517 bus buffer or the PCA9546A switch. Multiplexors, Switches, and Repeaters are doing the same function, dividing the bus into segments, as repeaters.
- Switched pull-up circuit (see Section 2.4.3): A switched pull-up circuit can be used to accelerate rising edges by switching a low value pull-up alternately in and out when needed.

2.2.1 Reduced f_{SCL}

Operating above the maximum allowable bus capacitance through reducing the bus speed, f_{SCL}, is possible without affecting the performance of the system. This is due to two factors:

a. In a synchronous serial transmission like an RS-232 communication, the clock speed is predetermined. Since the receiver does not get an explicit clock signal, it has to rely on the accuracy of the sender's timing. Any deviation is almost fatal as it may cause errors sooner or later. This is not the case with I2C.

   In case of I2C bus, which is also a synchronous serial transmission, the situation is much more relaxed. The clock is transmitted by the sender and the receiver is always able to synchronize with that clock.

b. I2C specification defines several speed modes: 100 kbps for standard mode, 400 kbps for full speed, 1 Mbps for fast mode and 3.2 Mbps for high speed. The mentioned speed rates are maximum ratings. Compliant hardware guarantees that it can handle transmission speed up to the maximum clock rate specified by the mode. This does not imply that a transmission may not take place at any lower speed or even at a somewhat variable bit rate.

   In fact, a bus master does not even have full control over the actual timing. The reason for this is simple. The I2C bus uses open drain technology. The bus is kept on a high level and writing to the bus means to pull its level to ground. Depending on bus termination, serial resistors, capacitance, cable length, bus voltage and other factors, this process of pulling down the level and releasing it takes some time. It is therefore good practice for any device to read back the logic level of a modified line (both clock and data) before proceeding with further actions.

   A master will write a zero on the bus by first enabling the circuit to pull the bus low, then read back the state of the line and then proceed with the next step.

   As a consequence, a master set to clock at 100 kHz will most likely produce a lower speed on the bus. As the impact of electrical latency increases with higher clock rates, this effect shows more with higher speeds.

   In conclusion, since the timing can never be determined exactly and the transmitted information is often short, the accuracy of the bus clock is of very little relevance in most applications: it can be adjusted based on the application. Here, we consider the application when it is required to work with bus capacitance above that defined by I2C specification.

For the designer to determine a lower allowable bus operating frequency, he begins by finding the t_{LOW} and t_{HIGH} of the most limiting device on the bus. Refer to individual component data sheets for these values. Actual rise time (t_r) depends on the RC time constant. The most limiting fall time (t_f) depends on the lowest output drive on the bus. Be sure to allow for any devices that have a minimum t_r or t_f (Refer Figure 2.11). Equation (2.17) defines the resulting f_{max}.

$$f_{max} = \frac{1}{t_{LOW(min.)} + t_{HIGH(min.)} + t_{r(actual.)} + t_{f(actual.)}} \qquad (2.17)$$

Remark: Very long buses must also account for time of flight of signals. Actual results are slower, as real parts do not tend to control t_{LOW} and t_{HIGH} to the minimum from 30% to 70% or 70% to 30%, respectively.

2.2.2 Higher Drive Outputs

As mentioned above, t_r depends on the RC time constant, while t_f depends on the output drive power. When using higher output drives as PCA96xx Fast-mode Plus or the P82B bus buffers, the higher strength output drivers sink more current which results in considerably faster edge rates, mainly reducing t_f. Considering equation (2.3), reducing t_f allows increasing t_r for the same f_{max}. This, in turn, allows a higher bus capacitance.

An example of the drivers is P82B715. The P82B715 is a bidirectional logic driver that increases the allowable total I2C system wiring capacitance and requires no external directional control. It uses unidirectional analog current amplification to increase the current sink capability of I2C devices by a factor 10 and the theoretical load capacitance to 4000 pF (see Application Note AN255 for more examples).

We refer the reader to Phillips Application Note AN255 for more examples of such drivers.

Example 2.2: Use of Drivers with Higher drive outputs to Increase the Total Allowable Bus Capacitance

Normally the drivers are used only to reduce the effects of WIRING capacitance without changing the total I2C bus capacitance limit, which is 400 pF. In case if there is no large wiring capacitance component in the system, the drivers can be used to crease the total I2C bus system capacitance. Figure 2.23 gives a general block diagram for such method; using drivers with higher drive output.

In this method, the system capacitance is divided into a number of segments with roughly the same capacitance per segment. The effective pull-up

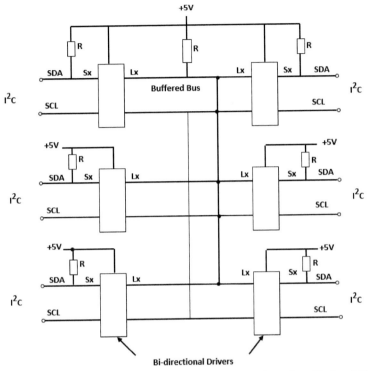

Figure 2.23 Example of using Bidirectional drivers to reduce the effects of WIRING capacitance.

load on any one of the I2C buses will be function: the local pull-up resistance, all other pull-ups and the buffer action.

Total load = Local pull-up resistance in parallel with

[(all other pull-ups connected in parallel) × (buffer action)]

In case of the circuit of Figure 2.23, we have 6 buses. Accordingly, besides the local R, we have 6 resistance connected in parallel. The 6 resistance in parallel have an equivalent resistance of R/6. The total load will be:

R in parallel with [(R/6) × (buffer factor)]

As an example, if we are using the driver P82B715, the buffer action is 10. Substituting in the above equation, the total load resistance = 2.7 kΩ.

For a bus rise time of 1.18 μs, the allowable capacitance with 2.7 kΩ is 430 pF. But the I2C specification limit is 400 pF, so in this example with 7 segments, this system allows a total capacitance of 7 × 400 pF = 2800 pF.

Using P82B715 driver with this arrangement is useful in systems that include TTL logic levels because P82B715 does not affect any of the logic thresholds in the system.

In the configuration of Figure 2.23, the effect of the failure of any VCC supply to any of the devices depends on internal structure of the driver. For example if we are using P82B715, the failure of any VCC will let the system hold the bus low because P82B715 has internal diodes from all I/Os to VCC. Other buffers/extenders (e.g., PCA9511/12/13/14/15/16/18 and P82B96) open circuit their I/Os if their supply is removed.

2.3 Managing the Delays in Large Systems: Use of Buffers, Repeaters, Multiplexers and Switches

The delays introduced at the Master and Slave need to be adjusted to avoid the following possible conditions:

1. At the START, the Master drives SDA LOW while SCL is HIGH, and then, after a required set-up time, it drives SCL LOW. If the Master is Fast-mode Plus compatible, then the set-up time generated could be just 260 ns. In a system with, say, 100 series buffers, it is possible that there will be some small 'skew' between the delays through the SDA and the SCL buffers. A skew of just 3 ns per buffer could mean that the Master SCL falling edge reaches a distant Slave before the SDA falling edge and it will not receive the START condition. If a hardware delay of around 10 ns per series buffer is added to the SCL falling edge at the Master, relative to the Master SDA falling edge during the START, that will ensure, after allowing for 10 ns skew/buffer, a correct START is delivered to even the most remote Slave. (As will be shown, the actual delays need to be slightly different because the hardware adds its own minimum delays.)

2. When the Master addresses a Slave, it is allowed to provide the data bits on SDA with minimal set-up or hold time relative to the falling edges of its SCL. If the Master is changing a data bit, then it can do that at essentially the same time as its SCL falls.

   This is especially the case if the bus is bit-bashed, and anyway for Fast-mode Plus (Fm+), this delay may be as short as 120 ns.

   If the Master data bit is changing from a HIGH to a LOW, then the situation is very similar to a START. Both Master SCL and SDA are driven LOW at nearly the same time, and "skew" might cause their intended relative timing to reverse after passing through many buffers.

To ensure the correct timing of data falling edges, the master hardware will delay all SDA falling edges by around 10 ns/series buffer, relative to the already delayed Master's SCL falling edges.

When the Master's transmitted data bit is changing from a LOW to a HIGH, then it can release the SDA line at essentially the same time as it drives its SCL LOW. The slew on the SCL falling edge means the SCL will be delayed by 70 ns/buffer more than the reach a distant Slave 7 μs before the Master's SCL falling edge. The consequent rising of SDA, while SCL is still HIGH, represents a STOP condition for that Slave.

In this example, the hardware at the Master must therefore delay the Master's SDA rising edges by more than 70 ns/series buffer (in addition to the 10 ns delay of the Master's SCL falling edges) to ensure its SCL falling edges reach every Slave before the rise of its SDA signal. For 100 series buffers, the delay becomes 8 μs, and it becomes clear that this can be a significant factor limiting the possible bus speed.

3. In a large system, a Slave located close to the Master will generate data bits on its bus section with closely the same timing as the Master would generate. Its data edges will be delayed only by the Slave's t_{VD} response delay relative to the received Master's SCL falling edges. For Fm+ parts that is typically only 200 ns. It is therefore necessary to delay the Slave's SDA signals in closely the same way as the Master's SDA above. The Slave SDA falling edges need to be delayed by $>$ 10 ns/series buffer in the system, and its rising edges need to be delayed 100 ns/series buffer.

Delaying the Slave's "output" data bit timing, while allowing undelayed input signals, requires the inclusion of a buffer with the capability to "split" the bidirectional SDA signals into their unidirectional components. Possible buffers are P82B96 and PCA9600 and either may be used at the Master, but because only one half of a buffer is required at a Slave, only P82B96 should be used at Slaves because only its Sx/Sy characteristic will allow its remaining half to be used to generate the delays.

As example for the hardware that can be used to generate delays at the master and at the slave, we introduce here the hardware of Figures 2.24 and 2.25 given in Application note NXP AN11084. Figure 2.24 shows the hardware for generating the delays at the Master while Figure 2.25 shows the hardware for the Slaves. While no delay would actually be needed at the slaves that have the largest number of series buffers, that is, are furthest from the Master, and mid-system Slaves could have smaller delays, the most practical system solution, for service and maintenance, will be to fit the same delay hardware at every Slave.

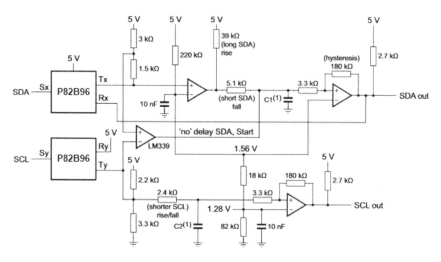

(1)   C1 = C2 = 5.6 pF / buffer (**minimum** = 180 pF)

Figure 2.24   Schematic for SCL/SDA delay generation to be fitted at the master.

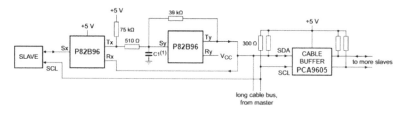

(1)   C1 = 10 pF / cable buffer.

Figure 2.25   Schematic for SDA delay generation fitted at each slave.

2.4 Bus Buffers, Repeaters, Multiplexers, and Switches

Another approach to coping with excess bus capacitance is to divide the bus into smaller segments using bus buffers, multiplexers, or switches. Using repeaters on the I2C is dividing, also, the bus into segments, that is, allowing working with excess capacitance.

2.4.1 Use of Buffers/Repeaters

I2C buffers/repeaters provide capacitive isolation from the I2C bus, so they are used to go beyond the maximum capacitive load allowed in the bus 400 pF (specified for standard and fast I2C mode).

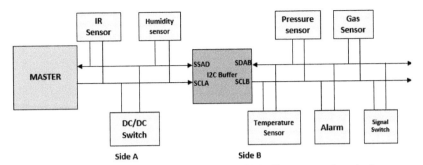

Figure 2.26 I2C bus utilizing I2C buffer, which allows more slave devices.

Buffers/repeaters allow for more slave devices to be added to the I2C bus in applications that have a heavily loaded I2C bus. In applications utilizing the I2C bus, slave devices and trace length are major contributors to the total capacitance of the I2C bus. A quick way to estimate the amount of bus capacitance is to account about 10 pF/slave device. If the total capacitance nears the maximum load of 400 pF, there could be issues with the rise time falling out of the I2C specification.

Figure 2.26 shows an example of a bus that uses a PCA9515 buffer to deal with high bus capacitance. Each segment is then allowed to have the maximum capacitance so the total bus can have twice the maximum capacitance. Keep in mind that adding a buffer always adds delays–a buffer delay plus an additional transition time to each edge, which reduces the maximum operating frequency and may also introduce special V_{IL} and V_{OL} considerations.

2.4.2 Use of Multiplexors and Switches

The I2C multiplexor splits, exactly as buffers are doing, the I2C bus into I2C bus sub-branches or subsections. I2C multiplexors allow the I2C master to communicate with only one of the multiple devices connected. I2C multiplexors are used to solve address conflicts, besides using them to reduce the load capacitance. They isolate devices that are not needed to reduce the overall system loading. When one of the connected devices is selected, the I2C multiplexor acts as a wire. The cumulative capacitive loading of the main I2C bus and the other active I2C sub-branches must be considered; so, care must be taken that each branch does not exceed the 400 pF specified (beyond this, the rising and falling times specifications would be violated).

I2C switches are like I2C multiplexors but more than one device can be selected simultaneously.

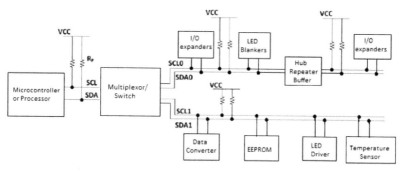

Figure 2.27 Use of multiplexor/switches and repeater.

Figure 2.27 gives a general I2C application in which multiplexer/switches are used to solve the capacitance load.

Not only multiplexors, switches, and buffers are used to solve the problem of bus capacitance but also they have other applications; one of them is to solve the address conflict that may happen when using many devices on the bus. Multiplexors, switches, and buffers are discussed again in Section 2.5.

Refer to application notes AN255, I2C / SMBus Repeaters, Hubs and Expanders; and AN262, PCA954x Family of I2C / SMBus Multiplexers and Switches for more details on this subject and the devices available from NXP Semiconductors.

2.4.3 Switched Pull-up Circuit

The supply voltage (V_{cc}) and the maximum output LOW level determine the minimum value of pull-up resistor R_p. For example, with a supply voltage of $V_{cc} = 5$ V $\pm$ 10% and $V_{OL}(max) = 0.4$ V at 3 mA, $R_p(min) = (5.5–0.4) / 0.003 = 1.7$ kΩ. As shown in Figure 2.28, this value of R_p limits the maximum bus capacitance to about 200 pF to meet the maximum t_r requirement of 300 ns. If the bus has a higher capacitance than this, a switched pull-up circuit (as shown in Figure 2.28) can be used.

The switched pull-up circuit in Figure 2.28 is for a supply voltage of V_{cc} =5 V $\pm$ 10% and a maximum capacitive load of 400 pF. Since it is controlled by the bus levels, it needs no additional switching control signals. During the rising/falling edges, the bilateral switch in the HCT4066 switches pull-up resistor R_{p2} on/off at bus levels between 0.8 V and 2.0 V. Combined resistors R_{p1} and R_{p2} can pull up the bus line within the maximum specified rise time (t_r) of 300 ns. Series resistors R_s are optional. They protect the I/O stages of the I2C bus devices from high-voltage spikes on the bus lines and minimize

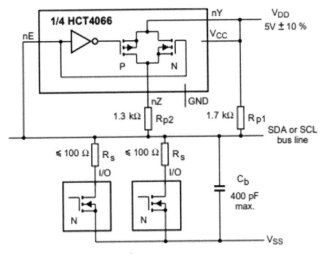

FAST - MODE I^{2}C BUS DEVICES

Figure 2.28 Switched pull-up circuit.

crosstalk and undershoot of the bus line signals. The maximum value of R_s is determined by the maximum permitted voltage drop across this resistor when the bus line is switched to the LOW level in order to switch off R_{p2}. Additionally, some bus buffers contain integral rise time accelerators. Standalone rise time accelerators are also available.

Static Voltage Offset: Certain devices within the I2C buffer/repeater portfolio, such as TCA9517/A and TCA9617/A/B, contain a static voltage offset. The type of buffer design on the B-side prevents these devices from being used in series with devices that use a static voltage offset. When a static voltage offset is present on the B-side of these TCA devices, they will be unable to recognize buffered low signals as a valid low and do not propagate it as a buffered low again. An example of this would include taking two TCA9517A devices and placing them in an A-B B-A configuration. By having both B-sides in series, the design runs the risk of miscommunicating low signals as the device will be unable to recognize this and result in transmitting a high.

Level Shifting: Many of the devices in the I2C buffer/repeater portfolio have the ability to level-translate from master voltages to slave voltages to ensure communication between the voltage mismatch. For example, the TCA9517/A is an example of a level-translating buffer that has two voltage rails which allows for the mismatched master and slave voltages to appropriately communicate to each other through the I2C bus.

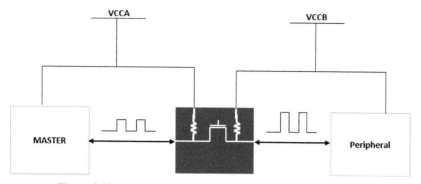

Figure 2.29   Operation of an I2C voltage translator/level shifter.

I2C Voltage Translation: While TI's I2C buffers/repeaters offer level shifting for master and slave voltage mismatches, TI also offers level shifting on the I2C bus if buffering is not needed. PCA9306 and TCA9406 are two devices in TI's portfolio that offer level shifting. The internal structure of both devices is similar to that of a switch which allows the bidirectional open-drain bus to pass through while level-shifting to ensure proper communication through I2C. The operation of the I2C level-shifting is the same as the voltage translation performed by I2C buffers/repeaters (Figure 2.29).

2.5 Use of Multiplexor/Switches to Solve Address Conflict

2.5.1 Use of Multiplexor to Solve Address Conflict

The multiplexor is used in Figure 2.27 to solve the problem capacitance load. As a matter of fact, multiplexors can be used to solve another serious problem that may happen on I2C bus: address conflict. Address conflict occurs when more than one device with the same slave address is connected to the same I2C bus.

Some devices have slave address that has one part fixed/hardware and one part programmable. Examples of such devices are the Texas devices CDCE(L)949, CDCE(L)937, CDCE(L)925, and CDCE(L)913. Those devices belong to the same family of products with the same performance but having different outputs.

CDCE(L)949 and CDCE(L)937 share the same hardwired address. The same situation applies to CDCE(L)925 and CDCE(L)913. For this reason, if the slave address programmable part changes within each of the defined groups, it is possible to reprogram them to have the same slave address. That is, CDCE(L)949 can be programmed to have the same address as that of the CDCE(L)937. But in this situation, when connected to the same I2C bus, a conflict arises.

To avoid such address conflict, some devices can re-program the last few bits (normally the last two) of their default slave address. For example, CDCE(L)949, CDCE(L)937, CDCE(L)925, and CDCE(L)913 can reprogram the last two bits of their default slave address. Therefore, up to four devices of each can be plugged into the I2C bus.

The default setups of each of these devices are set to have different slave address. Therefore, with the default setup, each of them has to be able to share the same I2C bus with no conflict.

Some other devices offer the possibility to fix the slave address with control signals. Thus, the same device can overwrite its slave address with control pins, and it is unnecessary to preprogram the device.

If the slave device has a fixed address or if the preceding options are not preferred, then an I2C multiplexor or I2C buffers can be used.

I2C multiplexors split the I2C bus into several subbranches and allow the I2C master to select and address one of multiple identical devices, thus resolving address conflict. The multiplexor connects the main I2C bus to the selected slave device and removes electrically the nonselected devices. This device is programmable via I2C protocol, and therefore, no additional pins or control logics are required. TI's PCA9544A can multiplex up to four slave devices with the same slave address, as can be seen in Figure 2.30.

2.5.2 Use of Switches to Solve the Slave Address Conflict

I2C Switches: I2C switches are slave devices controlled by the I2C bus that fans out multiple I2C channels to enable more control over your bus.

Applications with slave devices that share the same address can use an I2C switch to prevent slave address conflicts. By having the same slave address, the I2C master will be unable to tell which slave device it is speaking to, and most importantly, it may communicate commands to the wrong slave device. A solution is to incorporate an I2C switch (Figures 2.31 and 2.32). An I2C switch has the capability of isolating slave devices with the same addresses to prevent corruption.

Switches are also used to isolate portions of an I2C bus which have legacy slave devices that cannot handle higher speed bus transactions. Also switches can be used to turn off portions of the bus which are not powered constantly, or are shut down temporarily, to save power.

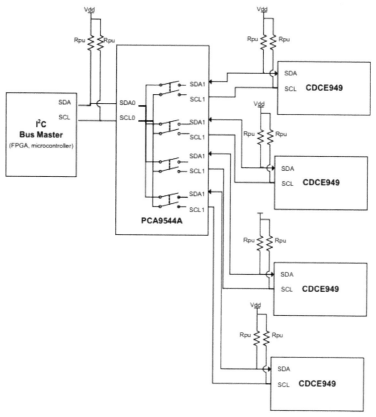

Figure 2.30   Slave address conflict solved with I2C multiplexor.

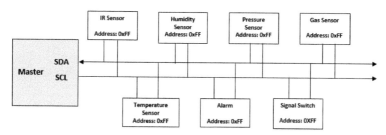

Figure 2.31   I2C bus with multiple slave devices with the same address.

A common misconception about I2C switches is that they are often confused as a multiplexer. The difference between a switch and multiplexer is that a multiplexer utilizes a selection bit to enable a specific channel while a switch can enable one or many channels.

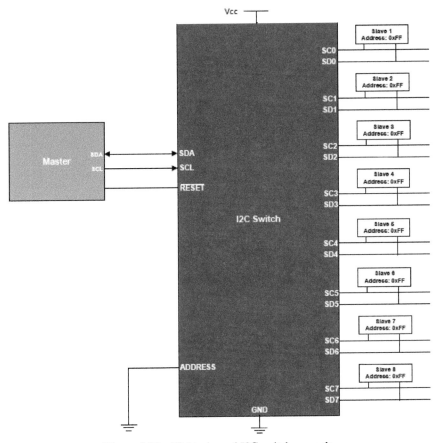

Figure 2.32   Eight-channel I2C switch example.

2.5.3 Control Register

I2C switches determine which channels are connected internally through their control register. Following the successful acknowledgment of the address byte, the bus master sends a command byte that is stored in the control register of the I2C switch. This register can be written and read via the I2C bus. Each bit in the command byte corresponds to an SCn or SDn (SCn/SDn) channel and a high (or 1) selects this channel. Multiple SCn/SDn channels may be selected at the same time. When a channel is selected, the channel becomes active after a stop condition has been placed on the I2C bus. This ensures that all SCn/SDn lines are in a high state when the channel is made active, so that no false conditions are generated at the time of connection. A stop condition always must occur immediately after the acknowledge cycle. If multiple bytes are received by the I2C switch, it saves the last byte received

2.5.4 Level Shifters

As mentioned before under **"Level Shifting,"** I2C buses are not limited to a defined voltage. In some applications, different I2C reference voltages are used for different ICs. Sometimes, it is necessary to have all of them sharing the same bus. To be able to recognize what a logical zero and logical one is, a level-shifter is necessary. Different to most interfaces, the I2C bus does not have a dedicated direction, for example, multiple devices may pull the SCL line to ground.

This causes serious problems building level shifters since they have to take care of moving a request from one side to the other and vice versa without introducing back coupling and oscillation. The basic idea to handle this is to limit the bandwidth on the bus. This gives poor transfer rates of course.

Many of the devices in the I2C buffer/repeater portfolio have the ability to level-translate from master voltages to slave voltages to ensure communication between the voltage mismatch. For example, the TCA9517/A is an example of a level-translating buffer that has two voltage rails which allows for the mismatched master and slave voltages to appropriately communicate to each other through the I2C bus. If the I2C buffer/repeater does not have the ability for level-shifting, a level shifter has to be used.

Commercial level shifters as Levii and Optii are using different approaches providing higher bandwidths. Optii has some additional outstanding feature: it separates the I2C buses galvanically, so two I2C buses can be logically connected but electrically insulated.

2.5.5 Switches and Level Shifting

I2C switches allow level shifting between the master side and slave side to ensure compatibility betweenI2C master and I2C slave as well as between multiple slave devices operating at different voltages.

Mismatched voltages are a concern with voltages of master and slave devices moving towards lower voltages. For example, it is possible for a 1.8-V master I2C bus to communicate with a 3.3-V slave device on another channel. All channels may be at different voltages, with the only requirement being that the VCC pin must be connected to the lowest bus supply voltage that the switch will see.

2.6 Advantages and Limitation of I2C Communication

a. Advantages of I2C Communication

I2C communication or protocol has a significant edge over its peers such as serial port communication and SPI. Let us have a look into the various advantages that renders the I2C protocol so effective for short distance intraboard communication.

1. Flexibility–The I2C protocol supports multi-master, multi-slave communication, which implies the possibility of adding a lot of functionality to any design. More than one master IC controlling and communicating with the slave ICs can speed things up and add functionalities to the embedded system.

ICs can be added or removed from the hardware without affecting any other circuits on bus.

It uses open collector bus design and hence there is flexibility to use voltage on the bus. Moreover slew rates are limited.

2. Addressing feature–The addressing is very simple and does not need any CS (Chip Select) lines to add extra devices like SPI. Adding any extra device is very easy.

3. Simplicity–I2C protocol doesn't complicate the design. It requires only two bidirectional signal lines to establish communication among multiple devices unlike SPI which requires four. Hence, it minimizes interconnections between ICs due to fewer pins and also fewer PCB tracks. This results into smaller and less expensive PCBs. Further, the pin count is low as well; I2C requires least number of pins (just two pins) to perform serial data transfer.

4. Better error handling mechanism–To improve the error detection and correction mechanism, the I2C protocol relies on ACK/NACK feature, which is a robust error correction feature. ACK stands for Acknowledgment, whereas NACK means No Acknowledgment. It is easy to perform diagnosis and debugging. Hence, it is easy to trace malfunctionalities.

5. Adaptable–The I2C protocol is adaptable in the sense that it can work well with both slow ICs and fast ICs.

b. Disadvantages and Limitation

I2C communication doesn't have too many disadvantages. The fact that the protocol has been in use for over 30 years highlights this fact. However, it suffers from a few minor limitations.

• Conflicts–Address Limitation

The assignment of slave addresses is one weakness of I2C. Seven bits is too few to prevent address collisions between the many thousands of available devices. What alleviates the issue of address collisions between different vendors and what allows to connect to several identical devices are that manufacturers dedicate pins that can be used to set the slave address to one of a few address options per device. Two or three pins are typical, and with many devices, there are three or more wiring options per address pin.

10-bit I2C addresses are not yet widely used, and many host operating systems do not support them. Neither is the complex SMBus "ARP" scheme for dynamically assigning addresses (other than for PCI cards with SMBus presence, for which it is required).

• Bus Configuration

Automatic bus configuration is a related issue. A given address may be used by a number of different protocol-incompatible devices in various systems, and hardly any device types can be detected at runtime. For example, 0x51 may be used by a 24LC02 or 24C32 EEPROM, with incompatible addressing, or by a PCF8563 RTC, which cannot reliably be distinguished from either (without changing device state, which might not be allowed). The only reliable configuration mechanisms available to hosts involve out-of-band mechanisms such as tables provided by system firmware, which list the available devices. Again, this issue can partially be addressed by ARP in SMBus systems, especially when vendor and product identifiers are used, but that has not really caught on. The rev. 03 version of the I2C specification adds a device ID mechanism.

• Slower Speeds–Speed Limitation

I2C protocol uses pull-up resistors rather than the push-pull ones used by its peers. Due to the open-drain design, the speed is limited.

Actually, I2C supports a limited range of speeds. Hosts supporting the multi-megabit speeds are rare. Support for the Fm+ 1 Mbps speed is more widespread, since its electronics are simple variants of what is used at lower speeds. Many devices do not support the 400 kbps speed (in part because SMBus does not yet support it). I2C nodes implemented in software (instead of dedicated hardware) may not even support the 100 kbps speed; so the whole range defined in the specification is rarely usable. All devices must at least partially support the highest speed used or they may spuriously detect their device address.

Devices are allowed to stretch clock cycles to suit their particular needs, which can starve bandwidth needed by faster devices and increase latencies when talking to other device addresses. Bus capacitance also places a limit on the transfer speed, especially when current sources are not used to decrease signal rise times. Meaning longer I2C bus traces will limit the bus speed. Additionally, the protocol is managed by software stack. This increases processing overheads on the μP and μC.

- **Potential Fault**

Because I2C is a shared bus, there is the potential for any device to have a fault and hang the entire bus. For example, if any device holds the SDA or SCL line low, it prevents the master from sending START or STOP commands to reset the bus. Thus, it is common for designs to include a reset signal that provides an external method of resetting the bus devices. However many devices do not have a dedicated reset pin, forcing the designer to put in circuitry to allow devices to be power cycled if they need to be reset.

Because of these limits (address management, bus configuration, potential faults, speed), few I2C bus segments have even a dozen devices. It is common for systems to have several such segments. One might be dedicated to use with high-speed devices, for low-latency power management. Another might be used to control a few devices where latency and throughput are not important issues; yet, another segment might be used only to read EEPROM chips describing add-on cards (such as the SPD standard used with DRAM sticks).

- **Debugging**

Since the bus is shared by many devices, debugging an I2C bus (detecting which device is misbehaving) for issues is pretty difficult.

- **Requires More Space**

Now, as an embedded system engineer, you know how valuable PCB real estate is. So, it isn't such a positive attribute that the I2C protocol requires so much space for its pull-up resistors. Added to that, the hardware complexity increases when number of master/slave devices are more in the circuit which may need more space.

One possible additional disadvantage is the fact that I2C is half duplex communication: At a time, data are transmitted only in one direction.

Despite these minor limitations, I2C is a robust and desirable protocol. It achieves exactly what it was meant to facilitate low-speed communication. So, if you are designing an entry-level embedded system, I2C communication may be the right choice for you.

2.7 Common Problems in I2C Bus Systems

Often, problems in I2C bus systems have a simple cause and can easily be verified by checking with simple equipment.

- *Is there a stable supply voltage Vcc?* Often, external I2C devices (like I2C masters or monitors) must be provided with Vcc.
- *Are appropriate termination resistances attached between SDA, SCL and Vcc?* The voltage level on SDA and SCL must be Vcc as long as the bus is idle and drop near GND if shorted to GND. [Note: Very few I2C masters exist which drive SCL high and low, that is, the SCL line is not open-drain. In this case, a termination resistor is not needed and SCL cannot be pulled low. These masters will not work together with other masters (as they have no multi-master support) and may not be used with devices which stretch SCL during transfers.]
- *Are SDA and SCL mixed up?* This may accidentally happen, for example, when connecting I2C buses with cables or connectors.
- *Do all I2C devices support the I2C supply voltage used on the bus?*
- *Do all I2C devices support the maximum SCL clock rate used on the bus?*
- If more than one I2C master is connected to the bus: *do all masters provide multi-master support?*

Further diagnostics is possible with Vcc, SCL and SDA connected to an oscilloscope and/or an appropriate I2C monitor device:

- *Are the high- and low-level voltages on SDA and SCL correct during I2C transfers?* The I2C standard defines the low-level threshold with 0.3 Vcc and the high-level threshold with 0.7 Vcc. Modifying the termination resistance Rp and the serial resistors Rs or lowering the SCL clock rate could help here.
- *Are there spikes or noise on SDA, SCL, or even Vcc?* They may result from interferences from other components or because the capacitances Cp and/or Cc are too high. The effects can often be reduced by using shorter interconnections.

2.7.1 Obscure Problems in Systems

Additional problems may arise if I2C devices which do not fully conform to the I2C specification are used on the bus. They are often hard to find without appropriate equipment like sophisticated I2C monitors or logic analyzers, especially if they are not easily reproducible. Potential sources for non-obvious problems are:

- Some I2C masters (especially if implemented in software) have no multi-master support and/or cannot be used with certain devices which stretch SCL during transfers. Note that fully functioning multi-master support in software-implemented I2C masters is not trivial.
- Very simple I2C master implementations may not even recognize error conditions like negative acknowledges from slave devices.
- Some I2C devices do not use the correct voltage thresholds to recognize high and low levels on SDA and SCL, leaving them especially vulnerable to wrong combinations of serial and termination resistors.
- Some I2C devices do not have Schmitt trigger inputs and/or spike filters, which makes them especially susceptible to crosstalk and noise.
- Some I2C devices cancel an I2C transfer after a certain period of inactivity on the I2C bus and consider the bus idle afterwards. The I2C specification does not specify any timeout conditions–an I2C device can occupy the bus for an arbitrary time period.

2.7.1.1 Analyzing obscure problems

As already mentioned, the causes for some problems are hard to track down. In general, these problems only occur occasionally and often are difficult to reproduce and analyze without adequate tools like I2C monitors.

Tracii XL 2.0 is an I2C monitor which can detect I2C error conditions that are very hard or even nearly impossible to find with usual oscilloscopes; it helps to analyze them by triggering oscilloscopes or recording digital and even analog traces of SDA and SCL.

2.7.2 Typical Problems on the I2C Bus

In this section, some examples of typical problems on the I2C bus are discussed. The problems are:

- Blocked I2C Bus
- No Acknowledge From I2C Slave
- Master Reports Arbitration Lost
- Data Bytes From Slave Are 0xff

2.7.2.1 Blocked I2C Bus

Blocked I2C bus is the case when there is no communication after reset.

There are many reasons for a microcontroller to go through a reset condition. Watchdog triggering, user events and intentional reconfiguration are only some examples.

For systems which employ an I2C bus, such reset of a microcontroller may have rather unexpected side effects if other components continue to operate during the reset phase.

I2C slaves work as a state machine. After they have been addressed, they keep receiving bytes until they see another start condition or a stop condition sent by the I2C bus master.

But what if the I2C master goes through a reset right in the middle of transmitting or receiving a byte from the I2C bus slave?

The slave is not aware of the reset and since I2C doesn't really define timeouts, it may well wait for the next clock event to send or receive a bit, not listening to any start condition which is likely to occur after a reset as startup sequence.

In other words, the bus is stuck, and your I2C bus is blocked.

One rather clumsy but easy to implement solution is to toggle the clock line multiple (16) times before doing any I2C operation after power-up of the microcontroller, that is, after it has possibly gone through reset. This sequence can be followed by a stop condition.

In many cases, this will advance the state machine of any blocked slave to a point where it accepts the next start condition again.

A safer but more complex solution is to power cycle all I2C devices after a reset condition.

The key point to keep in mind is that an I2C slave is not stateless and therefore may not function properly out of the box after reset or–equally common–a spike caused by EMC influence interpreted as clock signal.

2.7.2.2 No acknowledge from I2C slave

During a master transmitter transfer, the slave suddenly does not acknowledge the bytes from the master anymore. This may happen directly after the address byte or later during the data transfer stage. Possible reasons are as follows:

- The I2C slave could not correctly interpret the data on SDA because the SDA high- or low-level voltages do not reach its appropriate input thresholds.
- The I2C slave missed an SCL cycle because the SCL high- or low-level voltages do not reach its appropriate input thresholds.
- The I2C slave accidently interpreted a spike as an SCL cycle.

With adequate serial resistors between master and slave, an analog shot of the signals at the slave's SDA and SCL pins provides a clue whether the slave acknowledges and to which SCL clock pulse. The different SDA low levels due to the serial resistor make it possible to distinguish acknowledges from the slave from data bits from the master.

2.7.2.3 Master reports arbitration lost

Sometimes during an I2C transfer, the master reports "arbitration lost" or something similar and cancels the transfer, although there is no other active master on the bus.

Possible reasons are the same as the ones described in "No Acknowledge from I2C Slave," but here they provoke the slave to pull down SDA when it should not. The master detects this as an arbitration lost event and stops the transfer.

2.7.2.4 Data bytes from slave are 0xff

A slave addressed by a master receiver transfer acknowledges the address byte and starts to send its data correctly, but suddenly all bytes sent by the slave are 0xff. Possible reasons are as follows:

- The I2C slave missed an SCL cycle because the SCL high- or low-level voltages do not reach its appropriate input thresholds.
- The I2C slave accidently interpreted a spike as an SCL cycle.

After having received the last byte during a master receiver transfer, the master leaves SDA high during the acknowledge clock cycle–this causes the slave to not send data anymore, regardless whether a stop condition follows or not. Here, the slave is not synchronized anymore with the master and interprets a high SDA as a "not acknowledge" from the master.

2.8 Case Study: I2C in AVR

This section discusses I2C in AVR microcontroller, taking Atmega 32 as reference. In AVR, this I2C module is called two-wire interface (TWI). The TWI module in the AVR is composed of four submodules (Figure 2.33):

- Bit rate generation unit
- Bus interface unit
- Address match unit
- Control unit

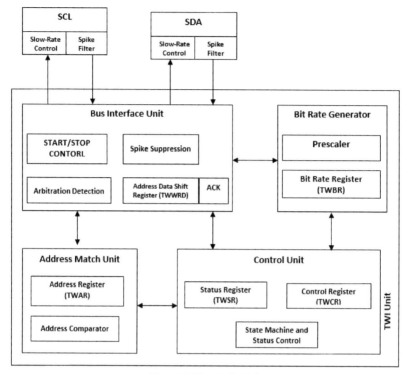

Figure 2.33   AVR I2C.

2.8.1 AVR Atmega32 I2C Registers

In the AVR microcontroller, five major registers are associated with the TWI:

- TWBR (TWI Bit rate Register)
- TWCR (TWI Control Register)
- TWSR (TWI Status Register)
- TWAR (TWI Address Register)
- TWDR (TWI Data Register)

2.8.1.1 TWBR (TWI bit rate register)

This register is used in master mode to set the division factor for bit rate generator (SCL clock frequency). The bit rate generator unit controls the time period of SCL. The SCL clock frequency is decided by the Bit Rate Register (TWBR) and the prescaler bits of TWSR register (explained next).

TWBR

7	6	5	4	3	2	1	0
TWBR7	TWBR6	TWBR5	TWBR4	TWBR3	TWBR2	TWBR1	TWBR0

Bits [7:0]–Bit Rate

It selects the division factor for the bit rate generator. The bit rate generator is basically a frequency divider. It generates the SCL clock frequency in the Master modes.

2.8.1.2 TWSR (TWI status register)

TWSR

7	6	5	4	3	2	1	0
TWS7	TWS6	TWS5	TWS4	TWS3	-	TWPS1	TWPS0

- **Bits [7:3]–TWS: TWI Status**

These five bits reflect the status of the TWI logic and the two-wire serial bus.

Bit 2–Reserved Bit

- This bit is reserved and will always read as zero.

Bits [1:0]–TWPS: TWI Prescaler Bits

These bits are used to set the prescaler of the "Bit Rate generator unit. These bits can be read and written.

Value of prescaler according to TWPS bits		
TWPS1	TWPS0	Prescaler Value
0	0	1
0	1	4
1	0	16
1	1	64

The formula for calculating SCL clock frequency is as follows:

$$SCL\ frequency = \frac{CPU\ clock\ frequency}{16 + 2(TWBR).(4^{TWPS})}$$

Example 2.3

Calculate the SCL frequency if the value of TWPS bits in TWSR is 01 (decimal 1) and the value of TWBR is 0010 0110 (38 decimal). Assume that CPU clock frequency is 8 MHz.

Solution

Using the above formula and the given values, the SCL frequency will be:

$$8 \text{ MHz}/((16 + 2(38) \times 4^1) = 25 \text{ kHz}$$

2.8.1.3 TWCR (TWI control register)

This register is used to control all TWI operations.

TWCR

7	6	5	4	3	2	1	0
TWINT	TWEA	TWSTA	TWSTO	TWWC	TWEN	–	TWIE
R/W	R/W	R/W	R/W	R	R/W	R	R/W
0	0	0	0	0	0	0	0

- **Bit 7–TWINT: TWI Interrupt Flag**

This bit is used to monitor the current status of the TWI. This bit is set by hardware when the TWI has finished its current job and expects application software response. SCL will be low till the time TWINT bit is set. It is not automatically cleared by hardware when executing the interrupt routine. So, you must clear TWINT Flag by writing a logic one to it.

Also note that clearing this flag starts the operation of the TWI. So before clearing this flag, all accesses to the other TWI registers must be complete.

- **Bit 6–TWEA: TWI Enable Acknowledge Bit**

It controls the generation of the acknowledge pulse. If the TWEA bit is written to one, the ACK pulse is generated on the TWI bus if the following conditions are met:

1. The device's own slave address has been received.
2. A general call has been received, while the TWGCE bit in the TWAR is set.
3. A data byte has been received in Master Receiver or Slave Receiver mode.

On the other hand, if TWEA bit is zero, the device can be virtually disconnected from the two wire serial bus temporarily. And to resume address recognition write one to it.

Note: The TWI status values are different when data are transferred on the basis of receive the ACK (TWEA is set) or the NOT ACK (TWEA is clear) pulse.

- **Bit 5–TWSTA: TWI START Condition Bit**

By writing one to this bit, the device becomes master controller on the Two-wire Serial Bus. It checks if the bus is available (bus is not busy in transmission for other masters) and then generates the start condition on the bus. If the bus is not free, the TWI waits until a STOP condition is detected, and then generates a new START condition to claim for being the master of the bus.

- **Bit 4–TWSTO: TWI STOP Condition Bit**

In Master mode, write the one TWSTO bit to generate a STOP condition. This bit is cleared automatically when the STOP condition is executed on the bus.

In slave mode, setting the TWSTO bit can be used to recover from an error condition.

- **Bit 3–TWWC: TWI Write Collision Flag**

This bit is used to avoid the collision in writing to TWDR. The TWWC bit will get set when TWDR is attempting to write (TWINT flag is low). The TWWC flag is cleared by writing TWDR when TWINT is high.

- **Bit 2–TWEN: TWI Enable Bit**

The TWEN bit is used to enable/disable the TWI interface. When this bit is written to one, the TWI operation is enabled. If this bit is written to zero, the TWI is switched off and the TWI transmission gets terminated.

- **Bit 1–Reserved Bit**

This bit is a reserved bit and will always read as zero.

- **Bit 0–TWIE: TWI Interrupt Enable Bit**

When this bit is written to one, and the I-bit in SREG is set, the TWI interrupt request will be activated for as long as the TWINT Flag is high.

2.8.1.4 TWDR (TWI data register)

In Transmit mode, this register contains the next byte to be transmitted in transmission mode, and in case of receive mode, it has last byte received. Note that, it is writable only when the TWI is not in the process of shifting a byte.

TWDR

7	6	5	4	3	2	1	0
TWD7	TWD6	TWD5	TWD4	TWD3	TWDD2	TWD1	TWD0

2.8.1.5 TWAR [TWI (slave) address register]

7	6	5	4	3	2	1	0
TWA6	TWA5	TWA4	TWA3	TWA2	TWA1	TWA0	TWCE
R/W	R/W	R/W	R/W	R/W	R/W	R/W	R/W
1	1	1	1	1	1	1	0

- **Bits 7.1–TWA: TWI (slave) Address Bits**

These bits are used to store 7-bit slave address to which TWI responds when programmed as Slave Transmitter or Receiver.

- **Bit 0–TWGCA: TWI General call Recognition Enable Bit**

This bit is used to enable the slave to acknowledge for the general call address (0x00).

The general call addresses all devices on the bus using the I^2C address 0.

If a device does not need the information provided, it simply does nothing. Devices processing the message acknowledge this address and behave as slave receiver.

The master cannot detect how many devices are using the message. The second byte contains a command.

2.8.2 Programming of the AVR TWI in Master Operating Mode

- We must be able to
    - Initialize the TWI
    - Transmit a START condition
    - Send or receive data
    - Transmit a STOP condition.

2.8.2.1 Initialize the TWI in master operating mode

- Set the TWI module clock frequency by setting the values of the TWBR register and the TWPS bits in the TWSR register
- Enable the TWI module by setting the TWEN bit in the TWCR register to one

2.8.2.2 Transmit START Condition

- Set the TWSTA bit to one to tell the TWI to initiate a START condition when the bus is free

- Set the TWINT bit to one to clear the interrupt flag to initiate operation of the TWI module to transmit the START condition
- Poll the TWINT flag in the TWCR register to see whether the START condition is transmitted completely

2.8.2.3 Send Data
- Copy the data byte to the TWDR
- Set the TWEN and TWINT bits of the TWCR register to one to start sending the byte
- Poll the TWINT flag in the TWCR register to see whether the byte is transmitted completely

2.8.2.4 Receive Data
- Set the TWEN and TWINT bits of the TWCR register to one to start receiving a byte
- If you want to return ACK after receiving data you should also set the TWEA bit of the TWCR register to one
- Poll the TWINT flag in the TWCR register to see whether a byte has been received completely
- Copy the received byte from the TWDR to another register to save it

2.8.2.5 Transmit STOP Condition
- Set the TWEN, TWSTO, and TWINT bits of the TWCR register to one

2.8.3 Use I2C/TWI (Two-wire Interface) in AVR ATmega32

This section explores the TWI interfacing between two ATmega32 controllers. As discussed before, I2C, and accordingly, TWI works in four modes:

1. MASTER as a transmitter.
2. MASTER as a receiver.
3. SLAVE as a receiver.
4. SLAVE as a transmitter.

The objective of this section is to establish the communication between two ATmega32 using TWI/I2C interface. Two cases will be considered:

- Master is transmitting and slave is receiving.
- Master is receiving and slave is transmitting.

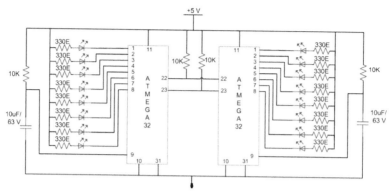

Figure 2.34   Circuit diagram.

First the Master starts by sending data, then the slave transmits complement of the received data to the master. When the Master receives the
complemented data, it shifts the original data to left. This process of transmitting and receiving continues. As the data value reaches 0x80, the whole
process is repeated. At the starting, value of the original data is 0x01. The
received value is displayed on PORTB at both the ends.

2.8.3.1 Circuit diagram
The circuit is shown in Figure 2.34.

2.8.3.2 Code explanation for MASTER controller
Step 1: Initialization of master
Initialization of MASTER means to set the TWI clock frequency (SCL). It is
done by setting bit rate in TWBR and prescaler bits in TWSR.

$$SCL\ freq = \frac{CPU\ clock\ freq.}{16 + 2(TWBR).(4^{TWPS})} \qquad (2.18)$$

Equation Of TWI Clock Frequency to initialize the Master in AVR

```
*********************************************************************

void TWI_init_master(void) // Function to initialize master
{
    TWBR=0x01;      // Bit rate
    TWSR=(0<<TWPS1) | (0<<TWPS0);  // Setting prescaler bits
        // SCL freq= F_CPU/(16+2(TWBR).4^TWPS)
}

*********************************************************************
```

Step 2: Send start condition

The start condition in TWI is explained before. The AVR microcontroller has in-built registers which makes this job much easier.

1. Clear TWINT flag by writing a logic one to it.
2. Set TWSTA bit to send start condition.
3. Set TWEN bit to initialize the TWI.
4. Monitor the status of TWINT flag.
5. Check the ACK byte (using while condition since the SCL frequency is very small as compared to micro controller clock frequency). The ACK byte can be compared by monitoring the status of TWSR.

```
*******************************************************************
void TWI_start(void)
{
    // Clear TWI interrupt flag, Put start condition on SDA,Enable TWI
    TWCR= (1<<TWINT) | (1<<TWSTA) | (1<<TWEN);
    while(!(TWCR & (1<<TWINT)));// Wait till start condition is
    transmitted
    while((TWSR & 0xF8)!= 0x08); // Check for the acknowledgment
}

*******************************************************************
```

Step 3: Send the slave address, data direction bit (write) and wait for the ACK signal

START condition	7 bit slave address	Data direction bit	Slave ACK

Start condition bit in TWI

These three processes are controlled by AVR's TWI registers:

1. Put the seven bit slave address and the direction control bit in TWDR.
2. Clear TWINT flag.
3. Enable TWI by writing logic one to TWEN bit.
4. Monitor the status of TWINT flag, the TWINT flag will get cleared when the data in TWDR is been transmitted.
5. Check for the correct acknowledgment.

```
*******************************************************************
void TWI_read_address(unsigned char data)
{
    TWDR=data;    // Address and read instruction
```

```
TWCR=(1<<TWINT) | (1<<TWEN); // Clear TWI interrupt flag,Enable TWI
while (!(TWCR & (1<<TWINT))); // Wait till complete TWDR byte
received
while((TWSR & 0xF8)!= 0x40);  // Check for the acknowledgment
}
```

* *

Step 4: Send the 8-bit data and wait for the ACK

START condition	7 bit slave address	Data direction bit	Slave ACK	8 bit data transmission	Receiver ACK

Data transfer in TWI of AVR

1. Put the 8 bit data in TWDR.
   8 bits = 7 bit slave address + Data direction bit (write = 0).
2. Clear TWINT flag.
3. Set TWEN bit to enable TWI.
4. Monitor the status of TWINT flag to get data transmission completed.
5. Check for the acknowledgment.

* *

```
void TWI_write_data(unsigned char data)
{
    TWDR=data;    // put data in TWDR
    TWCR=(1<<TWINT) | (1<<TWEN); // Clear TWI interrupt flag,Enable TWI
    while (!(TWCR & (1<<TWINT)));// Wait till complete TWDR byte
    transmitted
    while((TWSR & 0xF8) != 0x28); // Check for the acknowledgment
}
```

* *

Step 5: Send the STOP condition

START condition	7 bit slave address	Data direction bit	Slave ACK	8 bit data transmission	Receiver ACK	STOP condition

TWI STOP condition bit

To send the stop condition use TWSTO:

1. Clear TWINT flag.
2. Set TWEN bit.

3. Write logic one to TWSTO bit so send STOP condition on SDA and SCL line.
4. Monitor the status of TWSTO bit, as TWSO bit get cleared means the stop condition has been transmitted.

****

```
void TWI_stop(void)
{
    // Clear TWI interrupt flag, Put stop condition on SDA,Enable TWI
    TWCR= (1<<TWINT) | (1<<TWEN) | (1<<TWSTO);
    while(!(TWCR & (1<<TWSTO)));  // Wait till stop condition is
    transmitted
}
```

****

Up till here the data transmission from slave side is complete, the MASTER is working in mode one. As per the objective, the data received by MASTER is displayed on PORTB. The flow chart for MASTER as a Transmitter (mode one) is given below.

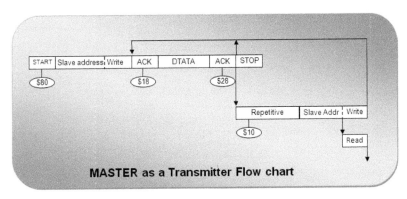

MASTER as a Transmitter Flow chart

Flow chart of MASTER as transmitter in TWI interfacing using AVR

From here, the MASTER would be working in mode two, that is, MASTER becomes a receiver. The AVR TWI works in mode 2.

Step 6: Send the START condition on bus lines
This step is as similar to the previous one.
Note: In Step 6, the START condition is sent after the STOP condition. If one more start condition is sent before the STOP condition in–between, then it is called as repetitive start condition. The repetitive start condition is same as the

START condition but the only difference is between the acknowledgments. For more details about repetitive start refer to the data sheet. If the data are sent continuously in the same direction, then there would be no need of start condition, repetitive start or stop condition in-between. The second data can be transmitted just after receiving the acknowledgment of first data byte (as shown in the above flow chart).

Step 7: Send the slave address and data direction bit (read) and wait for the ACK signal

1. Put the 8 bit data in TWDR.
   8 bits = 7 bit slave address + Data direction bit (read = 1).
2. Clear TWINT flag.
3. Set TWEN bit to enable TWI.
4. Monitor the status of TWINT flag to get data transmission completed.
5. Check acknowledgment.

```
*******************************************************************

void TWI_read_address(unsigned char data)
{
    TWDR=data;     // Address and read instruction
    TWCR=(1<<TWINT) | (1<<TWEN);     // Clear TWI interrupt flag,
    Enable TWI
    while (!(TWCR & (1<<TWINT))); // Wait till complete TWDR byte
    received
    while((TWSR & 0xF8)!= 0x40);   // Check for the acknowledgment
}


*******************************************************************
```

Step 8: Read the data from SDA bus

1. Clear TWINT flag.
2. Set TWEN bit to enable TWI.
3. Monitor the status of TWINT flag, as the TIWNT flag get set indicates that the value in TWDR has been received.
4. Check for the acknowledgment. If the master wants to receive the last byte from slave, the status of TWSR register will be 0x58. After receiving the last byte either a repetitive start condition is issued by the master to continue the communication or a stop condition must be given by the master to stop the communication process. Else if the master wants to keep on receiving more byte from slave, the status of TWSR register will be 0x50.

To acknowledge slave about last byte, TWEA bit is used while transmitting the data. If TWEA bit is set, reception continues from the MASTER side. And if TWEA bit is low, MASTER orders slave to send the last byte.

5. Get the received data and send it on PORTB.

```
*****************************************************************

void TWI_read_data(void)
{
    TWCR=(1<<TWINT) | (1<<TWEN);    // Clear TWI interrupt flag,
    Enable TWI
    while (!(TWCR & (1<<TWINT)));    // Wait till complete TWDR byte
    transmitted
    while((TWSR & 0xF8) != 0x58;     // Check for the acknowledgment
    recv_data=TWDR;
    PORTB=recv_data;
}

*****************************************************************
```

Step 9: Send STOP condition

The stop condition is already explained.

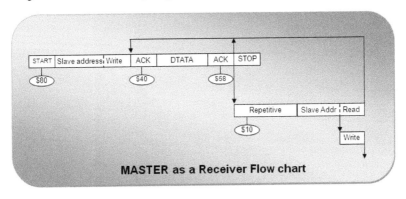

MASTER as a Receiver Flow chart

Flow chart of MASTER as receiver in TWI interfacing using AVR

2.8.3.3 Code explanation for slave controller
Step 1: Initialization of the slave controller

Initialization of the slave controller is done by assigning address to the slave. The seven bit slave address is filled in TWI Slave Address Register (TWAR). The LSB of TWAR, that is, TWGCE bit is used to enable the slave to acknowledge for the general call address (0x00).

```
*************************************************************************
void TWI_init_slave(void) // Function to initialize slave
{
    TWAR=0x20;      // Fill slave address to TWAR
}

*************************************************************************
```

Step 2: Check the status of TWSR register

If the value of TWSR is 0x60, it means the data sent by the master in the next step is meant to read by this particular slave only and the slave sends back the acknowledgment to the master corresponding to the read operation. If the TWSR status is 0x70, the SLAVE is requested to read the data at the general call (0x00). At this stage, the SLAVE acts as receiver. The AVR TWI is working in mode 3.

1. Clear TWIN flag.
2. Enable TWI.
3. Set TEWA to receive acknowledgment.
4. Monitor the status of TWINT flag.
5. Match the status of TWSR. If the status is 0x60 read data or else jump to (1).

```
*************************************************************************
void TWI_match_read_slave(void) //Function to match the slave address
                               and
                        //slave direction bit(read)
{
    while((TWSR & 0xF8)!= 0x60)  // Loop till correct acknowledgment
                                 // have been received
    {
        // Get acknowledgment, Enable TWI, Clear TWI interrupt flag
        TWCR=(1<<TWEA) | (1<<TWEN) | (1<<TWINT);
        while (!(TWCR & (1<<TWINT)));  // Wait for TWINT flag
    }
}

*************************************************************************
```

Step 3: Read data

Read the data sent by the MASTER.

1. Clear TWINT flag.
2. Enable TWI.
3. Set TWEA for receiving ACK.
4. Get the data form TWDR, display it on PORTB.

****

```
void TWI_read_slave(void)
{
    // Clear TWI interrupt flag,Get acknowledgment, Enable TWI
    TWCR= (1<<TWINT)|(1<<TWEA)|(1<<TWEN);
    while (!(TWCR & (1<<TWINT)));      // Wait for TWINT flag
    while((TWSR & 0xF8)!=0x80);            // Wait for acknowledgment
    recv_data=TWDR;                       // Get value from TWDR
    PORTB=recv_data;                  // send the receive value on
    PORTB
}
```

****

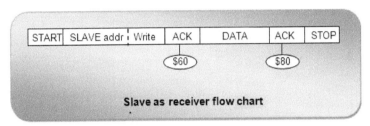

Slave as receiver flow chart

Slave as receiver in TWI interfacing using AVR

From here, the slave becomes a transmitter on the request of master to send data. The AVR TWI works in mode 4.

Step 4: Check the status of TWSR register

If the value of TWSR is 0xA8, it means the master wants to receive data from the particular slave and the slave sends back the acknowledgment to the master corresponding to the write operation.

1. Clear TWIN flag.
2. Enable TWI.
3. Set TEWA to receive acknowledgment.
4. Monitor the status of TWINT flag.
5. Match the status of TWSR. If the status is 0xA8, send data or else jump to (1).

****

```
void TWI_match_write_slave(void) //Function to match the slave address
                          and
                          // slave direction bit(write)
{
    while((TWSR & 0xF8)!= 0xA8)   // Loop till correct acknowledgment
                          // have been received
```

```
    {
        // Get acknowledgment, Enable TWI, Clear TWI interrupt flag
        TWCR=(1<<TWEA) | (1<<TWEN) | (1<<TWINT);
        while (!(TWCR & (1<<TWINT)));  // Wait for TWINT flag
    }
}
```

* *

Step 5: Write the data on SDA bus

1. Put the data in TWDR.
2. Enable TWI.
3. Cleat the TWINT flag.
4. Monitor the TWINT flag. As it get cleared signifies the data has been send.
5. Check for the ACK. Since the TWEA bit was not set during writing data on SDA bus, it signifies master that this is the last data to be sent, and in turn, it receives a NOT ACK, and the status of TWSR becomes 0xC0. And if the TWEA bit was set during the data transmission, it receives an ACK and the status of TWSR is becomes 0xB8. For more details refer to the data sheet.

* *

```
void TWI_write_slave(void) // Function to write data
{
    TWDR= write_data;          // Fill TWDR register with the data to be
    sent
    TWCR= (1<<TWEN) | (1<<TWINT);   // Enable TWI, Clear TWI interrupt
    flag
    while((TWSR & 0xF8) != 0xC0);   // Wait for the acknowledgment
}
```

* *

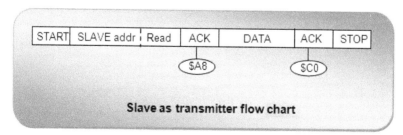

Slave as transmitter flow chart

Slave as transmitter in TWI interfacing using AVR

3

Improved Inter-integrated Circuits (I3C)

3.1 Derivative Technologies

Because of I2C simplicity and flexibility, I2C and its derivatives are widely employed by control applications to manage sensors. I2C is the basis of many platform-based applications such as system management and power management solutions and others. Most of these applications and system architectures have had to develop a protocol layer of their own on top of I2C to define a set of application-specific command to extend the use of I2C protocol to suit the specific needs of the function being performed. Some of such protocols are:

1. **ACCESS.bus**, or **A.b** for short, is a peripheral-interconnect computer bus developed by Philips and DEC in the early 1990s, based on Philips' I2C system. It is similar in purpose to USB, in that it allows low-speed devices to be added or removed from a computer on the fly. While it was made available earlier than USB, it never became popular as USB gained in popularity.
2. **System Management Bus (SMBus)** is an architecture developed based on the principles of I2C and is used as a control bus for system related tasks. SMBus is used to pass commands or messages to and from devices instead of tripping individual control lines.
3. The **Power Management Bus (PMBus)** is developed as a protocol over SMBus and is used as a control bus for power management tasks in a system. The system uses PMBus to communicate between power converters and a system host. It makes intelligent control of the power converters possible.
4. **Intelligent Platform Management Interface (IPMI)** is developed by IPMI based on the principles of I2C to define a standardized message-based interface for platform management hardware. IPMI

provides general system management functions such as automatic alerts, automatic system shutdown, and restart.

5. The VESA **Display Data channel (DDC)** is yet another protocol developed based on I2C and is used for communication between Host and Display to enable display control functions. Communication with graphics adapter in display devices (monitors, television, etc.) is managed by Display Data Channel circuitry.

6. The **Camera Command Interface (CCI)** is a command interface protocol for camera controllers that defines a communication protocol and physical layer interface based on I2C specification.

7. High-availability systems (AdvancedTCA, MicroTCA) use 2-way redundant I2C for shelf management. Multi-master I2C capability is a requirement in these systems.

8. **Improved Inter-integrated Circuit (I3C)**: This is the newest development in I2C. It is recently developed, 2017, by IPMI to fulfill the requirements of sensor networks. It is the subject of the next section.

Notes: The above-mentioned variants have differences in voltage and clock frequency ranges and may have interrupt lines.

TWI (Two-wire Interface) or TWSI (Two-wire Serial Interface): TWI (Two-wire Interface) or TWSI (Two-wire Serial Interface) is not a derivative of I2C; it is essentially the same bus implemented on various system-on-chip processors from Atmel and other vendors. Vendors use the name TWI, even though I2C is not a registered trademark as of 2014-11-07. Trademark protection only exists for the respective logo (see upper right corner), and patents on IšC have now lapsed.

In some cases, use of the term "two-wire interface" indicates incomplete implementation of the I2C specification. Not supporting arbitration or clock stretching is one common limitation, which is still useful for a single master communicating with simple slaves that never stretch the clock.

3.2 Improved Inter-integrated Circuits (I3C)

The I2C derivatives mentioned above are based on the assumptions that the volume of data flowing between devices is relatively low. This is a drawback if we consider the integration of sensors into almost all electronic devices of today and applications built around them, advent of 3D graphics, highly power aware devices, smart mobile devices, wearables, industrial control, automotive systems, smart home and health-care equipment.

Another drawback that has become apparent to designers and users of serial communication systems is the heterogeneous use of protocols such as SPI, UART along with I2C to create practical solutions.

These drawbacks became more serious as we consider the fact that the sensor world itself has been evolving from controlling discrete sensors to managing multiple sensors. The sensor technology is beginning to closely mimic the ultimate sensing machine–the human being. At the heart of this technology is a microcontroller (a "brain") that collates and fuses data collected from multiple sensors to derive a more holistic and meaningful interpretation of the collected data than what one would get by using the data from each discrete sensors in isolation. This is heading the direction of "Sensor Fusion" where in the whole system is much greater than the sum of its individual parts.

The use of the existing I2C-based architectures with such growing world of sensors faces many challenges, and some of them are mentioned next.

3.2.1 Challenges with Existing I2C-based Architectures

Scalability: With the increase in the numbers of sensors that are employed by applications, the number of logic pins used for sensor communication and control has increased in direct proportion. It requires application processors and/or sensor hubs to interact with these multitude of sensors. In a typical application as shown in Figure 3.1, multiple digital communication interfaces

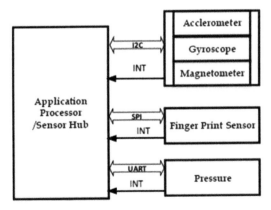

Figure 3.1 Typical application processor or sensor hub system with various digital interfaces.

are used along with supporting logic lines for dedicated interrupt and sleep signals.

- **Integration challenges with heterogeneous interfaces:** There is no consistent method to interface with sensors as the digital interfaces are fragmented with I2C, SPI, and UART interfaces. This also causes considerable integration challenges. Further, the digital interfaces do not have the scalability and flexibility to support advanced sensors features such as batching (buffering sensor samples in sensor hardware and delivering them in a batches instead of delivering continuously), power management, and so on.
- **Application-specific extensions to I2C:** Furthermore, I2C bus is used as the underlying communication protocol for system architectures like SMBus, PMBus, IPMI, and CCI. These architectures have defined their own extensions and modifications to physical layer interface of I2C protocol to build application-specific control management systems. For example:
- Address Resolution Protocol introduced in SMBus protocol which allows the devices to be "hot-plugged" and used immediately, without restarting the system. The devices are recognized automatically and assigned unique addresses. This advantage results in a plug-and-play user interface.
- Group command protocol introduced in PMBus specification is used to send commands to more than one device. This helps the Master to instruct all the targeted devices in one transfer.

The challenges, drawbacks, and limitations of I2C necessitate finding an improved version of I2C that can handle the sensor networks and other systems that need large amount of data to flow between devices at the same time to be compatible with I2C.

The most recent solution has been introduced by MIPI on 2017 and is called "**Improved Inter-integrated Circuits (I3C).**" MIPI I3C is the digital control and data communication interface. The purpose of developing MIPI I3C Interface is to standardize the sensor and control communication, reducing the number of physical pins, and to support low-power, high-speed interfaces which are lacking in the existing standard digital communication interfaces.

MIPI I3C standard retains the attributes of I2C while improving upon its capabilities and performance. It also provides a scalable interface architecture. The I3C Specification anticipates the needs of sensors and control interface solutions that embedded-system industry might need in the future. The provisions in I3C standard allow for heterogeneous solutions to be unified under a single protocol. The standardized Host Controller Interface as defined by the I3C standard also holds a great deal of promise in enabling easier adoption of I3C-based solutions. Because of these merits, I3C seems to emerge as a natural successor to I2C.

3.2.2 MIPI I3C Interface

I3C–an Overview

Improved Inter-integrated Circuit, I3C, is the next generation of the I2C. I3C is under development, hosted by the Sensor Working Group at the MIPI Alliance. The Improved Inter-integrated Circuit (I3C) (pronounced "Eye-three-See") is backward compatible with I2C but will be better suited for handling an abundance of sensors.

I3C was initially intended for mobile applications as a single interface that can be used for all digitally interfaced sensors. However, it is now intended for all mid-speed embedded and deeply embedded applications across sensors, actuators, power regulators, MCUs, FPGAs, etc. The interface is useful for other applications, as it offers high-speed data transfer at very low power levels while allowing multi-drop, which is highly desirable for any embedded system.

The MIPI I3C interface uses an I2C-like interface and supports multiple classes of devices including main master, secondary master, slave and I2C slave. I3C offers major improvements in terms of use, power consumption and performance. The Core uses just two pins and consumes a fraction of energy, reducing cost and complexity while allowing multiple sensors from different vendors to be easily interfaced to a controller or application processor. It gives developers unprecedented opportunity to craft innovative designs for any mobile product, from smartphones, to wearables, to safety systems in automobiles. MIPI I3C can integrate mechanical, motion, biometric and environmental, and any other type of sensor. It incorporates key attributes of the traditional I2C SPI, or UART interfaces to provide a new, unified, high-performing, very low power solution. It will be used to connect many components such as sensors, displays, always-on (low resolution) cameras, controllers, capacitive sensors, mobile applications, transducers, acoustics, and other peripherals. The technology is implemented on a standard CMOS I/O. It uses a two-wire interface, which reduces pin count and signal paths to offer system designers less complexity and more flexibility. It can also be used as a sideband interface to further reduce pin count. MIPI I3C supports a minimum data rate of 10 Mbps with options for higher performance high data rate modes, offering a substantial leap in performance and power efficiency compared with previous options.

Additional technical highlights include multi-master support, dynamic addressing, command-code compatibility, and a uniform approach for advanced power management features, such as sleep mode. It provides synchronous and asynchronous time-stamping to improve the accuracy of applications that use signals from various sensors. It can also batch and transmit data quickly to minimize energy consumption of the host processor.

The I3C interface, as mentioned before, uses an I2C-like interface with data line (SDA) and clock line (SCL). The open-drain SDA line allows for slaves to take control of the data bus and initiate interrupts. The push-pull SCL line is used by the master to clock the communication bus up to 12.5 MHz. The master can dynamically assign 7-bit addresses to all I3C devices while supporting the static addresses of legacy I2C devices. This ensures full compatibility between MIPI I3C and I2C. The Dynamic Address Assignment (DAA) is introduced in the physical layer, and the master can support the dynamic addresses of all MIPI I3C devices while supporting the static addresses of legacy I2C devices. This ensures backward compatibility between MIPI I3C and I2C.

MIPI I3C introduces the capabilities in the masters for request and handover of bus ownership between the masters in the system (a feature that is lacking in the I2C Protocol).

A set of Common Command Codes (CCCs) has been defined for standard operations like enabling and disabling events, managing MIPI I3C-specific features (dynamic addressing, timing control, and so on). These CCCs can be broadcasted (sent to all devices) or directed at a specific device on the bus. MIPI I3C specification also allows for defining user-specific common command codes based on the application needs.

A very useful feature allows MIPI I3C slaves to initiate in-band interrupts (which currently requires a dedicated signal line for both I2C and SPI devices). The in-band interrupt feature enables the slaves to interrupt the master whenever they require to communicate with the master.

The MIPI I3C interface provides several communication protocols including I2C-like single data rate (SDR) messaging mode (up to 12.5 Mbps), and high data rate (HDR) messaging modes (up to 33 Mbps). There are two main HDR modes, HDR-DDR (double data rate) and HDR-TSL/TSP (ternary symbol). These modes offer bit rates over 33 Mbps at a fraction of the per bit power of I2C (400 kHz, fast mode). Both SDR and HDR formats share a two-wire interface with a bidirectional data pin. The SDR format supports a mix of various message types like standard I2C messages, broadcast and CCC messages and slave-initiated requests (e.g., in-band interrupts or requests to assume the master role).

In conclusion, I3C introduces solutions for all the challenges that the current I2C faces and which mentioned in 9.28.1. Actually, the main purpose of MIPI I3C is threefold:

1. To standardize sensor communication;
2. To reduce the number of physical pins used in sensor system integration; and
3. To support low power, high speed, and other critical features that are currently covered by I2C and SPI.

In the following section, we will examine how some of the features currently used by legacy implementations can be made more efficient by transitioning to MIPI I3C.

3.2.3 I3C Bus Configuration and Device Roles

The I3C standard defines five device roles:

- Main master, which controls the I3C bus and function and includes bus ownership control and handoff to secondary masters.
- Secondary master, which takes temporary control of the I3C bus, needs permission from the main master, and passes control back to the main master once control tasks are exercised.
- Slave, which responds to either common or individual commands from the I3C master.
- Peer-to-peer slave, which can write to or read from another slave without the interaction of the master.
- I2C slave, for legacy I2C devices present in an I3C bus, to which I3C master devices can communicate but whose speed and capabilities are constrained.

The MIPI I3C specification defines different responsibilities for each type of device, such as managing SDA arbitration, dynamic address assignment, hot-join features, and HDR master and slave capability. Such, and other, capabilities will be discussed in detail in the next section. Summary of some of them is given here:

- *SDA arbitration*: It resolves bus ownership when multiple devices are transmitting at once. I3C uses the SDA line during the arbitration process and follows the common open-drain approach. The master typically manages the SDA arbitration.
- *Dynamic address assignment*: The I3C main master assigns each device a unique address, either when the bus is initialized or when a new device is connected to a configured I3C bus.
- *Hot-join feature*: Slaves don't have to be activated when the I3C bus is powered up, and could be connected but not activated, or added later on. Activating such slaves is known as hot-join and enables the master to assign a dynamic address to the slave when it asks for one.
- *HDR master and slave capability*: Masters and slaves capable of supporting high data rates of 16.84 Mbps and above are defined as HDR master/slave capability.

3.2.4 I3C Features

3.2.4.1 Dynamic addressing

I2C uses static addressing. One of the problems here is that a vendor might use the same address as another vendor for a product that uses the I2C bus, causing issues during integration. I3C removes this problem by providing addressing for devices on the I3C bus. Dynamic Addressing is a development of the "Dynamic Address Resolution (ARP)" introduced in the SMBus.

- **Dynamic Address Resolution (ARP)**

Dynamic Address Resolution (ARP) is a concept introduced in the SMBus that allows the bus devices to be "hot-plugged" and used immediately, without restarting the system.

A limitation in the I2C is the limited 7-bit address space. With all the different types of devices available, we are running out of address space that can be assigned to device categories. Also, to be able to have several identical instances of the same device on the bus I2C controller, implementations allow the user to set few address bits to obtain unique address for these identical instances. This use model further reduces the available address space and also creates a potential problem of conflicting address wherein two different device types have the same address. The standard method to resolve address conflicts is the use of Multiplexors/Switches to split the bus into several segments and enable one segment at a time under software control. Besides adding extra hardware, multiplexors/switches, this segmentation complicates the application firmware. The lack of an enumeration function makes it more involved to bring up an I2C system comprising of slaves from different vendors. SMBus has defined "Address Resolution Protocol" to address the challenges outlined above such that devices are recognized automatically and assigned unique addresses. This mimics Plug-and-Play capability.

This mechanism requires SMBus Master to generate a minimum of four transfers to assign dynamic address to a slave. The Master will issue "Prepare to ARP" as an initiator to prepare the slave for the address assignment procedure followed by "Reset Device" command to reset dynamic address if already assigned. The Master will then issue "Get UDID" command to receive the Unique identifier value from the device which is used for assigning the dynamic address in the following and final "Assign address" command. The SMBus Master repeats the outlined steps for every slave in the system to make sure all slaves have non-conflicting addresses.

Since ARP is built using I2C messaging primitives, it is not the most efficient way of handling address assignment. For example, the sequence of four consecutive commands that constitute ARP requires 445 ms at the highest speed (1 MHz) possible in SMBus to assign address to one slave. Once assigned, there is no easy mechanism to re-assign the dynamic address without performing address assignment procedure all over again for all slaves.

- **I3C Dynamic Addressing**

In contrast to ARP, I3C defines a set of commands to manage different scenarios under which address assignment can happen. The mechanism for Dynamic Address Assignment outlined by I3C protocol effectively achieves what is achieved by the four-step ARP mechanism of SMBus by a single Dynamic Address Assignment procedure. A single transfer is sufficient for assigning dynamic address to multiple devices. The analysis indicates that this can bring about an improvement of 99% (over the 445 ms of SMBus Protocol) in assigning a dynamic address. The I3C protocol also includes a command to assign dynamic address to devices that have a static address. This procedure consumes fewer cycles than Dynamic Address Assignment procedure. I3C also supports a command to Set New Dynamic Address directly to devices that already have a dynamic address without having to participate in Dynamic Address Assignment procedure all over again.

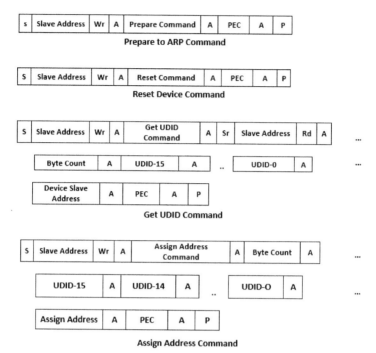

Figure 3.2 SMBus ARP commands.

Figure 3.3 Dynamic address assignment in I3C.

3.2.4.2 Provisional ID

During bus initialization, the I3C Master assigns a 7-bit dynamic address to each device on the bus. For this to happen, each Slave device must have a 48-bit Provisional ID (i.e., provisioned with its ID). The Provisional ID has multiple fields, including MIPI Manufacturer ID and a vendor-defined part number. The I3C Slave may also have a static address, which, if the Master knows it, allows for faster assignment of the dynamic address.

Concerning the 48-bit Provisional ID, it is divided into two parts:

- The first part of the ID contains a unique manufacturer ID. Companies do not have to be MIPI Alliance members to be assigned a unique manufacturer ID.
- The second part of the ID normally contains a part number (which is normally divided up into general and specific part info for that vendor), as

well as possibly an instance number which allows for multiple instances of the same device on the same bus. The instance ID is usually fed from a pin-strap, or fuse(s), or non-volatile memory (NVM).

A random number may be used for the part number, although normally only for test mode, as set by the Master using the ENTTM (Enter Test Mode) CCC. When a device that supports random values enters the test mode, the PID[31:0] bits are randomized. When the Master exits the test mode, the devices reset bits PID[31:0] to their default value. The use of a random number allows for many instances of the same device to be attached to a gang programmer/tester, relying on the random number to uniquely give each a dynamic address.

3.2.4.3 Possibility of collision during dynamic address assignment (DAA)

With most configurations, collision is not possible; each device will have its own manufacturer ID and part number, so no collisions are possible. If more than one instance of the same device is used on the same bus, then each instance must have a separate instance ID; otherwise, there would be a collision. Likewise, if any device is using a random number for its part number, then multiple instances from that manufacturer could collide (i.e., could have the same random value that time).

If the Master knows the number of devices on the bus, then it can detect this condition, since the number of dynamic addresses assigned would be less than the number of devices. Once detected, the I3C Master can take steps to resolve such collisions, for example, by resetting DAA and restarting the process, or by declaring a system error after a set maximum number (e.g., 3) of such attempts fail.

3.2.4.4 Rule of CCC in dynamic address assignment

The CCCs are the commands that an I3C Master uses to communicate to some or all of the Slaves on the I3C bus. The CCCs are sent to the I3C broadcast address (which is 7'h7E) so as not to interfere with normal messages sent to a Slave. The CCCs are used for standard operations like enabling/disabling events, managing I3C-specific features and other bus operations. CCCs can be either broadcasted (sent to all the devices on the bus) or directed at specific devices on the bus. All CCCs (i.e., the command numbers) are allocated by MIPI Alliance, with some reserved for specific purposes that include MIPI Alliance enhancements and vendor extensions.

All I3C devices must be able to process broadcast CCCs at any time, whether or not they have been assigned a dynamic address (DA). For example, an I3C device may act as an I2C device before it gets its DA assigned. However, it is expected to ACK the START with 7′h7E; the only exception would be if this device choosing to remain only as an I2C device, in which case, it would leave any 50 ns spike filter enabled. For those devices that do recognize START and address 7′h7E, those may see any CCC and not just ENTDAA (Enter DA Assignment) or SETDASA (Set Dynamic Address from Static Address). See the I3C specification for the required CCCs. The device can determine the effect of each CCC based on whether or not it has the DA assigned (e.g., the RSTDAA CCC [Reset DA Address] will probably have no effect before the DA has been assigned).

3.2.4.5 Hot-join

Hot-join describes an event where I3C slaves are allowed to join the I3C bus after the bus has been configured, enabling sensors to be shut down until needed.

Hot-join is also a concept used in SMBus. SMBus requires a separate side band ("SMBALERT") from all such hot-join aspiring slaves to notify the host when it attempts to hot-join. This adds to the hardware complexity and is not a very scalable solution. Devices that do not have the sideband signal implemented are expected to join as a Master and then issue "Host-notify" protocol to the Host to announce its intent to join the system as a slave. The Host will in turn assign the dynamic address to the device through ARP. The reduced complexity of not having a sideband implementation is compensated by increased complexity in having to support the Master role of operation by Slaves.

3.2.4.6 In-band Interrupt (IBI)

The 2-wire I3C interface uses a push–pull clock line (SCL) and an open-drain data line (SDA). The SDA allows slaves to take control of the data line and initiate interrupts. (SPI and I2C can only provide slave interrupts by using a third dedicated wire.) I3C slaves can request an interrupt when the bus is idle. I3C arbitrates conflicts from multiple slaves by allowing the lowest assigned address to win. An I3C slave can issue the IBI in the following two ways:

- Following a START (but not a Repeated START)
- If no START is forthcoming within the bus available condition, then an I3C slave can issue a START request by pulling the SDA line low. The

I3C Master would then complete the START condition by pulling the SCL clock line low and taking over the SDA.

When the device joins system dynamically and requires a dynamic address, the device generates a Hot-join request through In-band Interrupt message, and master will perform Dynamic Address Assignment procedure to assign the dynamic address for the newly joined device.

The inbuilt provisions of I3C protocol provides for better and efficient alternatives to ARP and hot-Join of SMBus and obviates the need for sideband implementation and reduces the overhead of Master functionality having to be available in the slave just to be able to hot-join.

3.2.4.6.1 Interrupts

I2C and SPI are typically used to support multiple sensors, but they both have drawbacks for sensor interconnections. Neither of them have a method to notify the master about a change in their state or to initiate a data transfer from the sensor to the master. These notifications are currently being performed by additional logic using general purpose input and output (GPIO) signals. A fallout of this mechanism is that Slaves cannot be easily re-deployed in newer systems without porting the Slave-specific "interrupt handling" capability of the Master over to the new system.

MIPI I3C has the potential to replace both I2C and SPI with a more power efficient two-wire interface. The in-band interrupts mechanism of MIPI I3C interface eliminates or reduces the need for external GPIOs. By being able to interrupt the Master based on sensor functionality, I3C slaves can potentially operate in a more "Sensor Aware" manner and reduce the overhead of polling and servicing interrupts from the Master. This translates to reduced complexity of Master implementation and reduction in power consumption.

The slave initiates an interrupt via an in-band message. The payload of the message may be utilized to specify slave-specific information/action to be performed by the Master or interrupt data. Introduction of in-band interrupts in a multi-drop environment enables sensors to be pro-active in it their interaction with the Master rather than having to depend on Master to poll and service them. Polling mechanism burdens a Master with the responsibility to having to service interrupts from any slave; this makes the implementation inflexible. In-band interrupts are broadcast messages which can be serviced by "any" Master capable of doing so. The slaves can also be agnostic of the Master servicing the interrupt. This allows for a more flexible

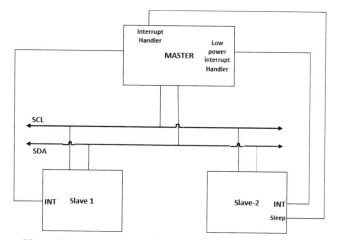

Figure 3.4   I2C system with external sideband GPIO signals.

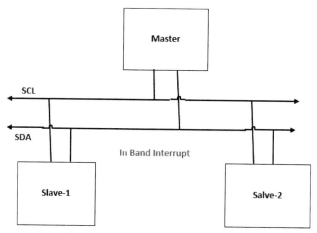

Figure 3.5   I3C system with in-band interrupt support.

implementation that does not call for a tight coupling between the Master and Slave w.r.t interrupt handling that existed in legacy systems. Thus the use of in-band interrupts results in a simpler and scalable implementation.

3.2.5 Multiple Masters

I3C allows for multiple masters on the same bus. I3C has one Main Master that initially configures the bus and act as an initial current Master. Optionally, the bus can have multiple Secondary Master devices that initially act as

Slaves. Any Secondary Master device can request to become the current Master. Once the current Master agrees, and transfers the current Master control to a given Secondary Master device, then that device becomes the current Master.

How Can Masters and Slaves Communicate on the I3C Bus?

The basic byte-based messaging schemes of I2C and SPI map easily onto I3C. Additionally, a set of common command codes (CCCs) has been defined for standard operations like enabling and disabling events, managing I3C-specific features (dynamic addressing, timing control, etc.) and others. CCCs can be either broadcasted (sent to all devices on the bus) or directed at a specific device on the bus.

CCCs do not interfere with, and do not use up any of the message space of, normal Master-to-Slave communications. I3C provides a separate namespace for CCCs.

CCCs (Common Command Codes)

The CCCs are the commands that an I3C Master uses to communicate to some or all of the Slaves on the I3C bus. The CCCs are sent to the I3C broadcast address (which is 7'h7E) so as not to interfere with normal messages sent to a Slave. The CCCs are used for standard operations like enabling/disabling events, managing I3C-specific features and other bus operations. CCCs can be either broadcasted (sent to all the devices on the bus) or directed at specific devices on the bus. All CCCs (i.e., the command numbers) are allocated by MIPI Alliance, with some reserved for specific purposes that include MIPI Alliance enhancements and vendor extensions.

3.2.6 Increased Throughput

The I3C protocol supports higher-speed transfers compared to I2C with the same two-wire interface. The master clock uses the SCL to clock the communication bus to 12.5 MHz. There are two main HDR modes, HDR-DDR (double data rate) and HDR-TSL/TSP (ternary symbol). These HDR modes support speeds of up to 33 Mbps at a fraction of the per bit power of I2C (400 kHz, fast mode). As a matter of fact I3C offers four data transfer modes that effectively provide data bit rates at 11.1, 20, 23.5, and 33.3 Mbps, including I3C bit rate error detection. The increased speeds of operation of I3C allows I3C-based implementations to replace SPI-based legacy systems due to its ability to evacuate accumulated data quickly in batches.

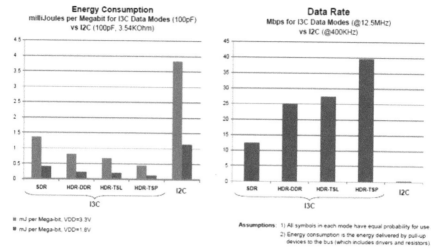

Figure 3.6   Energy consumption and raw data rate: I3C versus I2C (figure courtesy from https://www.synopsys.com/dw/doc.php/wp/mipi_i3c_wp.pdf).

3.2.7 Power Consumption

Increased speed of operation of MIPI I3C interface has resulted in significant reduction of per bit power consumption. As a fact, the power consumption per bit transfer in all I3C modes is more efficient than I2C, due to the use of push–pull (vs. open-drain) and strong pull-up signaling.

Further, I3C can save considerable device power through higher data rates (because the device can be put back to sleep sooner), built-in configuration and control (without intruding on the main communication protocols), in-band interrupt (IBI) as a low-cost wake mechanism, and the ability for Slaves to shut down all internal clocks while still operating correctly on the I3C bus.

The bar charts in Figure 3.6 compare the energy consumption (per bit) of the various MIPI I3C modes with I2C (left) and the corresponding data rates (right). These numbers conclusively demonstrate that I3C is a more power efficient interface as compared to I2C.

3.2.7.1 Pull-up resistor

In general, I3C does not need pull-up resistor. I3C Masters control an active pull-up resistance on SDA, which they can enable and disable. This may be a board-level resistor controlled by a pin, or it may be internal to the Master.

Much of the activity on the I3C bus is in push–pull mode (i.e., with the pull-up resistor disabled) in order to achieve higher data rate. However, for some bus management activities, and for backwards compatibility with I2C, pull-up-resistor-based open-drain mode is enabled, for example, arbitration during dynamic address assignment, and in-band interrupt. Also, the ACK/NACK during the 9th bit is done using pull-up resistor. With few exceptions, it is the responsibility of the I3C Master to provide open-drain class pull-up resistor when the bus is in the open drain mode.

3.2.7.2 High-keeper

A high-keeper is used for Master-to-Slave and Slave-to-Master bus hand-off, as well as optionally when the bus is idle. The high-keeper may be a passive weak pull-up resistor on the bus, or an active weak pull-up or equivalent in the Master. The high-keeper only has to be strong enough to prevent system-leakage from pulling the bus low. At the same time, the high-keeper has to be weak enough that a Slave with a normal IOL driver is able to pull the bus line low within the minimum period.

3.2.8 I3C Bus Activity States

The I3C bus activity states provide a mechanism for the Master to inform the Slaves about the expected upcoming levels of activity or inactivity on the bus, in order to help Slaves better manage their internal states (e.g., to save power). I3C bus has four activity states.

The four activity states (and their expected activity interval) are:

- **Activity State 0:** Normal activity.
- **Activity State 1:** Expect quiet for at least 100 μs.
- **Activity State 2:** Expect quiet for at least 2 ms.
- **Activity State 3:** Expect quiet for at least 50 ms.

3.2.9 I3C Bus Conditions When the Bus Is Considered Inactive

In addition to open drain, pull-up, and high-keeper, the I3C bus has three distinct conditions under which the bus is considered inactive: Bus Free, Bus Available, and Bus Idle.

- **Bus Free** condition is defined as a period occurring after a STOP and before a START and for a given duration (e.g., t_{CAS} and t_{BUF} timing).
- **Bus Available** condition is defined as a Bus Free condition for at least t_{AVAL} duration. A Slave may only issue a START request (e.g., for In-Band Interrupt or Master Handoff) after a bus available condition.
- **Bus Idle** condition is defined to help ensure bus stability during hot-join events. This condition is defined as a period during which the Bus

Available condition is sustained continuously for a duration of at least t_{IDLE}.

3.2.10 Time-stamping Capability Defined in the I3C Bus

The I3C bus supports an optional time control mechanism. One mode is synchronous (from the synchronized timing reference) and four modes are asynchronous (Slave provides timestamp data). All I3C Masters are expected to support at least Async Mode 0.

- **Synchronous:** The Master emits a periodic time sync that allows Slaves to set their sampling time relative to this sync. This may be used in conjunction with one of the Asynchronous modes.
- **Asynchronous:** The Slaves apply their own timestamp to the data at the time they acquire samples, permitting the Master to time-correlate samples received from multiple different Slaves or sensors.

There are four types of asynchronous time controls:

- **Async Mode 0:** Basic mode that assumes that a Slave has access to a reasonably accurate and stable clock source to drive the time stamping– at least accurate for the duration of the time it has to measure (i.e., from event to IBI). A set of counters, in conjunction with IBI, are used to communicate time-stamping information to the Master.
- **Async Mode 1:** Advanced mode extends the basic mode by using some mutually identifiable bus events like I3C START.
- **Async Mode 2:** High precision mode that uses SCL falling edges (for SDR and HDR-DDR modes) as a common timing reference for Master and Slave. A burst oscillator is used to interpolate the time between a detected event and next SCL falling edge. For HDR-TSL and HDR-TSP modes, the mode uses both SDA transitions and SCL transitions as a timing reference.
- **Async Mode 3:** Highest-precision triggerable mode that supports precise time triggering and measurement across multiple transducers applications like beam forming.

3.2.11 Error Detection and Recovery Methods in I3C

The I3C bus has elaborate error detection and recovery methods. Seven Slave error types (S0 to S6) and three Master error types (M0 to M2) are defined for the SDR mode, along with suggested recovery methods. Similarly, a set of errors are defined for each of the HDR modes.

3.2.12 Maximum Capacitance Load Allowed on the I3C Bus

The I3C specification lists the maximum per device capacitance on SCL and SDA, but the goal is that most or all devices will be well below that. Capacitance alone is not sufficient to determine maximum frequency on the I3C bus (as with any bus). It is important to consider maximum propagation length, effect of stubs, internal clock-to-data (t_{SCO}) of the Slaves, as well as capacitive load.

3.2.13 Maximum Wire Length for I3C Communication

The maximum wire length would be a function of speed, as all the reflections and bus turnaround must complete within one cycle. Larger distances can be achieved at the lower speeds than at the higher ones. For example at 1 meter (between Master and Slave), the maximum effective speed is around 6 MHz for read, to allow for clock propagation time to Slave and SDA return time to Master.

It is not, directly, possible to use I2C repeaters with I3C bus to solve the problem of bus length limitation, for the following two reasons: (1) The I3C bus works with push–pull modes (in addition to the open drain for some transfers) and (2) with much higher speeds. Most of such devices are quite limited in speed, because of the lag effect of changing states on SCL and SDA due to both series resistance and assumptions about open drain.

3.2.14 Repeaters and I3C

The I2C repeaters cannot be used directly for I3C for a couple of reasons: (1) The I3C bus works with push–pull modes (in addition to the open drain for some transfers) and (2) with much higher speeds. Most of such devices are quite limited in speed, because of the lag effect of changing states on SCL and SDA due to both series resistance and assumptions about open drain. Long-wire approaches are being evaluated for a future version of the I3C specification.

3.2.15 Resolving Communication Conflicts on the I3C Bus

The I3C slaves are only allowed to drive the bus under certain situations. Besides during a read and when ACKing their own address, they may also drive after a START (but not Repeated START). After a START, the I3C bus reverts back to open-drain pull-up resistor mode, and thus, the slave that drives a low value (logic 0) would win.

3.2.16 Can I3C Devices Cause the Communication Bus to "Hang"?

Unlike I2C, there is no natural way to "hang" the bus. In I2C, clock stretching (where the slave holds the clock low, stopping it from operating) often causes serious problems with no fix: there is simply no way to get the Slave's attention if it has hung the bus. By contrast, in I3C only the Master drives the clock, and so the Slave performs all actions on SDA relative to that clock, thereby eliminating the normal causes of such hangs.

Further, since I3C is designed to ensure that I3C Slaves can operate their back-end I3C peripheral off the SCL clock (vs. oversampling), problems elsewhere in the Slave will not translate into bus hangs.

If a system implementer is highly concerned about a Slave accidently locking itself, then a separate hard-reset line could be used. For the next revision of the I3C specification, a feature called in-band hardware reset (IBHR) is under consideration to reset non-responsive I3C slaves (i.e., if a Master emits a certain pattern that does not occur during regular communication, then the devices on the bus would treat it just like a hardwired reset line).

3.2.17 MIPI I3C Specifications and Software Development

MIPI specifications can help the software designers to develop their application software. The following MIPI specifications are expected to help with software development:

- MIPI specification for I3C Host Controller Interface (I3C HCI), v1.0 **[MIPI02]** (In development) creates a standard definition that allows a single OS driver (aka "in-box driver") to support I3C hardware from several vendors, while also allowing vendor-specific extensions or improvements. The target audience of the HCI specification is application processor host controllers, in particular developers of host controller (i.e., I3C main Master) hardware and developers of I3C host controller software.
- MIPI specification for Discovery and Configuration (DisCo), v1.0 **[MIPI03]** describes a standardized device discovery and configuration mechanism for interfaces based on MIPI specifications, which can simplify component design and system integration, also oriented to application processors.
- MIPI DisCoSM specification for I3CSM, v1.0 **[MIPI04]** (In development) allows operating system software to use ACPI (Advanced Configuration and Power Interface) structures to discover and configure

the I3C host controller and attached devices in ACPI-compliant systems, also oriented to application processors.
* In addition, an *MIPI Application Note for I3CSM [MIPI05]* is being developed to help ASIC hardware developers, system designers, and others working in the more deeply embedded I3C devices.

We note here that core I3C infrastructure is being added to the Linux Kernel via patchwork.kernel.org.

3.2.18 Expected New Features for MIPI I3C

While not finalized yet, the following features are under consideration:
* Grouped addressing
* Additional slave error detection/recovery
* CCC support in HDR-DDR/TSP
* HDR-DDR end write
* HDR-TSP end transfer
* Clock-to-data refinement
* Timing control disable
* New minimum tIDLE
* In-band hardware reset (IBHR)
* Multi-lane, for speed
* HDR-DDR-end CRC

Summary

* I3C simplifies system design to a true two-wire interface
* Backward compatible to existing I2C devices (supports legacy I2C messaging)
* Reduces device die size
* Lowers power consumption
* Supports in-bound error checking and CRC
* Supports peer-to-peer slave communication
* Supports hot-plug capability
* Dynamic addressing while supporting static addressing for legacy I2C devices
* Supports I2C-like SDR messaging and optional HDR messaging (up to 30 Mbps)
* Supports multi-master and multi-drop capabilities

References

[MIPI01]   *MIPI Alliance Specification for I3CSM (Improved Inter Integrated Circuit)*, version 1.0, MIPI Alliance, Inc., MIPI Board adopted 31 December 2016.

[MIPI02]   *MIPI Alliance Specification for I3C Host Controller Interface (I3C HCISM)*, version 1.0, MIPI Alliance, Inc., In press.

[MIPI03]   *MIPI Alliance Specification for Discovery and Configuration (Dis-CoSM)*, version 1.0, MIPI Alliance, Inc., MIPI Board adopted 28 December 2016.

[MIPI04]   *MIPI Alliance DisCoSM Specification for I3CSM*, version 1.0, MIPI Alliance, Inc., In press.

[MIPI05]   *MIPI Alliance Application Note for I3CSM*, version 1.0, MIPI Alliance, Inc., In press.

[MIPI06]   *MIPI Alliance Specification for Camera Serial Interface 2 (CSI-2SM)*, version 2.1, MIPI Alliance, Inc., In press.

[MIPI07]   *MIPI Alliance Specification for Debug for I3CSM*, version 1.0, MIPI Alliance, Inc., In press.

[MIPI08]   *MIPI Alliance Specification for Virtual GPIO Interface (VGISM)*, version 1.0, MIPI Alliance, Inc., In press.

[MIPI09]   *MIPI Alliance Conformance Test Suite (CTS) for I3CSM v1.0*, CTS version 1.0, MIPI Alliance, Inc., In press.

Section II

Controller Area Network (CAN)

This section deals with "Controller Area Network". The section consists of two chapters:

Chapter 4: Controller Area Network (CAN)
Chapter 5: CAN Data Link

The two chapters give enough knowledge for the reader to let him design CAN bus.

4

Controller Area Network (CAN)

There are a number of methods of networking microcontrollers. These include the I2C (or Two-wire Serial Interface (TWI)), the Controller Area Network (CAN) protocol, and the ZigBee wireless interface.

Inter Integrated Circuit (IIC or I2C) protocol considered in Section 1 of this book (Chapters 1 to 3). This section (Chapters 4 and 5) handles the Controller Area Network (CAN) protocol.

4.1 Introduction to CAN: Overview

4.1.1 What is CAN?

The *controller area network* (CAN) bus is a broadcast serial communication standard for real-time applications that allows multiple processors in a system, possibly connected over a twisted pair multidrop cable, to communicate efficiently with each other.

Controller Area Network (CAN bus) is developed by Robert Bosch in 1986 as a multi-master, message broadcast robust vehicle bus standard that allows microcontrollers and devices to communicate with each other in applications without a host computer. It is a *message-based* protocol, designed originally for multiplex electrical wiring within automobiles to save on copper, but is also used in many other contexts. Unlike a traditional network such as USB or Ethernet, CAN does not send large blocks of data point-to-point from node A to node B under the supervision of a central bus master. In a CAN network, many short messages like temperature or RPM are broadcast to the entire network, which provides for data consistency in every node of the system.

Vehicle Bus

A **vehicle bus** is a specialized internal communications network that interconnects components inside a vehicle (e.g., automobile, bus, train, industrial or agricultural vehicle, ship, or aircraft). Special requirements for vehicle control such as assurance of message delivery, of non-conflicting messages, of minimum time of delivery, of low cost, and of Electro Magnetic Field (EMF) noise resilience, as well as redundant routing and other characteristics mandate the use of less common networking protocols. Protocols include Controller Area Network (CAN), Local Interconnect Network (LIN) and others. Conventional computer networking technologies (such as Ethernet and TCP/IP are rarely used, except in aircraft, where implementations of the ARINC 664 such as the Avionics Full-Duplex Switched Ethernet are used. Aircraft that use AFDX include the B787, the A400M and the A380.

(Source: Wikipedia)

4.1.1.1 Meaning of Massage-based Protocol

(See also "Key Characteristics of CAN)

CAN as *message-based protocol* does not use the conventional address scheme for identifying the devices. Data messages transmitted from any node on a CAN bus do not contain addresses of either the transmitting node (data source), or of any intended receiving node (destination node). Instead, CAN broadcasts the messages to all the nodes in the network using a unique *identifier* for the contents of the message. The identifier is unique throughout the network. All the nodes receive the message and each node runs an acceptance test on the identifier to determine if the contents of the message are relevant to it or not. Any node finds that the contents are relevant to it, this node processes the message otherwise it ignores the message.

For example, an automotive airbag sensor can be connected via CAN to a safety system router node only. This router node takes in other safety system information and routes it to all other nodes on the safety system network. Then all the other nodes on the safety system network can receive the latest airbag sensor information from the router at the same time, acknowledge if the message was received properly, and decide whether to utilize this information or discard it.

CAN uses the identifier also to determine the priority of the message in terms of competition for bus access. The lower the numerical value of the identifier, the higher the priority of the message. CAN standards use the identifier to avoid collision on the bus. This technique is used by CAN to achieve what is called "***non-destructive arbitration***". It guarantees that in situations where two or more nodes attempt to transmit at the same time, the messages are sent in order of priority and that no messages are lost. In the following sections the principle of using identifier is given in detail.

4.1.2 The Benefits of CAN Bus in Automobile Industry

Robert Bosch, German automotive system supplier, released CAN protocol to the Society of Automotive Engineers congress in 1986. Bosch introduced CAN as a method for enabling robust serial communication. The goal was to make automobiles more reliable, safe and fuel-efficient while decreasing wiring harness weight and complexity. His company promoted the manufacture of the BMW 850 coupe that year as the first vehicle featuring the CAN bus system. Since the wiring of the electronics in the vehicle was drastically reduced because of using CAN bus, the coupe itself weighed over 100 pounds less than previous models of the automobile, and the sensors worked significantly faster: This achievement considered as the first benefit gained by using CAN bus (see Figure 4.1).

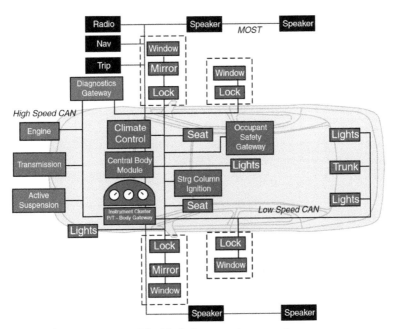

Figure 4.1 Typical automotive networks.

Since that time, the Society of Automotive Engineers has required that 100 percent of all automobiles sold in the United States be outfitted with the CAN bus protocol since 2008, and a similar mandate has been instituted in Europe.

In addition to creating lighter vehicles, the CAN bus protocol has many other significant benefits. By connecting all the modules in a vehicle to one central line, the control system is simplified and more easily regulated. This design allows any connected module to alert the main controller to an event that as it occurs which will cause the rest of the system to respond accordingly. The shared data line allows multiple modules to be attached with less invasive efforts, making the potential for error much lower.

If one module (node) on the CAN bus fails, it does not necessarily cause the failure of other modules. Unless the two systems are directly related and one cannot run without the other, the healthy modules will continue to function perfectly despite the loss of one bad module. This makes the entire system safer.

Additionally, diagnostics of the vehicle are simpler and more specific due to the specialized and self-contained nature of the modules. Diagnostics can now pinpoint the exact cause of the module failure with accuracy and speed. When one module needs maintenance, that module can simply be targeted and replaced rather than tearing apart the entire vehicle.

One of the most significant benefits of CAN bus technology is the ratio it offers of price to performance. Currently CAN bus is one of the most affordable network systems in addition to being one of the most reliable, making it the main cost-effective choice of automobile manufacturers around the world.

CAN bus offers a simple interface for attaching an additional node (module). The node can be used for diagnostics, reprogramming, monitoring, and more. Many tools have been created to tie into the CAN bus. Currently, there are solutions to do USB to CAN, wireless to CAN, PCIx to CAN and even a CAN bus logger.

4.1.3 CAN History

Controller area network was first developed by Robert Bosch in 1986. Bosch published several versions of the CAN specification and the latest is CAN 2.0 published in 1991. This specification has two parts; part A is for the standard format with an 11-bit identifier, and part B is for the extended format with a 29-bit identifier. A CAN device that uses 11-bit identifiers is commonly called CAN 2.0A and a CAN device that uses 29-bit identifiers is commonly called CAN 2.0B. The CAN frame uses 11-bit identifier is called "Standard Frame Format" and that uses 29-bit is called "Extended Frame Format". In 1993, the International Organization for Standardization (ISO), documented it as CAN standard ISO 11898 (for applications up to 1 Mbps) and ISO 11519 (for applications up to 125 kbps).

ISO 11898 is actually a family of specifications that cover two layers: The **Physical Layer** and the **Data Link Layer**.

ISO 11898 covers the following link and physical layers standards:

- ISO 11898-1: CAN Data Link Layer and Physical Signaling
- ISO 11898-2: CAN High-Speed Medium Access Unit
- ISO 11898-3: CAN Low-Speed, Fault-Tolerant, Medium-Dependent Interface
- ISO 11898-4: CAN Time-Triggered Communication
- ISO 11898-5: CAN High-Speed Medium Access Unit with Low-Power Mode
- ISO 11898-6: CAN High-speed medium access unit with selective wake-up functionality

Other standards are:

- ISO 11992-1: CAN fault-tolerant for truck/trailer communication
- ISO 11783-2: 250 kbps, Agricultural Standard
- SAE J1939-11: 250 kbps, Shielded Twisted Pair (STP)
- SAE J1939-15: 250 kbps, Unshielded Twisted Pair (UTP) (reduced layer)
- SAE J2411: Single-wire CAN (SWC) ISO 11898-2 uses a two-wire balanced signaling scheme. It is the most used physical layer in car powertrain applications and industrial control networks.

Notes:

- The physical layer standards ISO 11898-2 and ISO 11898-3 are not part of the Bosch CAN 2.0 specification. These standards may be purchased from the ISO.
- **ISO 11898-2** uses a two-wire balanced signaling scheme. It is the most used physical layer in car powertrain applications and industrial control networks.
- ISO 11898-4 standard defines the time-triggered communication on CAN (TTCAN). It is based on the CAN data link layer protocol providing a system clock for the scheduling of messages.

- SAE J1939 standard uses a two-wire twisted pair, -11 has a shield around the pair while -15 does not.
- SAE 1939 is widely used in heavy-duty (truck) and autobus industry as well as in agricultural & construction equipment.
- ISO 11783-2 uses four unshielded twisted wires: two for CAN and two for terminating bias circuit (TBC) power and ground. This bus is used on agricultural tractors. This bus is intended to provide interconnectivity with any implementation adhering to the standard.

4.1.3.1 Higher layer implementations: Higher layer protocols (HLPs)

As the CAN standard does not include tasks of application layer protocols, such as flow control, device addressing, and transportation of data blocks larger than one message, and above all, application data, many implementations of higher layer protocols were created. Among these are:

- DeviceNet (used for industrial automation)
- CANopen (used for industrial automation)
- Smart Distributed System (SDS)
- CANaerospace (for the aviation industry)
- SAE J1939 (heavy road vehicles)
- ISO 15765-4
- ISO 11783 or ISOBUS (agriculture)
- SmartCraft
- NMEA 2000 (marine industry)
- CAN Kingdom
- SafetyBUS p (used for industrial automation)
- EnergyBus (used for electrical vehicles)
- MilCAN
- RV-C (used for recreational vehicles)
- ARINC 825 (for the aviation industry)

For passenger cars, manufacturers have preferred to develop each their own standard. Examples:

- GMLAN (for General Motors)
- CCP/XCP
- ISO14229

In 2012, Bosch released CAN FD 1.0 or CAN with *Flexible Data Rate*. This specification uses a different frame format that allows a different data length as well as optionally switching to a faster bit rate after the arbitration is decided. CAN FD is compatible with existing CAN 2.0 networks so new CAN FD devices can coexist on the same network with existing CAN devices.

4.1.4 Error Detection and Security in CAN

CAN includes a number of error detection mechanisms that can be used to detect any error in the data on the message and on the bit level and force the message to be destroyed and retransmitted in case of detecting an error. The use of the detection mechanisms ensures that the message a node does receive is error free and contains valid data. The error detection mechanisms, also, guarantee that the nodes will not miss any message.

Concerning security, CAN as a low-level protocol does not support any security features intrinsically. Applications are expected to deploy their own security mechanisms, for example, to authenticate each other. Failure to do so may result in various sorts of attacks, if the opponent manages to insert messages on the bus. Password mechanisms exist for data transfer that can modify the control unit software, like software download or ignition key codes, but usually not for standard communication.

4.1.5 The Application of CAN Bus

Although the CAN bus protocol was introduced for automotive use, today the system is used in several other networking applications. With the simple system of networking individual modules into a single central line of control, many companies are taking advantage of the CAN bus protocol in assembly lines, industrial vehicles, medical equipment, test equipment mobile machines and any other machinery that uses multiple modules that must communicate with each other.

Autonomous industrial machinery, as well as mobile machines, have been increasingly integrating CAN bus protocol because of the embedded control in the system. Based on the reliability and efficiency of the protocol, medical and military equipment have also seen a rise in the application of CAN bus. As machinery becomes increasingly complex, the application of the CAN bus system is likely to rise with the technology.

Applications: Automotive

A modern automobile may have as many as 70 electronic control units (ECU) for various subsystems [3]. Typically the biggest processor is the engine control unit (also engine control module/ECM or Powertrain Control Module/PCM in automobiles); others are used for transmission, airbags, antilock braking/ABS, cruise control, electric power steering/EPS, audio systems, windows, doors, mirror adjustment, etc. Some of these form independent subsystems, but communications among others are essential. A subsystem may need to control actuators or receive feedback from sensors. The CAN standard was devised to fill this need. The CAN bus may be used in vehicles to connect engine control unit and transmission, or (on a different bus) to connect the door locks, climate control, seat control, etc. (see Figure 4.1). Today the CAN bus is also used as a fieldbus in general automation environments, primarily due to the low cost of some CAN Controllers and processors. Bosch holds patents on the technology, and manufacturers of CAN-compatible microprocessors pay license fees to Bosch, which are normally passed on to the customer in the price of the chip. Manufacturers of products with custom ASICs or FPGAs containing CAN-compatible modules may need to pay a fee for the CAN Protocol License.

4.2 CAN Standard

4.2.1 CAN Architecture: OSI 7 layer reference model and CAN model

Figure 4.2 shows the OSI/ISO 7 layer reference model. This model is a guideline for network developers that defines different levels that data should pass through to travel between two devices on the network. CAN covers the lowest two layers of the OSI/ISO model: Physical Layer and Data Link Layer as shown in Figure 4.2. Figure 4.3 gives the two layers in more detail. Figure 4.4 links the model layers with the components needed to be included in any CAN Node. Figure 4.4 is based on Figure 4.3 that reflects the **CAN BUS (ISO-11898-2)**.

The ISO 11898-1 doesn't cover HLPs which, generally, map the upper 5 layers of the 7 layer model (Figure 4.2). It also does not specify what transceivers or connectors should be used for a CAN network. Considering this fact, CAN protocol exists in the lower two layers (Data link and part of the Physical layer).

Many Higher Layer Protocols (HLPs) were developed for and standardized on CAN: Both proprietary and non-proprietary protocols exist. Some of high level protocols are given before in Section 4.1.3.1.

From the above discussions, it is possible to say that the CAN protocol can be described in terms of four levels of abstractions: application, object,

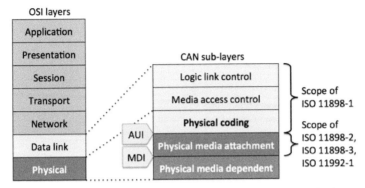

Figure 4.2 The OSI 7-layer model and the ISO 11898 standard.

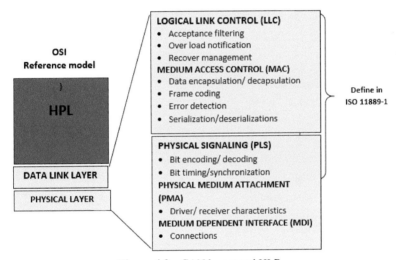

Figure 4.3 CAN layers and HLP.

data-link (or transfer) and physical levels. CAN specifications describe the physical and data-link layers while all the remaining layers are managed by the microprocessor in software as appropriate. The four layers are (see Figure 4.5):

1. **Application Layer:**

Many applications of CAN require services that are beyond the basic functionality specified by the Data-Link Layer but which may be implemented at the Application Layer. To meet such needs, several organizations have developed Application Layers.

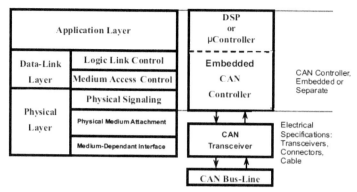

Figure 4.4   The layered ISO 11898 standard architecture and its reflection on CAN node requirements.

Application Layer
Object Layer
• Message Filtering • Message and Status Handling
Transfere Layer
• Fault Confinement • Error Detection and Signalling • Message Validation • Acknowledgment • Arbitration • Message Framing • Transfer Rate and Timing
Physical Layer
• Signal Level and Bit Representation • Transmission Medium

Figure 4.5   The four abstraction levels of CAN protocol.

2. **Object Layer:**

This includes the following functions:

- Message Filtering:

    i. Deciding which messages received by the transfer layer (Data-Link Layer) are actually to be used,

- Message and Status Handling,
- Providing an interface to application layer related hardware.

3. Data-Link Layer or Transfer Layer

This layer is the kernel of the protocol. It recognizes and understands the format of messages. It presents messages received to the object layer and accepts messages to be transmitted from the object layer. It constructs the messages to be sent to the Physical Layer, and decodes messages received from the Physical Layer. The data-link layer is responsible for:

- Fault Confinement
- Error Detection
- Message Validation
- Acknowledgment
- Arbitration
- Message Framing
- Transfer Rate and Bit Timing
- Information Routing

In CAN controllers, hardware is usually used to implement the data-link layer. The functions, operations and implementation of the data-link layer are considered latter.

4. Physical Layer

The physical layer is responsible for defining how the signals are actually transmitted. Some of the tasks of the physical layer are:

- **Transmission Medium**: The physical layer specifies the physical and electrical characteristics of the bus.
- **Signal Level and Bit Representation**: This includes the hardware required to convert the characters of a message into electrical signals for transmitted messages and electrical signals into characters for received messages.

The physical layer is all the time a "real" hardware. The physical layer and the data-link layer will be discussed after presenting some of the features and the characteristics of CAN BUS

Today, CAN is found in almost every market while development is still continuing in the area of HLPs to support the needs of existing and new developers. CAN continues to be widely popular mainly because it is an extremely robust, but also because the sheer volumes in the automotive market have driven the costs of a CAN node down. This has allowed the protocol to grow rapidly since 1985 while expanding into other Markets like marine, industrial, and medical, Agricultural and construction, Military, Factory and Building automation.

4.2.2 Higher Layer Protocols (HLPs)

As mentioned before (Section 4.1.2), CAN standard does not include tasks of application layer protocols, such as flow control, device addressing, and transportation of data blocks larger than one message, and above all, application data, many implementations of higher layer protocols were created, some of them are mentioned in Section 4.1.2.

Next, and after discussing the features and characteristics of the CAN bus, the details of the Physical and Data Link Layers are given.

4.3 CAN Features and Characteristics

4.3.1 Some Features of CAN Bus

Some of the features and uses of CAN bus are:

- To make the communication between different microprocessors more efficient without increasing the wiring needed, CAN does not use central control. Any processor in the system can send and receive messages directly without relying on any central control (central processor). This reduces the wiring, in some cases, by 90 percent. The non-destructive arbitration technique used by CAN guarantee the arrival of the messages without any collisions.
- Measurements needed by several controllers can be transmitted via the bus, thereby removing the need for each controller to have its own individual sensor.
- Flexibility and expansion: The use of message identifier technique instead of using addressing, simplifying the system configuration and gives its high degree of flexibility. The CAN specification does not specify the actual layout and structure of the physical bus itself. This gives the design engineers the possibility to reconfigure CAN systems by adding or removing network nodes easily. In the cases when the new nodes are purely receivers, and which need only existing transmitted data, it is possible to add them to the network without the need to make any changes to existing hardware or software. In general, any new node can be connected to CAN bus as it is able to transmit, or detect, on the physical bus, one of two signals called dominant or recessive. For

example, a dominant signal may be represented as logic '0' and recessive as logic '1' on a single data wire. Furthermore, the physical CAN bus must guarantee that if one of two devices asserts a dominant signal and another device simultaneously a recessive signal, the dominant signal prevails.

- Due to flexibility, expandability and content-oriented nature of CAN bus, it is easy to use by any project manager. Each unit in CAN system can be developed and tested separately. If all the individual functional units are proved to work independently, there is a very high probability that when all of them connected together to form a CAN system they will work correctly as a system.
- The individual nodes of a CAN system are connected together in a daisy chain. All the nodes are equal. Any processor can send a message to any other processor, and if any processor fails, the other systems in the machine will continue to work properly and communicate with each other.
- CAN messages are short, no more than eight bytes long. It is generally used for sending signals to trigger events, such as to lock seat belts during heavy braking, and measurement values, such as temperature and pressure readings. The content-oriented nature of CAN and the non-destructive arbitration technique used guarantee interrupt-free transmission. The message identifier defines the priority of the message which will prevent any conflicts. It also guarantees that the urgent messages (with the highest priority) will be delivered first.
- In order to increase the traffic reliability, CAN protocol include extensive error checking mechanisms. This is one of the features that lets CAN be extremely robust (see key characteristics of CAN)
- CAN system is able to transmit up to 7600 8-byte messages per seconds.
- The devices that are connected by a CAN network are typically sensors, actuators and control devices. A CAN message never reaches these devices directly, but instead a host-processor and a CAN Controller is needed between these devices and the bus.

4.3.2 Key characteristics of CAN

Some of the key characteristics of CAN protocol that will be discussed, with others, in detail in this chapter are as follows:

- CAN is an extremely robust serial communication protocol.
- CAN is Message based, not address base.

- Distributed control across the CAN network that allows peer to peer or master to slave style of communications.

a. **CAN is an extremely robust communication protocol**

Here are three examples why:

- Any CAN node on the BUS can detect errors in the message, and force the message to be destroyed and retransmitted. This feature helps to ensure that the message a node does receive contains valid data.
- The CAN Frame requires that every node "acknowledge" the message before it can be processed by that node. This acknowledge can only come after various error condition checks, including a 15-bit CRC on the message. If one CAN node finds an error with the message, the message is destroyed and retransmitted.
- The specification defines three different error states for a CAN node to be in, with each error state giving the CAN node different levels of bus access. That was designed to limit faulty nodes from permanently taking down the CAN bus.

CAN is Serial communication in which All nodes on the CAN bus are attach to common connection using the same bit rate.

The implementation of the above three examples will be considered in detail while discussing the data-link layer.

b. **CAN is message based, not address based**

This means that Messages are NOT transmitted from one node to another node based on the address of a CAN Node, instead a CAN node will broadcast its message to all nodes on the bus (Broadcast Communication concept), and it is up to the receiving node to determine it should act on that message.

The CAN concept of Broadcast Communication is shown in Figure 4.6. In Figure 4.6, station 2 is the transmitting station. After receiving the frame, it is the task of every node to decide if the message has to be accepted or not. So Acceptance Filtering has to be implemented in every CAN node (part of the data link layer discussion). The CAN Broadcast Communication can be compared with a radio station transmitting information about traffic accumulation for vehicle drivers. Every driver has to decide if the messages are important for him dependent on the motorway he wants to use.

It is possible to add new nodes to a CAN bus without having to update all of nodes with addressing information.

c. **CAN allows for distributed control across a Network:** This is because of the reliability of the data. This allows designers the flexibility to setup Master/Slave or Peer/Peer also called Consumer producer/networks.

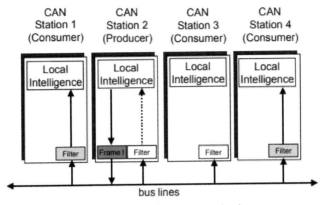

Figure 4.6 Broadcast communication.

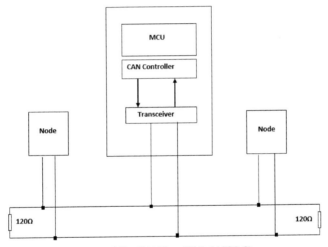

Figure 4.7 CAN bus (ISO-11898-2).

4.4 CAN Physical Layer

CAN physical layer is described by ISO-11898. ISO 11898-2 specifies the PMA and MDA (see Figure 4.2) sublayers of the physical layer. ISO 11898 and ISO 11898-2 are the basis of the discussions in this section.

4.4.1 CAN Bus Description

CAN Network is made of a group of nodes. Figure 4.7 gives a representation of a common CAN node/bus as described by ISO-11898.The smallest working CAN network contains at least two active nodes to allow that one to act as transmitting Node and the second to be receiving Node that will "ack" a

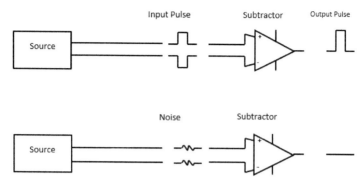

Figure 4.8   Differential signal: Advantage of using differential balanced 2-wire interface in noise elimination.

transmitted message. The complexity of the node can range from a simple I/O device up to an embedded computer with a CAN interface and sophisticated software. The node may also be a gateway allowing a standard computer to communicate over a USB or Ethernet port to the devices on a CAN network. The main types of CAN nodes are given in Section 4.4.2.1.

All nodes are connected to each other through a two wire bus. The wires are a twisted pair with a 120 Ω (nominal) characteristic impedance.

The CAN bus is a differential balanced 2-wire interface running over either a shielded twisted pair (STP), unshielded twisted pair (UTP), or ribbon cable. Each node uses a Male 9-pin D connector (see Section 4.4.1.1.10). The use of differential signal tends to eliminate the effect of the noise generated on the bus (Figure 4.8). For data communication on the differential wire bus, CAN uses for bit encoding the "Non Return to Zero (NRZ)" encoding technique (with bit-stuffing). NRZ encoding is used to ensure compact messages that have a minimum number of transitions and high resilience to external disturbance.

Figure 4.9 shows the configuration of CAN basic system. In this figure, CAN network is made of a number of nodes (n nodes) that are connected to a physical layer, which is a differential bus with termination R_t. Even though we are using a differential bus in this figure, but we must keep in mind that other possible physical layers exist. Unlike the 12C bus, the CAN bus does not require that one of the nodes is a master. Nodes can be added to the network at any time; it is not necessary to power the system down. The standard specifies that up to 30 nodes can be added to the bus.

The communication between nodes in Controller Area Network protocol is a synchronous. This means that all the nodes connected to the bus while

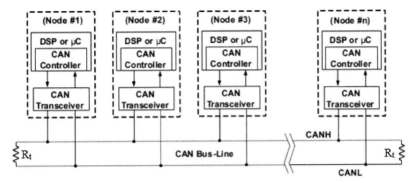

Figure 4.9 Basic CAN bus connection.

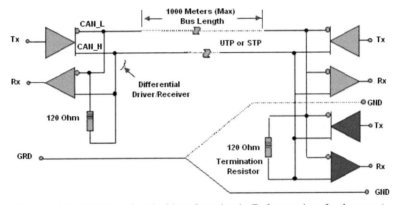

Figure 4.10 CAN bus electrical interface circuit (Ref. www.interfacebus.com).

receiving or transmitting must use the same clock rate and that all the clock rates in the network are based on the same reference (single) point. The use of synchronous serial communication results in more efficient data transmission. On the other hand, to keep any two clocks synchronized for long time represents a difficult task and needs some sort of fixed reference signal. With time, it is very common that the clocks lose their synchronization due to oscillator drift, propagation delays, and phase errors (see Section 4.4.1.2).

4.4.1.1 Electrical consideration

Electrically, the CAN bus utilizes balanced differential signaling; the current in each signal line is equal and opposite. Such a scheme significantly enhances noise immunity, increases common mode noise rejection, and improves fault tolerance over single-ended drive models. For bus interfacing, CAN uses the circuit shown in Figure 4.10.

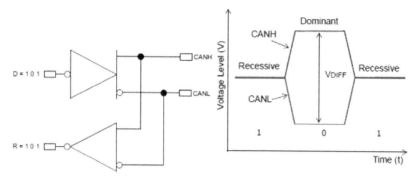

Figure 4.11   The inverted logic of a CAN bus.

The two serial lines that make up the CAN Bus are designated CANH and CANL. The bus signaling protocol defines two states, *dominant* and *recessive*. Similar to the wired AND configuration in the I2C bus, *dominant* will override *recessive*. When the transmitted data, TXD, is a logical 0, the bus is in the dominate state; the CANH signal line can be taken to voltage between 2.75 and 4.5 V (e.g., 3.5 V), and the CANL is taken to voltage between 0.5 and 2.25 V (e.g., 1.5 V), giving normally a difference of 2.0 V. When the transmitted data is a logical 1, the bus is in the *recessive* (quiescent) state, which is set to 2.5 V.

4.4.1.1.1 Inverted logic of CAN bus

Inverted logic of CAN bus is a fundamental CAN characteristic. Inverted logic shown in Figure 4.11 represents the opposite logic state between the bus, and the driver input and receiver output. Normally, a logic-high is associated with a one, and a logic-low is associated with a zero–but not so on a CAN bus. This is why, for example, Texas Instrument CAN (TI CAN) transceivers have the driver input and receiver output pins passively pulled high internally, so that in the absence of any input, the device automatically defaults to a recessive bus state on all input and output pins.

In the recessive state (i.e., logic '1' on the input), the differential voltage on CANH and CANL is less than the minimum threshold (<0.5 V receiver input or <1.5 V transmitter output) (see Figure 4.12).

In the dominant state (i.e., logic '0' on the input), the differential voltage on CANH and CANL is greater than the minimum threshold. A dominant bit overdrives a recessive bit on the bus to achieve nondestructive bitwise arbitration.

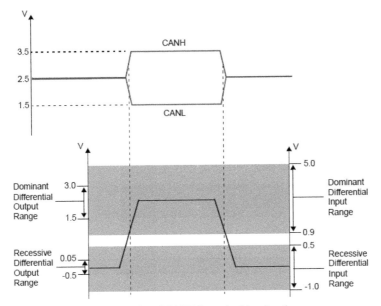

Figure 4.12 ISO11898 nominal bus levels.

4.4.1.1.2 Robustness

The ISO11898-2 specification requires that a compliant or compatible transceiver must meet a number of electrical specifications. Some of these specifications are intended to ensure the transceiver can survive harsh electrical conditions, thereby protecting the communications of the CAN node. The transceiver must survive short circuits on the CAN bus inputs from −3 V to +32 V and transient voltages from −150 V to +100 V.

The robustness of CAN may also be attributed in part to its abundant error checking procedures. CAN achieves utmost safety of data transfer. The error detection techniques used guarantee that the residual error probability for undetected corrupted message is less than:

$$\text{message error rate} * 4.7 * 10^{-11}$$

Also the system guarantees that any node will not miss (or loose) any message.

4.4.1.1.3 Bus lengths

ISO11898 specifies that a transceiver must be able to drive a 40 m bus at 1 Mbps. A longer bus length can be achieved by slowing the data rate. The biggest limitation to bus length is the transceiver's propagation delay.

4.4.1.1.4 **Bit rates and bus lengths**

Many factors are affecting the allowed bit rate to be used on a CAN Bus. The factors include:

Bus Length: The rate of data transmission depends on the total overall length of the bus (the total actual length of the wiring in the network) and the delays associated with the transceivers. This is a physical limitation and not set by CAN protocol. The length of the bus is limited by two factors:

- The propagation delay time: which is the time period necessary for a signal to go from one end of the bus to the other back again before the next signal is transmitted. This can be explained as follows: The basics of arbitration require that the front wave of the first bit of a message travel to the most remote node on a network and back again before the bit is designated by the receiver of the sending node as dominant or recessive (typically this sample is made at about two–thirds the bit width). With this limitation, the maximum bus length and signaling rate are determined by network parameters.
- Time needed by the electronic circuitry to transmit and receive these signals.

Increasing the length of the bus increases the sum of the propagation delay and the time needed by the transmitting and receiving devices. This, in turn, increases the nominal bit time and, accordingly, decreases the possible transmission bit rate. As a result, every CAN system must trade bus length for bit speed. For example, to have 1 Mbps speed, the maximum possible bus length is specified as 25 meters, for longer bus lengths it is necessary to reduce the bit rate. Table 4.1 gives some indication of the bit rates and the corresponding maximum bus length.

Table 4.1   Maximum bit rate versus bus length

Bit Rate	Length
1 Mbps	25 m
800 kbps	50 m
500 kbps	100 m
250 kbps	250 m
125 kbps	500 m
50 kbps	1000 m
20 kbps	2500 m
10 kbps	5000 m

The bit rate can always be slower than the maximum possible speed for a given bus length. Conversely, the bus length can be shorter than the maximum possible bus length for a given transmission speed. The CAN system designer needs to keep in mind that transmissions tend to become more reliable with a slower bit speed and shorter bus lengths.

Cable Bandwidth Limitation: Cable bandwidth limitations, which degrade the signal transition time and introduce inter-symbol interference (ISI), are primary factors reducing the achievable signaling rate when transmission distance is increased.

System Delay: For a CAN bus, the signaling rate is also determined from the total system delay–down and back between the two most distant nodes of a system and the sum of the delays into and out of the nodes on a bus with the typical 5ns/m prop delay of a twisted-pair cable.

Signal Amplitude Loss: Consideration must be given to the signal amplitude loss due to resistance of the cable and the input resistance of the transceivers. Under strict analysis, skin effects, proximity to other circuitry, dielectric loss, and radiation loss effects all act to influence the primary line parameters and degrade the signal.

A conservative rule of thumb for bus lengths over 100 m is derived from the product of the signaling rate in Mbps and the bus length in meters, which should be less than or equal to 50:

$$\text{Signaling rate (Mbps)} \times \text{Bus length (m)} \leq 50$$

If a 1000 m bus is required by an application, then by this approximation a 50 kbps signaling rate may safely be used. Lab experiments show that an actual safe signaling rate with 1000 m of 120-W characteristic impedance twisted-pair cable is indeed approximately 50 kbps. By providing this extra margin of safety, a lot of system variation can still take place without disruptions in communication.

A long cable length with a higher number of nodes than the Standard's recommended 30 nodes may require the use of higher cable quality, a CAN bus repeater, and tighter operating tolerances such as a 5% voltage supply regulation. In practice, however, almost any type of cable works to a certain degree, even a cheap phone line for short distances.

4.4.1.1.5 Propagation delay

The CAN protocol has defined a recessive (logic '1') and dominant (logic '0') state to implement a nondestructive bit-wise arbitration scheme. It is this

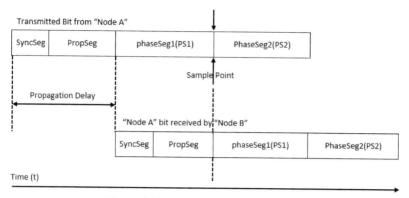

Figure 4.13   One-way propagation delay.

Note: Propagation delay will be discussed in detail in Section 4.4.3.1 while discussing "Bit Timing."

arbitration methodology that is affected most by propagation delays. Each node involved with arbitration must be able to sample each bit level within the same bit time.

For example, if two nodes at opposite ends of the bus start to transmit their messages at the same time, they must arbitrate for control of the bus. This arbitration is only effective if both nodes are able to sample during the same bit time. Figure 4.13 (see also Figure 4.24) shows a one-way propagation delay between two nodes. Extreme propagation delays (beyond the sample point) will result in invalid arbitration. This implies that bus lengths are limited at given CAN data rates. In Section 4.4.3.1.1, bit timing is considered in more detail.

A CAN system's propagation delay is calculated as being a signal's round-trip time on the physical bus (t_{bus}), the output driver delay (t_{drv}) and the input comparator delay (t_{cmp}). Assuming all nodes in the system have similar component delays, the propagation delay is explained mathematically by equation (4.1).

$$t_{prop} = 2 \cdot (t_{bus} + t_{cmp} + t_{drv}) \qquad (4.1)$$

4.4.1.1.6 Cables

Although unshielded 120-W cable is used in many applications, data transmission circuits employing CAN transceivers are used for jobs requiring a rugged interconnection with a wide common-mode voltage range. Therefore, shielded cable such as Belden Cable 3105A is recommended in these electronically harsh environments. Shielded cable and the Standard's 2 V to 7 V common-mode range of tolerable ground off-set, help to ensure data integrity.

Note that the HVD1050 CAN transceiver has an extended common-mode range of -12 V to 12 V.

While prefabricated cables for CAN applications are more expensive, they are more easily installed and verified, and may, therefore, reduce overall installation cost and time-to-market.

4.4.1.1.7 Shield termination

If a shield must be used, it is recommended that a short pig-tail be crimped to the shield end at each connector and then brought through a separate connector pin to a ground pin located as close to the connector as possible. Note that the network should be grounded at a single point at the source location.

This prevents parasitic currents from flowing in the shield between ground connections. If individual shielding of the signal pairs is used, use the same terminating technique as for the overall shield.

4.4.1.1.8 Grounding

There should be only one path for return current between the host and receiving nodes. This follows the same discussion in the shielding section. If a network is grounded in more than one location, parasitic current will flow. By grounding a network only at the source, potentially hazardous ground loops are avoided. The use of digital isolators such as the ISO721 (SLLS629) is recommended if it becomes necessary to connect the grounds of different sources.

Unused pins in connectors as well as unused wires in cables should be single-point grounded at the connector. Unused wires should be grounded at alternate ends to nearby ground pins.

4.4.1.1.9 Bus termination–line termination

Bus termination is used to minimize signal reflection on the bus. ISO-11898 requires that the CAN bus have a nominal characteristic line impedance of 120Ω. Therefore, the typical terminating resistor value for each end of the bus is 120Ω. There are a few different termination methods used to help increase EMC performance (see Figure 4.14).

1. Standard Termination
2. Split Termination
3. Biased Split Termination

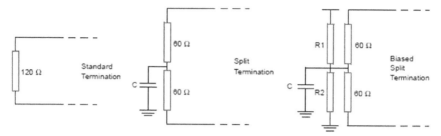

Figure 4.14　Termination converts.

Standard Termination: As the name implies, this termination uses a single 120 resistor at each end of the bus. This method is acceptable in many CAN systems.

Split Termination: Split termination is a concept that is growing in popularity because emission reduction can be achieved very easily. Split termination is a modified standard termination in which the single 120Ω resistor on each end of the bus is split into two 60Ω resistors, with a bypass capacitor tied between the resistors and to ground. The two resistors should match as close as possible.

4.4.1.1.10 Connectors

ISO-11898-2 does not specify the mechanical wires and connectors. Connectors, while not specified by the Standard, should have a characteristic impedance matching that of the bus line and terminators, and it should not affect standard operating parameters such as the minimum VOD.

The higher layer protocols such as CANopen and DeviceNet define the specific hardware required for implementation, including bus wire and connectors. Recommended products may be found on organization web-sites such as the CiA's CAN-cia.com which list connector and pin-out specifications for use in CANopen applications. These include the 9-pin DSUB shown in Figure 4.15, Multipole, RJ10, RJ45, M12, the 5-pin mini-style in Figure 4.16 and more micro-style connectors in the CiA specification document DR 303-1, V1-3.

4.4.2 CAN Node

Consider Figure 4.2 again, **CAN BUS (ISO-11898-2)**. The figure shows the components needed for any CAN node. This is shown in more detail in Figure 4.17.

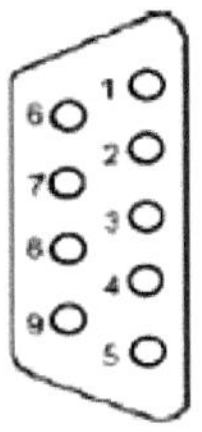

Pin		Description
1	Reserved	
2	CANL	CANL bus pin
3	V+	Optional 3.3-V or 5-V power supply for transceivers and digital isolators if required
4	Reserved	
5	CAN SHLD	Optional shield
6	V-	Ground return path/ 0V
7	CANH	CANH bus bin
8	Reserved	
9	V+	Optional 3.3-V or 5-V power supply for transceivers and digital isolators if required

Figure 4.15   CANopen DSUB connector.

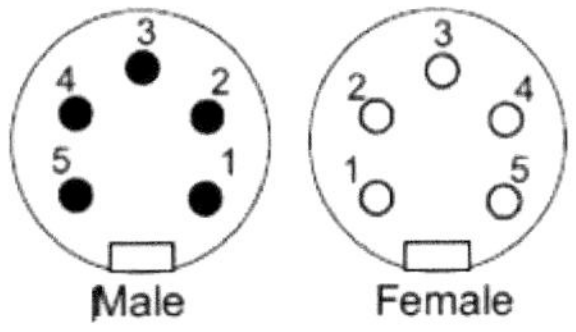

Pin		Description
1	CAN SHLD	Optional shield
2	V+	Optional 3.3-V or 5-V power supply for transceivers and digital isolators if required
3	V-	Ground return path/ 0V
4	CANH	CANH bus bin
5	CANL	CANL bus pin

Male Female

Figure 4.16   5-Pin mini-connector (ANSI/B.93.55M-1981).

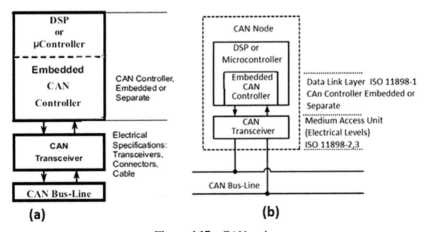

Figure 4.17   CAN node.

From Figures 4.7 and 4.17, each CAN node requires a:

- **Central processing unit, microprocessor, or host processor:**
  - The host processor decides what the received messages mean and what messages it wants to transmit.
  - Sensors, actuators and control devices can be connected to the host processor.
- **CAN controller**: Often an integral part of the microcontroller
  - Receiving: the CAN controller stores the received serial bits from the bus until an entire message is available, which can then be fetched by the host processor (usually by the CAN controller triggering an interrupt) (see Section 4.4.2.1).
  - Sending: the host processor sends the transmit message(s) to a CAN controller, which transmits the bits serially onto the bus when the bus is free.
- **Transceiver**: Defined by ISO 11898-2/3 Medium Access Unit [MAU] standards
  - Receiving: it converts the data stream from CANbus levels to levels that the CAN controller uses. It usually has protective circuitry to protect the CAN controller.
  - Transmitting: it converts the data stream from the CAN controller to CANbus levels.

Each node is able to send and receive messages, but not simultaneously. A message or Frame, as discussed latter, consists primarily of the ID (identifier), which represents the priority of the message, and up to eight data bytes. A CRC, acknowledge slot [ACK] and other overhead are also part of the message. The improved CAN FD extends the length of the data section to up to 64 bytes per frame. The message is transmitted serially onto the bus using a non-return-to-zero (NRZ) format and may be received by all nodes.

The devices that are connected by a CAN network are typically sensors, actuators, and other control devices. These devices are connected to the bus through a host processor, a CAN controller, and a CAN transceiver.

4.4.2.1 CAN nodes types

The analyses of Figures 4.8 and 4.17 and the discussions given in the above section show the possibility of three CAN Node types (Figure 4.18).

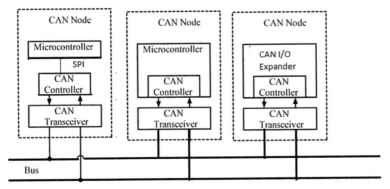

Figure 4.18 CAN node types.

1. **A node that does not have an integrated CAN module**:
   The node on the left of Figure 4.18 is an example of such a node: a microcontroller that does not have an integrated CAN module, but the application requires that it still needs to connect to the CAN bus.

   This is the case when the network designer needs that the application under consideration to be connected to the CAN bus, but, at the same time, the application requires a certain microcontroller that does not come with an integrated CAN module.

   The simplest solution in such cases is to use a standalone CAN module. The CAN module, now, replaces that which not integrated within the node.

2. **A node with integrated CAN module**:
   The node at the center of Figure 4.18 is an example: The node in this example is a microcontroller with an integrated CAN module. Such microcontroller allows for a faster internal connection between the CAN module and microcontroller. This is the most common type of nodes used on the CAN bus. Many microcontroller manufacturers provide microcontroller integrated within it CAN module. Microchip, for example, currently provides 8/16/32 bit PIC micros with the CAN module integrated.

3. **Node with CAN I/O expander**

   The node on the right side of Figure 4.18 is an example. This integrate circuit provides I/O expansion for a CAN network without a microcontroller. This device would have peripheral like GPIO, A2D, or PWM. Such Node can be configured with defaults to send out messages periodic or event driven on thresholds. Also other nodes can interact to read I/O or change Outputs. This node requires no microcontroller firmware to write or debug, but then lacks some flexibility with nonproprietary High Layer Protocols.

Such devices were designed for low cost simple sensor applications.

4.4.3 Sublayers of the CAN Physical Layer

Let us consider Figure 4.2 again. The figure shows that the CAN physical layer is divided into three parts (sublayers): The physical coding (PCS) implemented in the CAN controller chips, the physical media attachment (PMA) specifying the transceiver characteristics, and the physical media-dependent sublayers (PMS). The PMS is application-specific and is not generally standardized.

The figure shows also two interfaces: The AUI (attachment unit interface) is the interface between CAN controller and CAN transceiver chip; the MDI (medium-dependent interface) is the interface to physical bus lines.

The PCS comprises the bit encoding and bit decoding, the (re-)synchronization as well as the bit timing. This sublayer provides the attachment unit interface (AUI), which is the interface to the transceiver chips. The AUI are realized by means of the TxD and RxD pins. There are two implementations: the 5-V version and the 3.3-V version.

4.4.3.1 The physical coding sublayer: Overview

Controller Area Network (CAN) protocol, as mentioned before, is an asynchronous serial bus with Non-Return to Zero (NRZ) bit coding designed for fast, robust communications in harsh environments, such as automotive and industrial applications. The use of NRZ encoding, and in contrast to the Manchester coding, not every bit contains a falling or a rising edge. NRZ ensures a minimum number of transitions and high resilience to external disturbance. The two bits are encoded in medium states defined as "recessive" and "dominant". (0 is typically assumed as associated to the "dominant" state). The protocol allows multimaster access to the bus with deterministic collision resolution. At the very lowest level, this means that if multiple masters try to drive the bus state, the "dominant" configuration also prevails upon the "recessive".

During one bit, the voltage (0 V or +5 V resp. 3.3 V) is constant. In case of NRZ coding, the signal level can remain constant over a longer period of time if the transmitted bits have the same logical value. Therefore, measures must be taken to ensure that the maximum permissible interval between two signal edges is not exceeded. This is important for synchronization purposes. Hence, the CAN protocol uses so-called stuff-bits: After five bits of the same value, the transmitting CAN controller automatically includes a bit of the opposite value. The receiving CAN nodes de-stuff the bit sequence, meaning after five bits of the same value they automatically delete the following one. This way it is guaranteed that the original bit sequence is the same as the sequence forwarded to the host controller. Of course, these stuff-bits are visible on the bus lines.

If the bus is idle (recessive state for at least three bit-times), the first falling edge (recessive-to-dominant) is used to globally synchronize (hard synchronization) all CAN controllers. Every following recessive-to-dominant edge is used to locally synchronize (soft synchronization) the nodes. The re-synchronization is limited by the bit-timing settings.

On the bit-level–the OSI physical layer–Classical (standard) CAN uses synchronous bit transmission. This enhances the transmitting capacity but also means that a sophisticated method of bit synchronization is required. While in a character-oriented transmission (asynchronous) bit synchronization is performed upon the reception of the start bit available with each character, in a synchronous transmission protocol only one start bit is available at the beginning of a frame. To enable the receiver to correctly read the messages, continuous re-synchronization is required. Phase buffer segments are therefore inserted before and after the nominal sample point within a bit interval.

The details of bit timing, synchronization, bit rate, bus delay and other parameters are given next. The CAN protocol allows the user to program the bit timing parameters as the bit rate, the sample point of the bit, and the number of times the bit is sampled. With these features, the network can be optimized for a given application.

There are relationships between bit timing parameters, the physical bus propagation delays, and the oscillator tolerances throughout the system. This section of the chapter investigates these relationships as they are given in the original Bosh Standard and used in many CAN modules and how this assists in optimizing the bit timing for given physical system attributes.

4.4.3.1.1 Bit-timing setting for Standard CAN
4.4.3.1.1.1 The Can Bit Time

CAN has advanced features for coping with the time delays found in long bus lengths (in comparison to the bit rate) and coping with differences in clock Crystal frequencies for nodes on the bus.

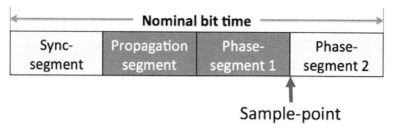

Sample-point

Figure 4.19 Bit timing.

The choice of bit timing is very important since it decides the bit rate, the sample point and the ability to resynchronize.

The CAN bit time is made up of non-overlapping segments. Each of these segments are made up of integer units called Time Quanta (TQ) (explained later). The Nominal Bit Rate (NBR) is defined in the CAN specification as the number of bits per second transmitted by an ideal transmitter with no resynchronization and can be described with the equation:

$$NBR = f_{bit} = \frac{1}{t_{bit}}$$

The CAN bit is logically divided into four nonoverlapping segments (Figure 4.19): synchronization segment, propagation segment, phase-segment 1, and phase-segment 2. Each segment consists of one or more time quanta (TQ). The synchronization segment has always a length of 1 TQ (one time quanta). The following propagation segment and phase-segment 1 are not distinguishable for the user. It is configured as one value. The forth part is the phase-segment 2. Between the two phase-segments, the sample-point is located. If the bit-timing settings move the sample-point to the end of the bit, the propagation segment is enlarged, so that you can realize longer networks. If the sample-point is configured in the other direction, the re-synchronization capability is increased.

The nominal bit time (NBT), accordingly, is the summation of the following four segments:

$$t_{bit} = t_{SyncSeg} + t_{PropSeg} + t_{PS1} + t_{PS2}$$

Associated with the NBT are the Sample Point, Synchronization Jump Width (SJW), and Information Processing Time (IPT), which are explained later, after defining the four segments.

(a) Synchronization segment (SynchSeg)

The synchronization segment (SyncSeg or Synch_Seg) is used to synchronize the various nodes on the bus. When a bit is sent on the bus, the leading edge is expected to be within this segment.

This segment is always one time quantum long.

(b) Propagation segment (PropSeg)

The Propagation Segment (PropSeg) exists to compensate for physical delays between nodes. The propagation delay is defined as twice the sum of the signal's propagation time on the bus line, including the delays associated with the bus driver. Equation (4.1) gives the relation. The PropSeg is programmable from 1–8 TQ.

(c) Phase Segment 1 (PhaseSeg1), Phase Segment 2 (PhaseSeg2)

The two phase segments, PS1 and PS2 are used to compensate for edge phase errors on the bus. PS1 can be lengthened or PS2 can be shortened by resynchronization. PS1 is programmable from 1TQ to 8TQ and PS2 is programmable from 2TQ to 8TQ.

4.4.3.1.1.2 Sample point

The sample point is the point in the bit time in which the logic level is read and interpreted. The sample point is located at the end of phase segment 1, that is, the edge between TSEG1 and TSEG2.. The exception to this rule is, if the sample mode is configured to sample three times per bit. In this case, the bit will be sampled three quanta in a row, with the last sample being taken in the edge between TSEG1 and TSEG2. In such case, the value of the bit is determined by a majority decision. Three samples should only be used for relatively slow baud rates.

The sample point should be the same in all connected CAN nodes, as shown in Figure 4.20. However, in practice it is sufficient if the sample points do not vary too much. The automotive industry has specified some recommendations for the bit timing settings in the SAE J2284 series. For CANopen networks, the recommendations given in CiA 301 should be used (CiA is abbreviation of "Controller in Automotive"). The sample points should be at 87,5 percent of the bit time. For most of the specified bit rates a range of 85 percent to 90 percent is allowed. Of course, the larger the tolerance range, the shorter the maximum possible length of the network.

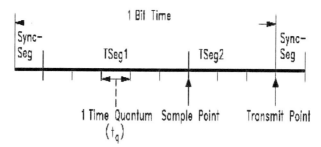

Figure 4.20   Bit segments.

4.4.3.1.1.3 Implementation of bit segments in practical CAN controllers

In most implementations of CAN controllers the segments seem to be implemented in a different way than described in the standard. The synch segment looks as in the standard and consists of one time quantum. The big difference is that the propagation segment and the phase segment 1 in the standard have been combined into one segment, TSEG1. Phase segment 2 is left untouched, but is renamed to TSEG2.

4.4.3.1.1.4 Information processing time (IPT)

The Information Processing Time (IPT) is the time required for the logic to determine the bit level of a sampled bit. The IPT begins at the sample point. It is measured in TQ and is fixed. For many CAN modules, for example, Microchip CAN module, it begins at 2TQ. Since phase segment 2 also begins at the sample point and is the last segment in the bit time, it is required that the minimum value of Phase Segment 2 (PS2) to be not less than the IPT. Therefore:

$$PS2_{min} = IPT = 2TQ$$

4.4.3.1.2 Bit timing setting for CAN FD

For each node, the distance (time) between the detected and the expected position of a falling or raising edge is called the phase error of that edge.

CAN FD uses two bit rates. The bit timing setting for the arbitration phase follows the same rules as in Classical CAN. In addition, the time quantum should be as short as possible, in order to reduce the quantization error in the data phase. This means a bit should be made of as many time quanta as possible. Some chipmakers provide a maximum of 385 tq for the arbitration phase.

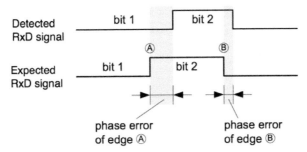

Figure 4.21 Bit-timing for CAN FD.

The data phase bit timing is independent of the arbitration phase bit timing. In the CAN FD data phase, the same parameters are to be set as in the arbitration phase (bit rate prescaler, length of propagation + phase-segment 1, phase-segment 2, and the synchronization jump width). Additionally, you need to configure the transmitter delay compensation (TDC) and its offset (up to two bits). This is necessary especially for higher bit rates. The CiA 601-1 document gives hints and guidelines to calculate appropriate values. There are bit rate depended effects to be considered as well as effects not dependent on the bit rate.

Because in the data phase the nodes are not synchronized, the time quanta in all nodes should be the same or multiple of the others. This is why only oscillator frequencies of 20 MHz, 40 MHz, or 80 MHz are allowed. The sample point should be at the same point in all nodes.

4.4.3.2 Calculation of baud-rate and sample point
(a) Calculation of baud rate
The baud rate of the bus can be calculated from:

$$\text{Baud rate} = f_{crystal}/(2 * n * (BRP + 1))$$

Where

– n is the number of time quanta for one bit and is defined as :

$$n = \text{SYNCHSEG+TSEG1+TSEG2}$$

– BRP is the value of the Baud Rate Prescaler.

Important Note: Some CAN controllers (like Intel 526) has another way of calculating the number of time quantas in a bit. In general, it is recommended for the designer to consult the user's manual of the CAN Controller that he/she is using in the application.

(b) Calculation of Sample point

$$Quanta_{before\ sample} = TSEG1 + 1$$
$$Quanta_{after\ sample} = TSEG2$$

Often the sample point is given in percent of the bit time. This is:

(TSEG1+1)/(TSEG1+1+TSEG2)

Important note: Some CAN controllers (like the C167CR) use another way of calculating $Quanta_{after\ sample}$. It is recommended for the user to consult the manual of the controller he/she is using.

Definition: Time quantum

Each of the four segments that make up a bit time is made up of integer units called Time Quanta (TQ). The length of each Time Quantum is based on the oscillator period (t_{OSC}). The base TQ equals twice the oscillator period. Figure 4.22 shows how the bit period is derived from TOSC and TQ. The TQ length equals one TQ Clock period (t_{BRPCLK}), which is programmable using a programmable prescaler named the Baud Rate Prescaler (BRP). This is shown in the following equation:

$$TQ = 2 \cdot BRP \cdot T_{OSC} = \frac{2 \cdot BRP}{f_{OSC}}$$

The value of BRP depends on the configuration as shown in Figure 4.22.

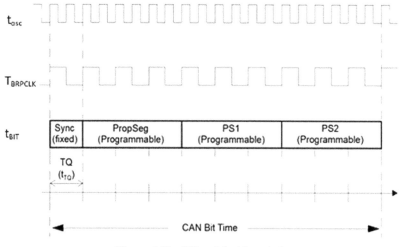

Figure 4.22   TQ and the bit period.

4.4.3.3 Synchronization jump width (SJW)

The Synchronization Jump Width (SJW) adjusts the bit clock as necessary by 1TQ–4TQ to maintain synchronization with the transmitted message.

The value of SJW is configured (adjusted) by the two bits SJW.0 and SJW.1 of one of the Bit timing Control Registers, SNF1, as follows:

<SJW.0, SJW.1>	00	01	10	11
Value of SJW in TQ	1 TQ	2 TQ	3 TQ	4 TQ

The calculation of the minimum value of SJW is given in Example 1 and the Bit Timing Control Register is given next.

4.4.3.4 Bit timing control registers

As mentioned before and as it is shown in Figure 4.22, the three segments PropSeg, PS1 and PS2 are programmable, that is, the number of quanta in each can be adjusted according to the application. This is achieved by using special register called "Bit Timing Control Register." The CAN Bit Timing Control (CNF) registers are the three registers that configure the CAN bit time. Figure 4.23 details the function of the CNF registers.

By adjusting the length of the TQ (t_{TQ}) and the number of TQs in each segment, both the nominal bit time and the sample point can easily be configured as desired.

Programming the Timing Segments: There are several requirements for programming the CAN bit timing segments.

1. PropSeg + PS1 = PS2
2. PropSeg + PS1 = t_{PROP}
3. PS2 > SJW

4.4.3.5 Synchronizing the bit time

All nodes on the CAN bus must have the same nominal bit rate. Noise, phase shifts, and oscillator drift create situations where the nominal bit rate does not equal the actual bit rate in a real system. Therefore, the nodes must have a method for achieving and maintaining synchronization with bus messages.

Nodes are synchronized on the bit edges so that every node agrees on the value of the bit currently transmitted on the bus. To do so, each node implements a synchronization protocol that keeps the receiver bit rate aligned with the actual rate of the transmitted bits. The synchronization protocol uses transition edges to resynchronize nodes. Hence, long sequences without bit transitions should be avoided to avoid drifts in the node bit clocks. This is

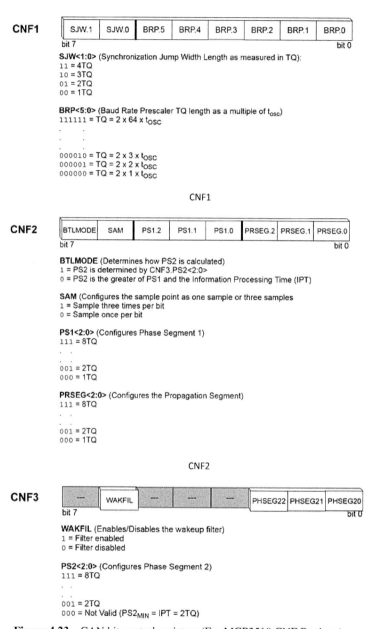

Figure 4.23   CAN bit control registers (For MCP2510 CNF Register).

the reason why the CAN protocol employs the so called "bit stuffing" or "bit padding" technique, which forces a complemented bit in the stream after 5 bits of the same type have been transmitted. Stuffing bits are automatically inserted by the transmission node and removed at the receiving side before processing the frame contents.

Synchronous bit transmission enables the CAN arbitration protocol and simplifies data-flow management, but also requires a sophisticated synchronization protocol. Bit synchronization is performed first upon the reception of the start bit available with each asynchronous transmission. Later, to enable the receiver(s) to correctly read the message content, continuous resynchronization is required. Other features of the protocol influence the definition of the bit timing. For the purpose of bus arbitration, message acknowledgment and error signaling, the protocol requires that nodes can change the status of a transmitted bit from recessive to dominant, with all the other nodes in the network being informed of the change in the bit status before the bit transmission ends. This means that the bit time must be at least large enough to accommodate the signal propagation from any sender to any receiver and back to the sender.

4.4.3.6 Oscillator tolerance

The bit timing for each node in a CAN system is derived from the reference frequency (fOSC) of its node. This creates a situation where phase shifting and oscillator drift will occur between nodes due to less than ideal oscillator tolerances between the nodes. The CAN specification indicates that the worst case oscillator tolerance is 1.58% and is only suitable for low bit rates (125 kbps or less). The details of the effect of oscillator tolerances can be found in some of the references of this chapter.

4.4.3.7 Propagation delay

The CAN protocol has defined a recessive (logic '1') and dominant (logic '0') state to implement a nondestructive bit-wise arbitration scheme. It is this arbitration methodology that is affected most by propagation delays. Each node involved with arbitration must be able to sample each bit level within the same bit time.

For example, if two nodes at opposite ends of the bus start to transmit their messages at the same time, they must arbitrate for control of the bus. This arbitration is only effective if both nodes are able to sample during the same bit time. Figures 4.24 and 4.25 show a one-way propagation delay between two nodes. Extreme propagation delays (beyond the sample point) will result

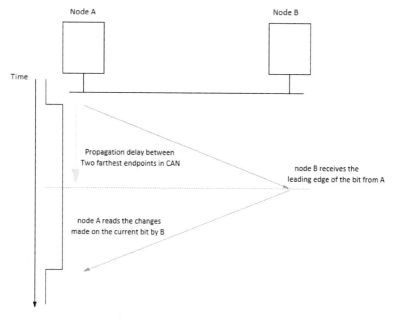

Figure 4.24   Bit propagation delay.

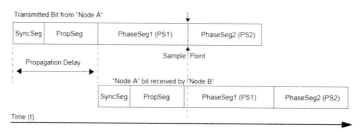

Figure 4.25   One-way propagation delay.

in invalid arbitration. This implies that bus lengths are limited at given CAN data rates. In Section 4.4.3.1.1, bit timing is considered.

A CAN system's propagation delay (t_{prop}) is calculated as being a signal's round-trip time on the physical bus (t_{bus}), the output driver delay (t_{drv}) and the input comparator delay (t_{cmp}). Assuming all nodes in the system have similar component delays, the propagation delay is explained mathematically by equation (4.1).

$$t_{prop} = 2 \cdot (t_{bus} + t_{cmp} + t_{drv})$$

4.4.3.8 Synchronization: Synchronization Mechanisms used in CAN

(a) Overview

All nodes on the CAN bus must have the same nominal bit rate. Noise, phase shifts, and oscillator drift create situations where the nominal bit rate does not equal the actual bit rate in a real system. Therefore, the nodes must have a method for achieving and maintaining synchronization with bus messages.

Nodes are synchronized on the bit edges so that every node agrees on the value of the bit currently transmitted on the bus. To do so, each node implements a synchronization protocol that keeps the receiver bit rate aligned with the actual rate of the transmitted bits. The synchronization protocol uses transition edges to resynchronize nodes. Hence, long sequences without bit transitions should be avoided to avoid drifts in the node bit clocks. This is the reason why the CAN protocol employs the so called "bit stuffing" or "bit padding" technique, which forces a complemented bit in the stream after 5 bits of the same type have been transmitted. Stuffing bits are automatically inserted by the transmission node and removed at the receiving side before processing the frame contents.

Synchronous bit transmission enables the CAN arbitration protocol and simplifies data-flow management, but also requires a sophisticated synchronization protocol. Bit synchronization is performed first upon the reception of the start bit available with each asynchronous transmission. Later, to enable the receiver(s) to correctly read the message content, continuous resynchronization is required. Other features of the protocol influence the definition of the bit timing. For the purpose of bus arbitration, message acknowledgment and error signaling, the protocol requires that nodes can change the status of a transmitted bit from recessive to dominant, with all the other nodes in the network being informed of the change in the bit status before the bit transmission ends. This means that the bit time must be at least large enough to accommodate the signal propagation from any sender to any receiver and back to the sender.

(b) Synchronization Mechanisms used in CAN

Any node receiving a frame has to be synchronized with the transmitter. The NRZ bit coding used in CAN does not encode a clock into the message. In other words, CAN system has no explicit clock signal that can be used by the receiver and the transmitter as reference to achieve synchronization. In spite of that, the receivers must synchronize to the transmitted data stream to insure messages are properly decoded. CAN nodes use two different mechanisms to synchronize their clocks, *"**hard synchronization**"* and "**re-synchronization**":

• *Hard synchronization*

Hard synchronization occurs only once during a message transmission. It only occurs on the first recessive-to-dominant (logic "1" to "0") edge during a bus idle condition, which indicates a Start-of-Frame (SOF) condition. The Start of Frame is the first bit of the frame and it is transmitted dominantly. Before it the CAN bus is in a recessive state. All the nodes connected to the bus automatically synchronize its clock using the transition created by the SOF bit. Hard synchronization causes the bit timing counter to be reset to the SyncSeg which causes the edge to lie within the SyncSeg. At this point, all of receivers will be synchronized to the transmitter.

Synchronization takes place during transmission. The phase segments are shortened or lengthened so that the following bit starts within the SyncSeg portion of the following bit time. In detail, PhaseSeg1 may be lengthened or PhaseSeg2 may be shortened. Damping is applied to the synchronization protocol. The amount of lengthening or shortening of the PHASE BUFFER SEGMENTs has an upper bound given by a programmable parameter Resynchronization Jump Width (RJW) (between 1 and min (4, PhaseSeg1) Time Quanta).

Synchronization information may be only be derived from transitions from one bit value to the other. Therefore, the possibility of resynchronizing a bus unit to the bit stream during a frame depends on the property that a maximum interval of time exists between any two bit transitions (enforced by the bit stuffing protocol).

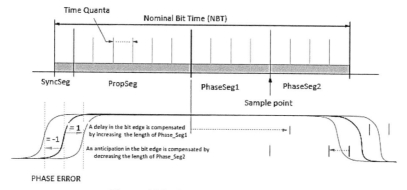

Figure 4.26   Sample point of CAN bit.

The device designer may program the bit-timing parameters in the CAN controller by means of the appropriate registers. We note here that depending on the size of the propagation delay segment the maximum possible bus length at a specific data rate (or the maximum possible data rate at a specific bus length) can be determined.

We note here that an SOF bit can be received both during Bus Idle, and also during Suspend Transmission and at the end of Interframe Space. This is why the Bosch CAN Reference Model enables Hard Synchronization not only for Bus Idle state, but also for Suspend state and the last bit of Interframe Space. Any node disables Hard Synchronization if it samples an edge from recessive to dominant or if it starts to send the dominant SOF bit.

Since synchronization on edges from dominant to recessive has become obsolete with the upgrade from CAN specification 1.1 to version 1.2, the Bosch CAN Reference Model does not support this kind of synchronization!

- *Resynchronization:*

Hard synchronization mechanism achieves the synchronization at the beginning of the message but it is not enough to compensate for oscillator drift, and phase differences between transmitter and receiver oscillators during the entire time of the frame. This makes the clocks to be not able to remain synchronized throughout the entire frame. To keep the clock continually resynchronize, CAN uses an additional synchronization mechanism: **"resynchronization"**

- Resynchronization occurs every time the bus transitions from recessive to dominant. The bit stuffing used in CAN (see latter)guarantee that a transition will occur in the bit stream even if the original message is a string of "0"s or "1"s. Bit stuffing does not allow the contents of the frame to has more than five consecutive identical bit levels. If it happened, the transmitter will automatically add (stuff) a bit of the opposite polarity into the bit stream.
- Resynchronization is automatically invoked if the system does not recognize the occurrence of an edge at the "Synchronization Segment" of any of the bit that follow the SOF bit of any received frame. These results in either shorten or lengthen the current bit time depending on where the edge occurs. The maximum amount by which the bit time is lengthened or shortened is determined by a user-programmable number of time quanta known as the *Synchronization Jump Width* (SJW).

Resynchronization is achieved by implementing a Digital Phase Lock Loop (DPLL) function which compares the actual position of a

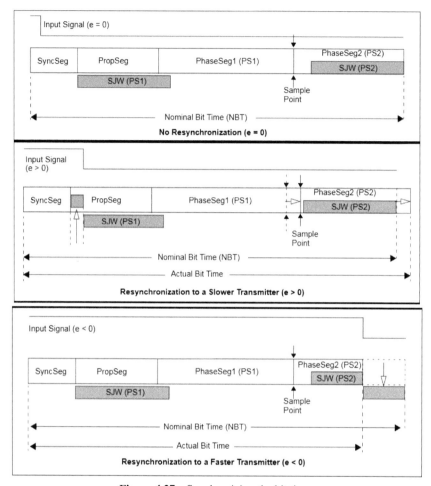

Figure 4.27   Synchronizing the bit time.

recessive-to-dominant edge on the bus to the position of the expected edge (within the SyncSeg) and adjusting the bit time as necessary. The phase error (e) of a bit is given by the position of the edge in relation to the SyncSeg, measured in TQ, and is defined as follows:

- e = 0; the edge lies within the SyncSeg.
- e > 0; the edge lies before the sample point. (TQ added to PS1).
- e < 0; the edge lies after the sample point of the previous bit. (TQ subtracted from PS2).

Figure 4.27 shows the synchronization of the Bit Time.

Table 4.2 CAN bit rate versus bus length

Bit Rate (kbps)	Bus Length
1000	30
500	100
250	250
125	500
62.5	1000

Figure 4.27 shows how phase errors, other than zero, cause the bit time to be lengthened or shortened. Synchronization Rules:

1. Only recessive-to-dominant edges will be used for synchronization.
2. Only one synchronization within one bit time is allowed.
3. An edge will be used for synchronization only if the value at the previous sample point differs from the bus value immediately after the edge.
4. A transmitting node will not resynchronize on a positive phase error $(e > 0)$. This implies that a transmitter will not resynchronize due to propagation delays of it is own transmitted message. The receivers will synchronize normally.
5. If the absolute magnitude of the phase error is greater than the SJW, then the appropriate phase segment will be adjusted by an amount equal to the SJW.

Putting It All Together: As indicated previously, the CAN protocol implements a non-destructive bitwise arbitration scheme that allows multiple nodes to arbitrate for control of the bus. Therefore, it is necessary for all the nodes to detect/sample the bits within the same bit time. The relationship between propagation delay and oscillator tolerance effects both the CAN data rate and the bus length. Table 4.2 shows some commonly accepted bus lengths versus data rates.

This application note does not cover all of the details for configuring the bit time for all scenarios; however, some general methodologies for configuring the CAN bit time are covered.

4.4.3.9 Calculating oscillator tolerance for SJW

The bit stuffing rule guarantees that no more than five like bits in a row will be transmitted during a message frame. The only exception is at the end of the message that includes ten recessive bits (one ACK delimiter, seven end-of-frame bits, and three inter frame space bits).

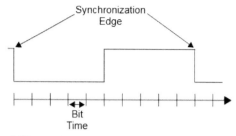

Figure 4.28   Maximum time between synchronization edges.

Resynchronization can only occur on recessive-to-dominant edges. This implies that there can be a maximum of ten bits between resynchronization due to bit stuffing (Figure 4.28).The oscillator tolerance between the slowest node and the fastest node can be used to determine the minimum SJW. Assuming Node A is the slow node (longest bit time) and Node B is the fast node (shortest bit time):

$$10\ t_{bit(A)} > 10\ t_{bit(B)} + t_{SJW(B)} \qquad (4.2)$$

Where

$$t_{bit(n)} = \text{bit time of node "n"}$$
$$t_{SJW(n)} = \text{SJW of node "n"}$$

4.4.3.10 Configuring the bit

In general, the longer the bus, the slower the maximum data rate due to propagation delays on the line. Increasing the oscillator tolerances between nodes can greatly amplify the relationship. CAN system designers must take this relationship into consideration when defining the network. The following examples demonstrate bit timings for achieving maximum oscillator tolerance or maximum bit rate.

Example 4.1: How to find Minimum SJW

Given the following:

    Nominal Bit Time = 1 μs
    Oscillator tolerance = 1.25%
    Number of TQ per bit = 8

Find SJW minimum

$$t_{bit(A)} = 1.01200\ \mu s$$

$$t_{bit(B)} = 0.98875 \; \mu s$$
$$TQ_{(A)} = 126.563 \; ns$$
$$TQ_{(B)} = 123.438 \; ns$$

Using Equation (4.2) we get

$$t_{SJW(B)} > 10 \, t_{bit(A)} - 10 \, t_{bit(B)} = 0.2325. \text{ We take it} = 0.250 \; \mu s$$

$$\text{Number } TQ_{SJW} > \frac{t_{SJW(B)}}{TQ_{(B)}} = \frac{250ns}{123.44ns} = 2.025$$

$$\text{Number } TQ_{SJW} = 3$$

Alternatively, the following equation can be used to maintain synchronization during normal bus operation: Solving for Oscillator Tolerance (Δf),

$$SJW > (2 \Delta f)(10 NBT)$$

Solving for Oscillator Tolerance Δf, we get

$$\Delta f < SJW/20 NBT$$

Example 4.2: Calculation of the Maximum Oscillator Tolerance.

The maximum oscillator tolerance for a maximum data rate is achieved when the phase segments 1 and 2 are equal to the maximum synchronization jump width (4 TQ). Also, the propagation segment is minimum, indicating a short bus and fast transceiver.

As mentioned before and given by equation (1), the propagation delay is twice the delays of the bus, the receiver circuitry, and the driver:

$$t_{prop} = 2 \cdot (t_{bus} + t_{cmp} + t_{drv})$$

Assume that the bus length is 50 m and the bus propagation delay is 5.5 ns/m:

$$t_{BUS} = 50 \text{ m} \times 5.5 \text{ ns/m} = 275 \text{ ns}$$
$$t_{CMP} = 40 \text{ ns}$$
$$t_{DRV} = 60 \text{ ns}$$

Then:

$$t_{prop} = 2 \cdot (t_{bus} + t_{cmp} + t_{drv}) = 750 \text{ ns}$$

Since the propagation segment is used to compensate for propagation delays and must be set to the minimum of 1QT, the implied time quantum $= t_{PROP} = 750$ ns

Figure 4.29 shows the bit timing in this case.

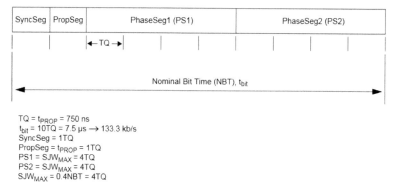

TQ = t$_{PROP}$ = 750 ns
t$_{bit}$ = 10TQ = 7.5 µs → 133.3 kb/s
SyncSeg = 1TQ
PropSeg = t$_{PROP}$ = 1TQ
PS1 = SJW$_{MAX}$ = 4TQ
PS2 = SJW$_{MAX}$ = 4TQ
SJW$_{MAX}$ = 0.4NBT = 4TQ

Figure 4.29   Bit timing for maximum oscillator tolerance.

Example 4.3: Calculation of Maximum Bit Rate.

The previous example showed that for a given bus length, the maximum data rate is inversely affected, due to oscillator tolerance (as oscillator tolerance goes up, the data rate goes down). To achieve the maximum bit rate for a given bus length, the emphasis is placed on configuring the bit time for the propagation delays (i.e., adjusting PropSeg to maximum). The oscillator tolerance must be minimized. Given the same delays as the previous example:

$$t_{BUS} = 50 \text{ m @ } 5.5 \text{ ns/m} = 275 \text{ ns}$$
$$t_{CMP} = 40 \text{ ns}$$
$$t_{DRV} = 60 \text{ ns}$$
$$t_{prop} = 2 \cdot (t_{bus} + t_{cmp} + t_{drv}) = 750 \text{ ns}$$

Since the oscillator tolerance is minimum, the phase segments and SJW can be set to the minimum. Assuming the bit time is 10TQ total, the PropSeg can be set to 6TQ which sets TQ = 125 ns. Figure 4.30 shows the bit timing for maximum bit rate.

4.4.3.11 The physical media attachment sublayer

The PMA sublayer is normally implemented in the transceiver chip. The inputs are the TxD and RxD signals from the CAN controller. The outputs drive the bus-lines. There are different solutions available:

- Most important is the CAN high-speed transmission as specified in ISO 11898-2.
- Increasingly in use are high-speed transceivers with low-power functionality also specified in ISO 11898-2 (formerly in ISO 11898-5).

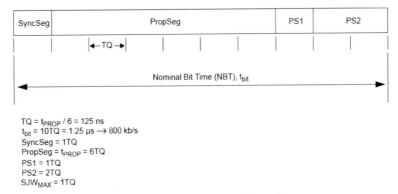

TQ = t$_{PROP}$ / 6 = 125 ns
t$_{bit}$ = 10TQ = 1.25 µs → 800 kb/s
SyncSeg = 1TQ
PropSeg = t$_{PROP}$ = 6TQ
PS1 = 1TQ
PS2 = 2TQ
SJW$_{MAX}$ = 1TQ

Figure 4.30 Bit-timing for maximum bit rate.

- In some automotive applications, high-speed transceivers with an additional selective wake-up capability as specified in ISO 11898-2 (formerly in ISO 11898-6) are used.
- The low-speed (up to 125 kbps), fault-tolerant, and low-power transceivers standardized in ISO 11898-3 will be increasingly substituted by high-speed transceivers with low-power functionality.
- The low-speed (up to 125 kbps) transceivers providing a higher voltage level as the others are required in Europe for truck-trailer point-to-point CAN communication (ISO 11992-1).
- Transceivers for single-wire communication (SWC) as specified in SAE J2411 provide a low-speed transmission and are rarely used.

The high-speed transceivers are qualified for different bit rates (e.g., 1 Mbps, 2 Mbps, 5 Mbps, and 8 Mbps). Components for Classical CAN transmissions up to 1 Mbps don't need to fulfill asymmetry requirements. Transceivers for bit rates above 1 Mbps do. Of course, you can use chips qualified for a higher bit rate than actually used in the network.

4.4.3.12 The medium-dependent sublayer

There are only a few general requirements for this sublayer. It is highly application-specific. Of course, if you use a high-speed transmission, a line topology is recommended (daisy chained or with very short stubs). Star (single or double) topologies with terminations in the center are often used for Classical CAN. For CAN FD networks they can't be recommended though. The same is true for hybrid topologies, made of a star and a line.

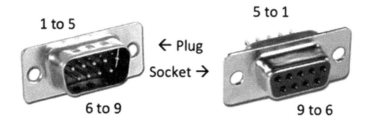

Pin	Signal	Description
1	-	Reserved for CIA
2	CAN_L	CAN bus-line dominant low
3	CAN_GRD	CAN ground
4	-	Reserved for CIA
5	(CAN_SHLD)	Optional CAN shield
6	(GND	Optional power ground
7	CAN_H	CAN bus-line dominant high
8	-	Reserved for CIA
9	(CAN_V+)	Optional (external) power supply

Figure 4.31   Pin description of 9-pin connector.

As early as 1992, CiA recommended a pinning for 9-pin D-sub connectors (DIN 41652) (Figure 4.31).

In case of high-speed transmissions, both ends are terminated nominally with 120 Ohm matching the impedance of all physical layer components (cable, connector, optional galvanic isolation circuitry). In order to achieve a maximum network length, the physical layer components should not cause delays. For the high-speed transmission cable, a delay of 5 ns/m is suitable.

Also, the standardization for pin-assignment for different connectors belongs to the medium-dependent sublayer. CiA has recommendations for the pinning of many connector types. The very first one for 9-pin D-sub connectors was already specified in 1992. Most companies use this pinning. Only CAN shields with a 9-pin D-sub connector for Raspberry have a different one.

For CAN open networks, the CiA 303-1 document provides more recommendations for the cabling and the other electro-mechanical components.

4.4.4 Maximum Number of Nodes

In practice, up to 64 nodes may be connected to a DeviceNet bus, 127 on a CANopen bus and up to 255 nodes may be connected together on a CANKingdom bus. When more than the standard 30 nodes are used on a bus, it is recommended that a transceiver with a high bus-input impedance, such as an HVD230 or HVD251 be used.

A problem may develop when too many transceivers source or sink current onto or from the bus. When a device begins to transmit a message, it has to sink or source all of the leakage current on the bus, plus drive the standard signal voltage levels across the termination resistance. If the current demand is too great, the device may be driven into thermal shut-down or destroyed.

To prevent transceiver damage, the voltage difference between reference grounds of the nodes on a bus should be held to a minimum. This is the common-mode voltage across the entire system and although many transceivers such as the HVD251 are designed to operate over an extended common-mode range, the cumulative current demand of too many devices at a common-mode voltage extreme may jeopardize network security. To enhance this common-mode security, most of the higher layer protocols like DeviceNet specify that power and ground wires be carried along with the signaling pair of wires. Several cable companies have developed 4-wire bundled cables specifically for these applications.

4.5 CAN Transceiver

In this section, the general features of CAN Transceiver are introduced. While explain the features refer will take place to SN65HVD251 as example and MCP2551 is taken as an example.

4.5.1 CAN Transceiver Features

4.5.1.1 Supply voltage

Most CAN transceivers like the SN65HVD251 require a 5-V power supply to reach the signal levels required by the ISO 11898 Standard. Other transceiver, for example, Texas Instrument SN65HVD23x family, requires 3.3 V. Shifting to 3.3 V allows designers to reduce total node power by 50% or more (Figure 4.32).

For applications using 3.3-V technology, such as the Texas Instruments TMS320C240x family of DSPs with integrated CAN controllers, the need

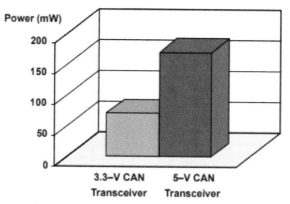

Figure 4.32 V-CAN transceiver power savings.

for a 5 V power supply can be eliminated. In addition to the inherent power savings of using a 3.3-V transceiver, this lowers the overall part count for the node, thereby reducing system cost and increasing system reliability.

4.5.1.2 High short-circuit protection

The CAN standard recommends that a transceiver survive bus wire short-circuits to each other, to the power supply, and to ground. This ensures that transceivers are not damaged by bus cable polarity reversals, cable crush, and accidental shorts to high power supplies. As an example, the 'HVD23x can survive short-circuits to voltages in the range of -4 V to 16 V while, the 'HVD251 provides an extended short-circuit protection to voltages in the -36 V to 36 V range.

4.5.1.3 High ESD protection

Static charge is an unbalanced electrical charge at rest, typically created by the physical contact of different materials. One surface gains electrons, while the other surface loses electrons. This results in an unbalanced electrical condition known as a static charge. When a static charge moves from one surface to another, it is referred to as an electrostatic discharge (ESD). It can occur only when the voltage differential between the two surfaces is sufficiently high to break down the dielectric strength of the medium separating the two surfaces.

In case if the CAN transceiver was not specifically designed to reject this coupled noise, the transceiver can respond to common-mode noise as if it were data on a bus and send meaningless data to a controller.

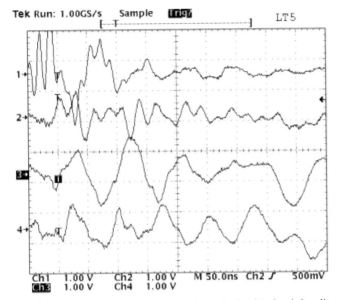

Figure 4.33 Common-mode noise coupled onto 4 twisted-pair bus lines.

The Texas Instrument transceivers, HVD23x and HVD251, for example, are specifically designed and tested for their ability to reject this common-mode noise.

4.5.1.4 High input impedance

A high-bus input impedance increases the number of nodes that can be added to a bus above the standard 30 nodes. The high impedance restricts the amount of current that a receiver sinks or sources onto a bus over common-mode voltage conditions. This ensures that a driver transmitting a message into such a condition is not required to sink or source an excessive amount of current from the sum of the receiver currents on the bus.

4.5.1.5 Controlled driver output transition times

Controlling driver output slew-rate dampens the rise-time of a dominant bit to improve signal quality and provides for longer stub lengths and a better bit-error rate.

4.5.1.6 Low current standby and sleep modes

Many applications are looking to lower-power opportunities as more electronics are added to designs. To achieve the low-power consumption, some transceivers, for example, "HVD230", has a standby mode. The standby mode is generally referred to as "listen only mode", since in this mode driver circuitry is switched off while the receiver continues to monitor bus activity. In the occurrence of a dominant bit on the bus, the receiver passes this information along to its DSP/CAN controller which in turn activates the circuits that are in standby. In some transceiver, for example, HVD231 both driver and receiver circuits can be switched off to create an extremely low-power sleep mode.

4.5.1.7 Thermal shutdown protection

Another safety feature that is recommended for transceiver is to poses "thermal shutdown circuitry." Another desirable safety feature for a CAN transceiver is the thermal shutdown circuitry. This safety feature exists in 'HVD23x and 'HVD251. This feature protects a device against the destructive currents and resulting heat that can occur in a short-circuit condition. Once thermal shutdown is activated, the device remains shut down until the circuitry is allowed to cool.

4.5.1.8 Glitch free power up and power down

This feature needed for hot-plugging onto a powered bus without disturbing the network. In case of TI transceiver, 'HVD23x's and 'HVD251's driver and receiver pins are passively pulled high internally, while the bus pins are biased internally to a high-impedance recessive state. This provides for a power-up into a known recessive condition without disturbing ongoing bus communication.

4.5.1.9 Unpowered node does not disturb the bus

Several CAN transceivers on the market today have very low output impedance when un–powered. This low impedance causes the device to sink any signal present on the bus and shuts down all data transmission. Other transceivers, for example, 'HVD23x and 'HVD251, have a very high output impedance in powered and unpowered conditions, and maintain the integrity of the bus when power or ground is removed from the circuit.

4.5.1.10 Reference voltage

Reference voltage on a CAN transceiver is the Vref pin (pin 5) of what is considered to be the standard. CAN transceiver footprint. This is the footprint of the first CAN transceiver to market, the NXP PCA82C250. When first

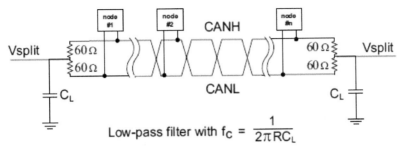

Figure 4.34 Split termination.

introduced, the Vcc/2 Vref pin served a particular NXP CAN controller as a voltage reference used to compare the bus voltage of a remaining single bus line in the event of an accident. If the voltages were the same, it was a recessive bit; if different, it was a dominant bit. Although some users consider it handy for use as an actual voltage reference at the node, it is typically unused.

4.5.1.11 V-Split

V-split is a fortified Vcc/2 Vref pin with the same ESD protection rating, short-circuit protection, and common-mode operating range as the bus pins. It is used to stabilize bus voltage at Vcc/2 and prevent it from drifting to a high common-mode voltage during periods of inactivity.

It also filters unwanted high-frequency noise from bus lines with the termination technique of Figure 4.34.

This is accomplished with a coupling capacitor between two ~60 O ±1% termination resistors to couple high-frequency noise to a solid ground potential. Care must be taken to match the two resistors carefully so as not to reduce the effective immunity. This technique improves the electromagnetic compatibility of a network. A typical value of CL for a high-speed CAN is 4.7 nF, which generates a 3-dB point at 1.1 Mbps. This, of course, is a signaling-rate-dependent value.

4.5.1.12 Loopback

This function places the bus input and output in a high-impedance state. The remaining transceiver circuitry remains active and available for driver-to-receiver loopback and self-diagnostic node functions without disturbing the bus.

4.5.1.13 Autobaud loopback

In autobaud loopback, the "bus-transmit" function of the transceiver is disabled, while the "bus-receive" function and all of the normal operating functions of the device remain intact. With the autobaud function engaged, normal bus activity can be monitored by the device.

Autobaud detection is best suited to applications that have a known selection of baud rates. For example, a popular industrial application has optional settings of 125 kbps, 250 kbps, or 500 kbps. Once a logic high has been applied to pin 5 (AB) of the HVD235, assume a baud rate such as 125 kbps, then wait for a message to be transmitted by another node on the bus. If the wrong baud rate has been selected, an error message is generated by the host CAN controller. However, because the "bus-transmit" function of the device has been disabled, no other nodes receive the error message of the controller.

This procedure makes use of the CAN controller's status register indications of message received and error warning status to signal if the current baud rate is correct or not. The warning status indicates that the CAN chip error counters have been incremented. A message-received status indicates that a good message has been received.

If an error is generated, reset the CAN controller with another baud rate and wait to receive another message. When an error-free message has been received, the correct baud rate has been detected.

4.5.2 CAN Transceiver Example: MCP2551 CAN TRANSCEIVER

The MCP2551 is a CAN Transceiver that implements the ISO-11898-2 physical layer specification. It supports a 1 Mbps data rate and is suitable for 12 V and 24 V systems. The MCP2551 provides short-circuit protection up to ±40 V and transient protection up to ±250 V.

In addition to being ISO-11898-2-compatible, the MCP2551 provides power-on reset and brown-out protection, as well as permanent dominant detection to ensure an unpowered or faulty node will not disturb the bus. The device implements configurable slope control on the bus pins to help reduce RFI emissions. Figure 4.35 shows the block diagram of the MCP2551.

4.5.2.1 General MCP2551 Operation
4.5.2.1.1 Transmit

The CAN protocol controller outputs a serial data stream to the logic TXD input of the MCP2551. The corresponding recessive or dominant state is output on the CANH and CANL pins.

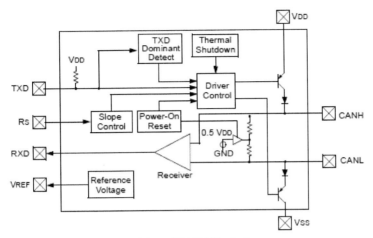

Figure 4.35 MCP2551 block diagram.

4.5.2.1.2 Receive
The MCP2551 receives dominant or recessive states on the same CANH and CANL pins as the transmit occurs. These states are output as logic levels on the RXD pin for the CAN protocol controller to receive CAN frames.

4.5.2.1.3 Recessive STATE
A logic '1' on the TXD input turns off the drivers to the CANH and CANL pins and the pins "float" to a nominal 2.5 V via biasing resistors.

4.5.2.1.4 Dominant STATE
A logic '0' on the TXD input turns on the CANH and CANL pin drivers. CANH drives ~1 V higher than the nominal 2.5 V recessive state to ~3.5 V. CANL drives ~1 V less than the nominal 2.5 V recessive state to ~1.5 V.

4.5.2.2 Modes of operation
There are three modes of operation that are externally controlled via the RS pin:

1. High Speed
2. Slope Control
3. Standby

4.5.2.2.1 HIGH SPEED
The high-speed mode is selected by connecting the RS pin to VSS. In this mode, the output drivers have fast rise and fall times that support the

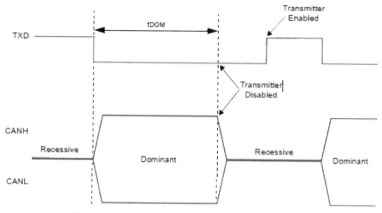

Figure 4.36   TXD permanent dominant detection.

higher bus rates up to 1 Mbps and/or maximum bus lengths by providing the minimum transceiver loop delays.

4.5.2.2.2 SLOPE CONTROL

If reduced EMI is required, the MCP2551 can be placed in slope control mode by connecting a resistor (REXT) from the RS pin to ground. In slope control mode, the single-ended slew rate (CANH or CANL) is basically proportional to the current out of the RS pin. The current must be in the range of 10 μA $<$ $-$IRS $<$ 200 μA, which corresponds to a voltage on the pin of 0.4 VDD $<$ VRS $<$ 0.6 VDD respectively (or 0.5 VDD typical).

The decreased slew rate implies a slower CAN data rate at a given bus length, or a reduced bus length at a given CAN data rate.

4.5.2.2.3 STANDBY

Standby (or sleep) mode is entered by connecting the RS pin to VDD. In sleep mode, the transmitter is switched off and the receiver operates in a reduced power mode. While the receive pin (RXD) is still functional, it will operate at a slower rate.

Standby mode can be used to place the device in low power mode and to turn off the transmitter in case the CAN controller malfunctions and sends unexpected data to the bus.

4.5.2.3 Permanent dominant detection on transmitter

The MCP2551 will turn off the transmitter to CANH and CANL if an extended dominant state is detected on the transmitter. This ability prevents a

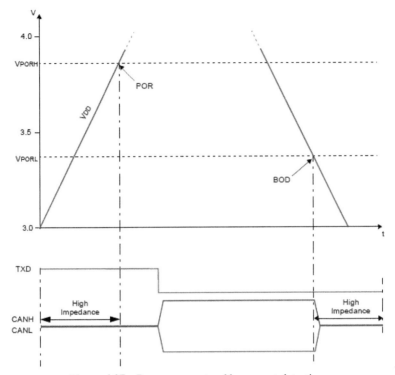

Figure 4.37 Power-on reset and brown-out detection.

faulty node (CAN controller or MCP2551) from permanently corrupting the CAN bus.

The drivers are disabled if TXD is low for more than ~1.25 ms (minimum). The drivers will remain disabled as long as TXD remains low. A rising edge on TXD will reset the timer logic and enable the drivers.

4.5.2.3.1 Power-on reset and brown-out
The MCP2551 incorporates both Power-On Reset (POR) and Brown-Out Detection (BOD) (see Figure 4.37).

4.5.2.3.2 Power-On Reset (POR)
When the MCP2551 is powered on, the CANH and CANL pins remain in the high impedance state until VDD reaches the POR high voltage (VPORH).

Additionally, if the TXD pin is low at power-up, the CANH and CANL pins will remain in high impedance until TXD goes high. After which, the drivers will function normally.

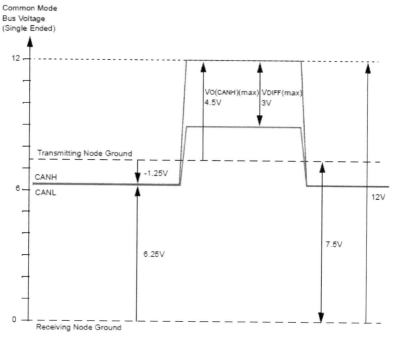

Figure 4.38   Receiving (node ground) below transmitting (node ground).

4.5.2.3.3 Brown-out detection (BOD)

BOD occurs when VDD goes below the power-on reset low voltage (VPORL). At this point, the CANH and CANL pins enter a high impedance state and will remain there until VPORH is reached.

4.5.2.3.4 Ground offsets

Since it is not required to provide a common ground between nodes, it is possible to have ground offsets between nodes. That is, each node may observe different single-ended bus voltages (common mode bus voltages) while maintaining the same differential voltage.

While the MCP2551 is specified to handle ground offsets from -12 V to $+12$ V, the ISO-11898 specification only requires -2 V to $+7$ V. Figure 4.38 and Figure 4.39 demonstrate how ground offsets appear between nodes.

Figure 4.39 shows the transmitting node with a positive ground offset with respect to the receiving node. The MCP2551 receiver can operate with CANH = $+12$ V.

The maximum CAN dominant output voltage (VO(CANH)) from the transmitting node is 4.5 V. Subtracting this maximum yields an actual ground offset (with respect to the receiving node) of 7.5 V for the transmitting node.

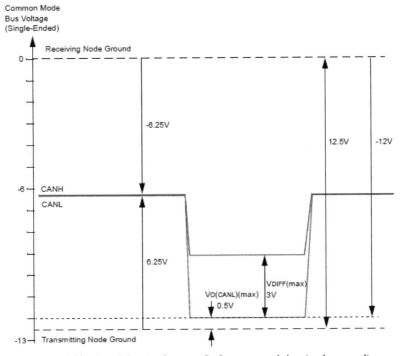

Figure 4.39 Receiving (node ground) above transmitting (node ground).

In the recessive state, each node attempts to pull the CANH and CANL pins to their biasing levels (2.5 V typical). However, the resulting common mode voltage in the recessive state becomes 6.25 V for the receiving node and −1.25 V for the transmitting node.

Figure 4.39 shows the transmitting node with a negative ground offset with respect to the receiving node. The MCP2551 receiver can operate with CANL = −12 V. The minimum CAN dominant output voltage (VO(CANL)) from the transmitting node is 0.5 V. Subtracting this minimum yields an actual ground offset, with respect to the receiving node, of −12.5 V. The common mode voltage for the recessive state is −6.25 V for the receiving node and 6.25 V for the transmitting node.

Since all nodes act as a transmitter for a portion of each message (i.e., each receiver must acknowledge (ACK) valid messages during the ACK slot), the largest ground offset allowed between nodes is 7.5 V, as shown in Figure 4.39.

Operating a CAN system with large ground offsets can lead to increased electromagnetic emissions. Steps must be taken to eliminate ground offsets if the system is sensitive to emissions.

5

CAN Data Link Layer

This Chapter is part of CAN bus. It concentrates on the Data Link Layer. The chapter will introduce the reader to many related subjects as: CAN Communication Services, Multi Buss Access, Standard and Extended CAN, CAN Messages and CAN bus Error Handling.

5.1 Data Link Layer

The CAN data link layer comprises two protocols: Classical CAN introduced in 1986 and implemented for the first time in 1988 and CAN FD launched in 2012 and internationally standardized in 2015 in ISO 11898-1. For a transitional period there are also non-ISO compliant implementations on the market. They are application-transparent, meaning they can be used for software development and designing prototype networks. However, CiA doesn't recommend using them for serial production.

The two CAN data link layer protocols have some common features. Any node has the right to request transmission rights at any time. The necessary bus arbitration method to avoid transmission conflicts is the same: Frames with the highest assigned identifier get bus access without delay. All frame types (data, remote, error, and overload frame) are transmitted in broadcast. The data frame structure comprising several fields is the same.

One of the unique features of the CAN data link layers is that all single-bit errors are detected. Multi-bit errors are detected with a high probability. In order to provide data consistency in all nodes, local errors are globalized. Additionally, the fault confinement implemented in the CAN data link layers precludes a single node from corrupting the communication of the others permanently.

197

Figure 5.1   CAN data frame.

Note: The structure of CAN data frames are the same for Classical CAN and CAN FD, just the field details are different.

The Classical CAN protocol uses just one bit rate in the arbitration and the data phase. The transmission speed is limited to 1 Mbps for short networks (theoretically up to 40 m). However, the achievable bit rate depends on the network length and the used physical layer elements such as cable, connector, and transceiver. The payload, the data field, is limited to 8 byte.

The CAN FD protocol allows payloads up to 64 byte. Additionally, it supports an optional second bit rate for the data-phase. The limitation of the speed in the arbitration phase is the same as for Classical CAN. In the data phase, the speed is limited by transceiver characteristic, the oscillator tolerance, and the topology (ringing). Data phase bit rates up to 8 Mbps are realistic when using a bus-line topology with very short, not terminated stubs.

The CAN data link layer is standardized in ISO 11898. The data link layer services are implemented in the Logical Link Control (LLC) and Medium Access Control (MAC) sublayers of a CAN controller (see Figure 4.3). The LLC provides acceptance filtering, overload notification and recovery management. The MAC is responsible for data encapsulation (de-capsulation), frame coding (stuffing/de-stuffing), medium access management, error detection, error signaling, acknowledgment, and serialization (de-serialization).

5.1.1 CAN Communication Services

5.1.1.1 Communication services

The CAN protocol provides two communication services:

- Write Object service: CAN Broadcast Communication.
- Read Object service: Remote Transmission Request (RTR)

CAN Broadcast Communication: Write Object Service

Write Object service transmits a Data Frame from one node (the producer) to one or more receiving nodes (consumers). This do not implies that one node will accept the message meaning that someone is interested in this

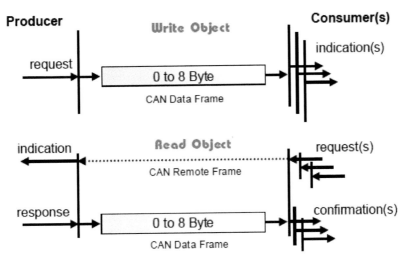

Figure 5.2 CAN communication services.

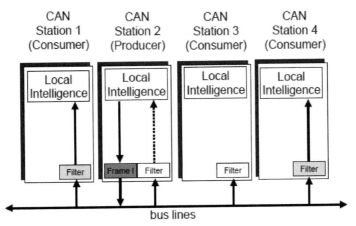

Figure 5.3 CAN broadcast communication.

information. This service is the classic CAN communication service: CAN Broadcast Communication (Figure 5.3).

The CAN concept of Broadcast Communication means that every station of the network can listen to the frames of the transmitting station (here: station 2, Figure 5.3). After receiving the frame it is the task of every node to decide if the message has to be accepted or not. So Acceptance Filtering has to be implemented in every CAN node.

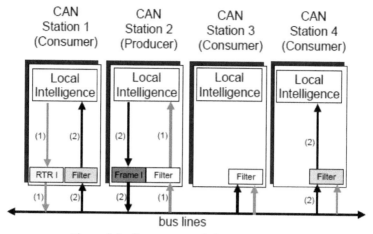

Figure 5.4   Remote transmission request (RTR).

The CAN Broadcast Communication can be compared with a radio station transmitting information about traffic accumulation for vehicle drivers.

Every driver has to decide if the messages are important for him dependent on the motorway he wants to use.

5.1.1.2 Remote transmission requests (RTR): Read object service

The second communication service is to request a specific message (Figure 5.4). This Read Object service is initiated by one or more consumers. Therefore these nodes will transmit a so-called Remote Frame (RTR). The node, which owns the requested information will transmit the corresponding Data Frame.

Remote Transmission Requests (RTR) are like questions. The node that has the answer will produce in a second communication the requested data. This data frame can be received also by other consumers, which are interested in this object Remote frames and data frames are identified by a specific field, called Identifier.

5.1.2 Multiple Bus Access

The CAN protocol allows simultaneous bus access from different nodes. If more than one node is accessing the bus, an arbitration is required. The bus access method used in CAN is a non-destructive, bit-wise arbitration, called Carrier Sense Multiple Access with Collision Detection and Arbitration on

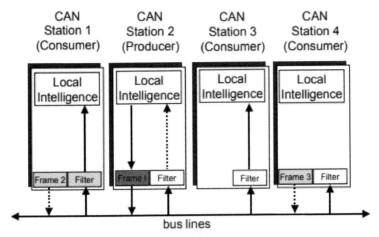

Figure 5.5 Multiple bus access.

Message Priority (CSMA/CD + AMP). CSMA means that each node on a bus must wait for a prescribed period of inactivity before attempting to send a message. CD+AMP means that collisions are resolved through a bit-wise arbitration, based upon a preprogrammed priority of each message in the identifier field of a message. The higher priority identifier always wins bus access.

The message priority is decoded in the CAN identifier.

When the bus is in Idle state, several nodes can start transmission of a frame. Every node reads back, bit by bit, from the bus during the complete message and compares the transmitted bit value with the received bit value.

Per definition the bits with a dominant value overwrites those with a recessive value (this has to be provided by the transceiver).

5.1.2.1 Meaning of CSMA/CD+AMP

Carrier Sense (CS): Every node must monitor bus for a period of no activity before sending a message. This period of no activity is called "inter-frame" space.

Multiple Access (MA): Once a period of no activity occurs, every node has an equal opportunity to transmit a message. This starts the arbitration period at the SOF.

Collision Detection (CD): If 2 nodes transmit at the same time, a collision occurs and is detected by the node transmitting a recessive bit.

Collision Resolution (CR): Collision resolution refers to "non-destructive bitwise arbitration."

- Messages remain intact even after collision occurs.
- All arbitration takes place without corruption or delay of the highest priority message.
- Any message that loses the CAN BUS in arbitration is automatically retransmitted at the next available time.

Nondestructive, accordingly, means that the node winning arbitration just continues on with the message, without the message being destroyed or corrupted by another node.

In other, words CSMA means that each node on a bus must wait for a pre-scribed period of inactivity before attempting to send a message. CD+AMP means that collisions are resolved through a bit-wise arbitration, based on a preprogrammed priority of each message in the identifier field of a message. The higher priority identifier always wins bus access. That is, the last logic-high in the identifier keeps on transmitting because it is the highest priority. Since every node on a bus takes part in writing every bit "as it is being written," an arbitrating node knows if it placed the logic-high bit on the bus.

Example: Consider Figure 5.6 that shows a bus connected to it three nodes. As mentioned before, the message priority is decoded in the CAN identifier. When the bus is in "Idle" state, several nodes can start transmission of a frame. Every node reads back, bit by bit, from the bus during the complete

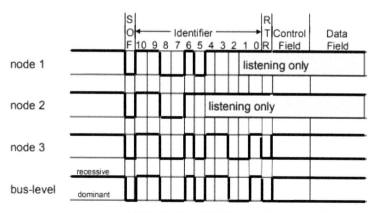

Node 3 wins arbitration and transmits his data.

Figure 5.6   Bus arbitration method.

message and compares the transmitted bit value with the received bit value. Per definition the bits with a dominant value overwrites those with a recessive value (this has to be provided by the transceiver).

5.1.2.2 Bus arbitration

Bus access is event-driven and takes place randomly. If two nodes try to occupy the bus simultaneously, access is implemented with a nondestructive, bit-wise arbitration. Nondestructive means that the node winning arbitration just continues on with the message, without the message being destroyed or corrupted by another node. Collision of the messages is avoided by the implemented CMSA/CA + AMP bus access method. Each node sends the bits of its message identifier and monitors the bus level. As long as the bits from all transmitters are identical nothing happens.

The allocation of priority to messages in the identifier is a feature of CAN that makes it particularly attractive for use within a real-time control environment. The lower the binary message identifier number, the higher its priority. An identifier consisting entirely of zeros is the highest priority message on a network since it holds the bus dominant the longest. Therefore, if two nodes begin to transmit simultaneously, the node that sends a zero (dominant) while the other nodes send a one (recessive) gets control of the CAN bus and goes on to complete its message. A dominant bit always overwrites a recessive bit on a CAN bus.

Figure 5.6 displays the arbitration process. At bit 5 nodes 1 and 3 send a dominant identifier bit. Node 2 sends a recessive identifier bit but reads back a dominant one. Node 2 loses bus arbitration and switches to listening only mode that is transmitting recessive bits. At bit 2 node 1 loses arbitration against node 3. This means that the message identifier of node 3 has a lower binary value and therefore a higher priority than the messages of nodes 1 and 2. In this way, the bus node with the highest priority message wins arbitration without losing time by having to repeat the message. Nodes 1 and 2 will send their messages after node 2 has finished his transmission.

The allocation of message priority is up to a system designer, but industry groups mutually agree upon the significance of certain messages. For example, a manufacturer of motor drives may specify that message 0010 is a winding current feedback signal from a motor on a CAN network and that 0011 is the tachometer speed. Since 0010 has the lowest binary identifier, messages relating to current values always have a higher priority on the bus than those concerned with tachometer readings.

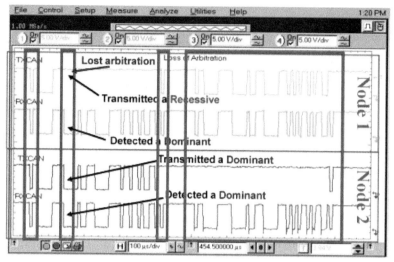

Figure 5.7   CSMA/CD-CR example: Case of two nodes.

CSMA/CD-CR: an Example

Figure 5.7 is a screenshot showing an example of two nodes attempting to transmit a message at the same time This screen shot shows the communication between the transceiver and the CAN module which is going to be digital.

The two signals on the top of the screen shot belong to Node 1, and they are the TX and RX digital lines between the transceiver and microcontroller. And the two signals on the bottom of the screen are the digital connections between Node 2's transceiver and microcontroller.

So starting at the left of the screen shot, we can see that both nodes started transmitting a 11-bit CAN frame with the dominant SOF. And if you can recall, the next portion of the CAN frame is the arbitration field in which the nodes will transmit out the ID for the data frame.

Both nodes will continue to transmit until there is a mismatch. We can see that Node 1 transmits a 1 (which is a recessive) while Node 2 transmits a 0 (which is a dominant). At this point, Node 2 "wins" arb and gains access to be the sole transmitter on the CAN bus, while Node 1, the losing node, stops transmitting and becomes a receiver.

Node 2 continues transmitting out the rest of the ID, Control field, Data (if there is any), and then the CRC field. At the ACK field, Node 2 will transmit out a recessive, and we can see that Node 1 transmits out a dominant which means that the CAN frame contained no errors.

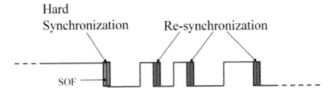

Figure 5.8 Synchronization and re-synchronization.

After the end of frame and a required 3-bit interframe space wait, Node 1 will transmit out its pending message, while Node 2 listens.

Each node has its own oscillator and internal CAN clock, so the question might arise, how is it possible that many nodes across a network can stay in synchronization without a clock line.

This is achieved by the receivers synchronizing on recessive to dominant edges.

CAN implements non-return to zero (NRZ) on the physical bus signal, which means there will be no edge between two like bits.

- Hard synchronization occurs at SOF and resets bit clock.
- Resynchronization occurs at recessive to dominant (1 to 0) edges and adjusts the bit clock as necessary.

5.1.3 Standard CAN or Extended CAN

As any serial communication system, the data exchange takes the form of frames. A CAN network can be configured to work with two different message (or "frame") formats: the standard or base frame format (described in CAN 2.0 A and CAN 2.0 B), and the extended frame format (only described by CAN 2.0 B). The only difference between the two formats is that the "CAN base frame" supports a length of 11 bits for the identifier, and the "CAN extended frame" supports a length of 29 bits for the identifier, made up of the 11-bit identifier ("base identifier") and an 18-bit extension ("identifier extension"). The distinction between CAN base frame format and CAN extended frame (Figure 5.9) format is made by using the IDE bit, which is transmitted as dominant in case of an 11-bit frame and transmitted as recessive in case of a 29-bit frame. CAN controllers that support extended frame format messages are also able to send and receive messages in CAN base frame format. All frames begin with a start-of-frame (SOF) bit that denotes the start of the frame transmission.

Standard Frame Format

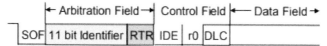

Extended Frame Format

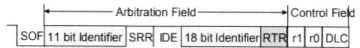

Figure 5.9   Arbitration field in case of standard and extended formats.

5.1.3.1 The bit fields of standard CAN and extended CAN

The bit fields of standard CAN and extended CAN are summarized in Figure 5.10.

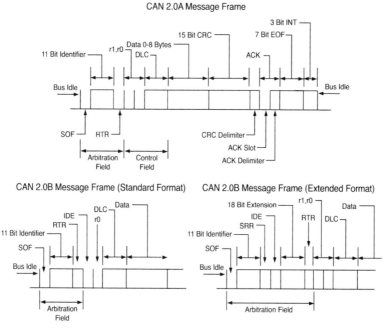

Figure 5.10   CAN frame.

5.1.3.2 Standard (base) CAN frame

The standard CAN Frame is shown in Figures 5.11 and 5.12. The frame consists of the following fields:

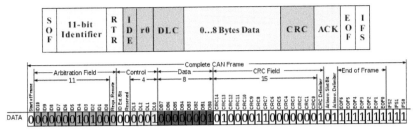

Figure 5.11 Base data frame format.

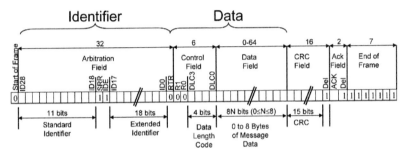

Figure 5.12 Standard CAN–The 11-bit identifier.

The standard data frame consists of the following fields:

• **SOF Field:** The data frame starts with a Start of Frame (SOF) which is a dominant bit. The SOF bit marks the start of a message and is used to synchronize the nodes on a bus after being idle.

• **Arbitration Field:** It contains 12 bits divided into two parts: ID Field of 11 bits and RTR of 1 bit.

   ○ **ID field** or Identifier Field: In Standard CAN, it contains 11 bits. The ID field is where the arbitration (arb) occurs and the node with the highest priority id (Lowest ID Number) wins the BUS. So in other words, a Node transmitting out a data frame with ID 0 will always "win" access to the BUS. A nondestructive bitwise arbitration method is used to allow multiple nodes to contend for the BUS. This process allows the highest priority message to remain intact even though there were collisions detected. And this process takes place without corruption or delay of the highest priority message.

   This is one reason why CAN works great in safety critical applications. (Arbitration will be discussed in more detail later.)

| RTR | IDE/r1 | r0 | DLC3 | DLC2 | DLC1 | DLC0 | Data/CRC |

No. of Data Bytes	Data Length Code (DLC)			
	DLC3	DLC2	DLC1	DLC0
0	d	d	d	d
1	d	d	d	r
2	d	d	r	d
3	d	d	r	r
4	d	r	d	d
5	d	r	d	r
6	d	r	r	d
7	d	r	r	r
8	r	d/r	d/r	d/r

Figure 5.13　Control field: DLC field (d, dominant; r, recessive).

The node that wins arb will continue to transmit while all other transmitters will stop transmitting and revert to receiving nodes. Every node on the bus is responsible to check the message for errors that range from bit stuffing to CRCs error.

○ **RTR:** The single remote transmission request (RTR) bit is dominant when information is required from another node. All nodes receive the request, but the identifier determines the specified node. The responding data are also received by all nodes and used by any node interested. In this way, all data being used in a system is uniform.

• **Control Field:** 6 bits (Figure 5.13). The Control Field in Standard Format includes four bits for the Data Length Code (DLC), and two reserved bits: IDE bit, which is transmitted dominant, and the reserved bit r0, which is also transmitted dominant.

○ IDE bit: Lets the receivers know if the message is standard or extended

○ r0: Reserved bit (for possible use by future standard amendment)

○ DLC: The 4-bit data length code (DLC) contains the number of bytes of data being transmitted. It lets the receivers know if the message has between 0 and 8 data bytes.

The admissible number of data bytes for a data frame ranges from 0 to 8. DLCs in the range of 0 to 7 indicate data filed length of 0 to 7 bytes. All other

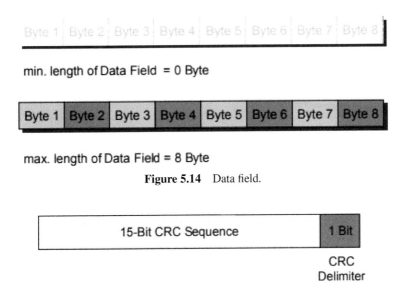

min. length of Data Field = 0 Byte

max. length of Data Field = 8 Byte

Figure 5.14 Data field.

Figure 5.15 CRC field.

Remark: The CRC Delimiter is a fixed formatted recessive bit.

DLCs indicate that the data filed is 8 bytes long. That means DLCs ranging from 9 to 15 may be used for application-specific purposes. The full range of possible DLCs is not required to be implemented.

- **Data Field:** Between 0 and 8 bytes (Figure 5.14): Next the transmitting node will transmit out the data.
  In some applications, it makes sense to transmit no data, for example, the case of indicating an event without any variable data. The event is identified by the identifier of the Data Frame, so it may be sufficient to transmit no data field.
- **CRC:** The CRC contains the 15-bit CRC sequence and the recessive 1-bit CRC Delimiter (Figure 5.15). The frame check sequence is derived from a cyclic redundancy code best suited for frames with bit counts less than 127 bit (BCH Code). The CRC is calculated from the SOF to the Data Field. The CRC code provides a Hamming distance of 6, meaning that 5 bit errors randomly distributed in the SOF, Arbitration, Control and Data Fields can be detected. In addition, burst errors up to a length of 15 bit can be detected.

CRC Generation

A generator polynomial is used to get the CRC. Linear Feedback Shift register is used for implementation.

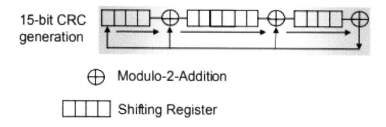

15-bit CRC generation

⊕ Modulo-2-Addition

▭ Shifting Register

$$x^{15} + x^{14} + x^{10} + x^{8} + x^{7} + x^{4} + x^{3} + 1$$

CRC Polynomial Generation

The receivers calculate the CRC in the same way as the transmitter as follows:

1. The message is regarded as polynomial and is divided by the generator polynomial: $x^{15} + x^{14} + x^{10} + x^{8} + x^{7} + x^{4} + x^{3} + 1$.
2. The division rest of this modulo 2 division is the CRC sequence which is transmitted together with the message.
3. The receiver divides the message inclusive the CRC sequence by the generator polynomial. A CRC error has to be detected, if the calculated result is not the same as that received in the CRC sequence. In this case, the receiver discards the message and transmits an Error Frame to request retransmission.

- **ACK** field (Figure 5.16): It contains 2 bits. The ACK Field is two bits long and contains the ACK Slot and the ACK Delimiter. The transmitter of a frame transmits both bits of the ACK Field recessive. A Receiver, which has received a valid message correctly, reports this to the Transmitter by sending a dominant bit during the ACK Slot. If the Transmitter detects a positive acknowledge, that is, a dominant

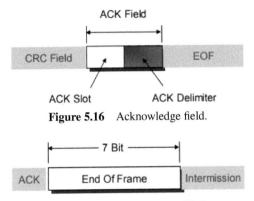

Figure 5.16 Acknowledge field.

Figure 5.17 End of frame (EOF).

ACK Slot, the Transmitter knows that in minimum one node has got his message correctly.

A receiving node may transmit a recessive to indicate that it did not receive a valid frame, but another node that did receive a valid frame may override this with a dominant. The transmitting node cannot know that the message has been received by all of the nodes on the CAN network.

- **EOF:** This is end-of-frame (EOF) (Figure 5.17).

  Each Data and Remote Frame is delimited by a flag sequence of seven recessive bits. The EOF field marks the end of a CAN frame (message) and disables bit stuffing, indicating a stuffing error when dominant. This EOF was introduced because an Error Frame caused by a global CRC failure should be transmitted within the length of Data or Remote Frame.

- **Interframe Spacing (IFS):** This 7-bit interframe space (IFS) contains the time required by the controller to move a correctly received frame to its proper position in a message buffer area (see Section 4.4).

Data frames and remote frames are separated from preceding frames by a bit field called interframe space. Interframe space consists of at least three consecutive recessive (1) bits. Following that, if a dominant bit is detected, it will be regarded as the "Start of frame" bit of the next frame. Overload frames and error frames are not preceded by an interframe space and multiple overload frames are not separated by an interframe space. Interframe space contains the bit fields intermission and bus idle, and suspends transmission for error passive stations, which have been transmitter of the previous message.

The standard frame fields are summarized in Table 5.1. The bit values are described for CAN-LO signal.

Table 5.1   Standard CAN frame

Field Name	Length (Bits)	Purpose
Start-of-frame	1	Denotes the start of frame transmission
Identifier	11	A (unique) identifier which also represents the message priority
Remote transmission request (RTR)	1	Must be dominant (0) for data frames and recessive (1) for remote request frames (see Remote Frame, below)
Identifier extension bit (IDE)	1	Must be dominant (0) for base frame format with 11-bit identifiers
Reserved bit (r0)	1	Reserved bit. Must be dominant (0), but accepted as either dominant or recessive.
Data length code (DLC)	4	Number of bytes of data (0–8 bytes)
Data field	0–64 (0-8 bytes)	Data to be transmitted (length in bytes dictated by DLC field)
CRC	15	Cyclic redundancy check
CRC delimiter	1	Must be recessive (1)
ACK slot	1	Transmitter sends recessive (1) and any receiver can assert a dominant (0)
ACK delimiter	1	Must be recessive (1)
End-of-frame (EOF)	7	Must be recessive (1)

Figure 5.18   Extended CAN: 29-bit identifier.

5.1.3.3 Extended frame format
Extended CAN

As shown in Figure 5.18, the Extended CAN message is the same as the Standard message with the addition of:

- SRR: The substitute remote request (SRR) bit replaces the RTR bit in the standard message location as a placeholder in the extended format.
- IDE: A recessive bit in the identifier extension (IDE) indicates that more identifier bits follow. The 18-bit extension follows IDE. The identifier in the extended format consists of two parts: A of 11 bits and B of 18 bits. The two parts are combined to form the 29 bits of the identifier.
- r0 and r1 bits: r1 following the RTR and r0 bits, an additional reserve bit has been included ahead of the DLC bit.

Table 5.2 Extended CAN format

Field Name	Length (Bits)	Purpose
Start-of-frame	1	Denotes the start of frame transmission
Identifier A (green)	11	First part of the (unique) identifier which also represents the message priority
Substitute remote request (SRR)	1	Must be recessive (1)
Identifier extension bit (IDE)	1	Must be recessive (1) for extended frame format with 29-bit identifiers
Identifier B	18	Second part of the (unique) identifier which also represents the message priority
Remote transmission request (RTR)	1	Must be dominant (0) for data frames and recessive (1) for remote request frames (see Remote Frame, below)
Reserved bits (r1, r0)	2	Reserved bits which must be set dominant (0), but accepted as either dominant or recessive
Data length code (DLC)	4	Number of bytes of data (0–8 bytes)[a]
Data field	0–64 (0-8 bytes)	Data to be transmitted (length dictated by DLC field)
CRC	15	Cyclic redundancy check
CRC delimiter	1	Must be recessive (1)
ACK slot	1	Transmitter sends recessive (1) and any receiver can assert a dominant (0)
ACK delimiter	1	Must be recessive (1)
End-of-frame (EOF)	7	Must be recessive (1)

The two identifier fields (A & B) combine to form a 29-bit identifier.
*It is physically possible for a value between 9 and 15 to be transmitted in the 4-bit DLC, although the data are still limited to 8 bytes. Certain controllers allow the transmission and/or reception of a DLC greater than 8, but the actual data length is always limited to 8 bytes.

The frame format is given in Table 5.2.

5.1.4 CAN Messages

5.1.4.1 CAN Frames

CAN bus has four frame types. If we consider the above discussions, CAN frames will be:

- **Data Frame** (Standard, Extended): A standard message used to transmit data over the network

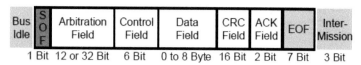

Figure 5.19   Data frame format.

- **Remote Frame** (Standard, Extended): A message sent by a receiver to request data from another node on the network: request the transmission of a specific identifier.
- **Error Frame** (Passive, Active): A frame transmitted by any node detecting an error. It is a message sent out by a receiver to destroy a frame that contains errors. The Error Frame tells the transmitter to send the message again.
- **Overload Frame:** An Overload Frame is similar to an error frame. A receiver would typically send out an Overload Frame to ask a transmitter to delay the next message sent. In other words, Overload Frame is used to inject a delay between data and/or remote frame.

5.1.4.1.1 Data frame

The data frame is the only frame for actual data transmission. There are two message formats. The major difference between the two formats is the Arbitration field. The two formats are:

- The standard (or Base) frame has an 11-bit ID in the arbitration field that allows up to 2048 unique messages onto the BUS.
- The extended frame has a 29-bit ID that allows up over 536 million unique messages onto the BUS.

The CAN standard requires that the implementation must accept the base frame format and may accept the extended frame format, but must tolerate the extended frame format.

Both standard (base) and extended data frames can coexist on the same CAN BUS. The standard 11 bit data frame will always have priority over the same extended 29-bit frame with identical 11-bit base ID.

It is recommended that no two nodes transmit the same data frame in terms of ID because this would break the arbitration process.

The Data Frame begins with a dominant Start of Frame (SOF) bit for hard synchronization of all nodes. The SOF bit is followed by the Arbitration Field reflecting content and priority of the message. The next field is the Control Field which specifies mainly the number of bytes of data contained

Table 5.3 CAN 2.0A message frame

Field	Length (Bits)	Description
Start of Frame (SOF)	1	Must be dominant
Identifier	11	Unique identifier indicates priority
Remote Transmission Request (RTR)	1	Dominant in data frames; recessive in remote frames
Reserved	2	Must be dominant
Data Length Code (DLC)	4	Number of data bytes (0–8)
Data Field	0–8 bytes	Length determined by DLC field
Cyclic Redundancy Check (CRC)	15	
CRC Delimiter	1	Must be recessive
Acknowledge (ACK)	1	Transmitter sends recessive; receiver asserts dominant
ACK Delimiter	1	Must be recessive
End of Frame (EOF)	7	Must be recessive

in the message. The Cyclic Redundancy Check (CRC) Field is used to detect possible transmission errors. It consists of a 15-bit CRC sequence completed by the recessive CRC delimiter bit. During the Acknowledgment (ACK) Field, the transmitting node sends out a recessive bit. Any node that has received an error-free frame acknowledges the correct reception of the frame by sending back a dominant bit. The recessive bits of the End of Frame end the Data Frame. Between two frames, there must be an recessive 3-bit Intermission field.

The standard and the extended frame formats are as shown in Tables 5.3 and 5.4.

5.1.4.1.2 Remote data frame

Generally data transmission is performed on an autonomous basis with the data source node (e.g., a sensor) sending out a Data Frame. It is also possible, however, for a destination node to request the data from the source by sending a Remote Frame. The intended purpose of the remote frame is to solicit the transmission of data from another node.

Remote Frames can come in either standard or extended formats and contain No data payload. Remote frames are used when one node needs to request data from another node. So for example Node A will transmit a remote frame request (RTR) onto the BUS with an Identifier that matches a

Table 5.4 CAN 2.0B message frame (extended frame format)

Field	Length (Bits)	Description
Start of Frame (SOF)	1	Must be dominant
Identifier–Standard and Extended Formats	11	Unique identifier corresponds to Base ID in Extended Format
Identifier–Extended Format	29	Comprised of 11-bit Base ID and 18-bit Extended ID
Remote Transmission Request (RTR)–Standard and Extended Formats	1	Dominant in data frames; recessive in remote frames. In Standard Format, the 11-bit identifier is followed by the RTR bit.
Substitute Remote Request (SRR)–Extended Format	1	Must be recessive. SRR is transmitted in Extended Frames at the position of the RTR bit in Standard Frames. In arbitration between standard and extended frames, recessive SRR guarantees the standard message frame prevails.
IDE–Standard and Extended Frames	1	Must be recessive for Extended Format; dominant for Standard Format.
Reserved r0–Standard Format	1	Must be dominant
Reserved r1, r0–Extended Format	2	Must be recessive
Data Length Code (DLC)	4	Number of data bytes (0–8)
Data Field	0–8 bytes	Length determined by DLC field
Cyclic Redundancy Check (CRC)	15	
CRC Delimiter	1	Must be recessive
Acknowledge (ACK)	1	Transmitter sends recessive; receiver asserts dominant
ACK Delimiter	1	Must be recessive
End of Frame (EOF)	7	Must be recessive

The two identifier fields (A & B) combine to form a 29-bit identifier.
Note: It is physically possible for a value between 9 and 15 to be transmitted in the 4-bit DLC although the data are still limited to 8 bytes. Certain controllers allow the transmission and/or reception of a DLC greater than 8, but the actual data length is always limited to 8 bytes.

data frame that typically sent by Node B. Upon seeing the RTR request Node B will transmit the data frame matching ID and data.

There are 2 differences between a Data Frame and a Remote Frame. First, the RTR bit is transmitted as a dominant bit in the Data Frame, and second,

● **11-bit Identifier field (Standard)**

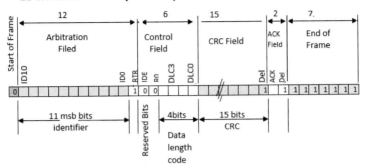

● **29-bit identifier field (extend)**

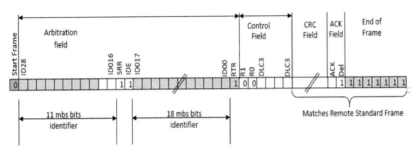

Figure 5.20 Remote data frame.

in the Remote Frame, there is no Data Field, that is,

RTR = 0; DOMINANT in data frame
RTR = 1; RECESSIVE in remote frame

In the very unlikely event of a Data Frame and a Remote Frame with the same identifier being transmitted at the same time, the Data Frame wins arbitration due to the dominant RTR bit following the identifier. In this way, the node that transmitted the Remote Frame receives the desired data immediately.

Remote frames are not typically used in Non-propriety HLP protocols, but can be found in some propriety protocols.

5.1.4.1.3 Error frame

The error frame is a special message that violates the formatting rules of a CAN message. It is transmitted when a node detects an error in a message, and causes all other nodes in the network to send an error frame as well. The

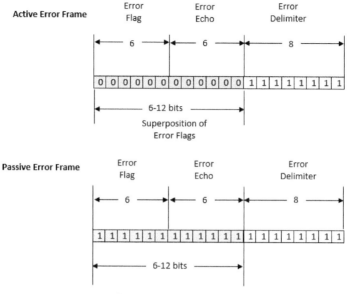

Figure 5.21   Error frame.

original transmitter then automatically retransmits the message. An elaborate system of error counters in the CAN controller ensures that a node cannot tie up a bus by repeatedly transmitting error frames.

Error frames are transmitted by any nodes when they detect an error with the CAN Data or Remote Frame. The error frame consists of two different fields:

- The first field is given by the superposition of ERROR FLAGS (6–12 dominant/recessive bits) contributed from different stations.
- The following second field is the ERROR DELIMITER (8 recessive bits).

There are two types of error flags: *Active error flag* and *Passive error flag*. If the node is in the error active state, it will transmit out an active error frame. If the node is in the passive error state, it will transmit out a passive error frame.

Active Error Flag

It contains six dominant bits transmitted by a node detecting an error on the network that is in error state "error active."

Passive Error Flag

It contains six recessive bits transmitted by a node detecting an active error frame on the network that is in error state "error passive."

The error frame can vary in length because it is possible for some noise localized next to a CAN node on the bus will flip a bit in data frame. If this occurs that node might see the bit flip while other nodes on the bus will not. Because of this "bit flip" the CRC calculation should be wrong, and it is possible that the bit stuffing rule may be violated. If that is the case that node will transmit out an error frame, and soon after the other nodes on the bus will start to see the bit stuffing rule being violated and they too will transmit out echoing error flags.

These error frames will "destroy" the current data or remote frame on the bus. The transmitting node will know that its message wasn't received properly by all nodes and will automatically attempt a retransmission at the next available quiet time on the bus.

How the node defines the error state: "Error Passive" or "Error Active"

There are two error counters in CAN:

1. Transmit error counter (TEC)
2. Receive error counter (REC)

- When TEC or REC is greater than 127 and lesser than 255, a Passive Error frame will be transmitted on the bus.
- When TEC and REC is lesser than 128, an Active Error frame will be transmitted on the bus.
- When TEC is greater than 255, then the node enters into Bus Off state, where no frames will be transmitted.

More details are provided in Section 5.2.

5.1.4.1.4 Overload frame

The Overload Frame is originally introduced to gain more time for processing the received data, today no CAN controller sends them actively. All CAN controllers are fast enough to store the received data.

The overload frame contains the two bit fields Overload Flag and Overload Delimiter. An overload frame is special version of the Error Frame that will not cause a retransmission of the last "destroyed" message. Instead the overload frame is used by a Node in two kinds of overload conditions:

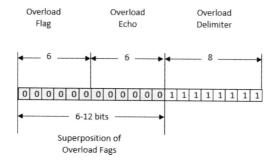

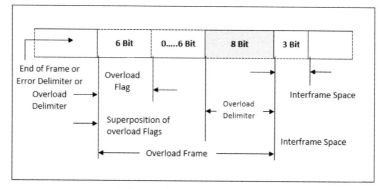

Figure 5.22   Overload frame.

1. To request a delay between data or remote frames during the "inter-frame" space or quiet time on the CAN bus. In other words, the internal condition of a receiver requires a delay of the next data frame or remote frame.

   If a node needs more time to process the message at the protocol level, it has the ability to transmit out up to two Overload frames before the next data or remote is present on the bus. What this does is request all nodes on the network to delay sending out the next Data or Remote Frames thus given that node more time to process the current message.

2. Detection of a dominant bit during intermission.

Another overload condition is that a message is valid for receivers, even when the last bit of EOF is received as dominant. Therefore, this dominant bit is not regarded as an error. On the other hand, the fixed-form bit filed EOF contains an illegal bit and the receiver of the dominant bit may have lost synchronization, which requires a reaction. The Reference CAN Model follows the example of a dominant bit as the last bit of Error or Overload Delimiter which are responded with an Overload Frame.

The start of an overload frame due to case 1 is only allowed to be started at the first bit time of an expected intermission, whereas overload frames due to case 2 start one bit after detecting the dominant bit. Overload Flag consists of six dominant bits. The overall form corresponds to that of the active error flag. The overload flag's form destroys the fixed form of the intermission field. As a consequence, all other stations also detect an overload condition and on their part start transmission of an overload flag. Overload Delimiter consists of eight recessive bits. The overload delimiter is of the same form as the error delimiter.

With an Overload Frame the transmitter is requested to delay the start of the next transmission.

The Overload Frame is identical to an Active Error Frame. The only difference is that an Overload Frame does not increase the error counters (see error confinement) and does not causes a retransmission of a frame. Every node may transmit consecutively only 2 Overload Frames.

5.1.4.1.5 Bit stuffing

To ensure enough transitions to maintain synchronization, a rule called "Bit Stuffing" is used. In Bit Stuffing, a bit of opposite polarity is inserted after five consecutive bits of the same polarity (see Figure 5.23). This practice is called bit stuffing and is necessary due to the non-return to zero (NRZ) coding used with CAN. This stuffed bit is added by the CAN node transmitting out the message at the protocol level and is expected and removed by the receivers at the protocol level so the user application is never aware of any stuffed bits.

All fields in the frame are stuffed with the exception of the CRC delimiter, ACK field and end of frame which are of fixed size and are not stuffed. In the fields where bit stuffing is used, six consecutive bits of the same polarity (111111 or 000000) are considered an error. An active error flag can be transmitted by a node when an error has been detected. The active error flag consists of six consecutive dominant bits and violates the rule of bit stuffing.

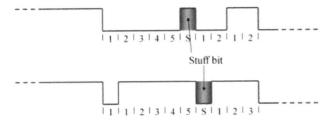

Figure 5.23 Bit stuffing rule.

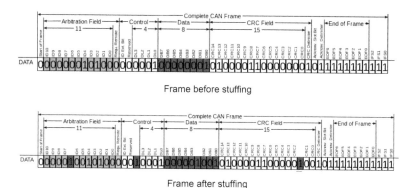

Figure 5.24   Complete frame before and after stuffing.

Bit stuffing means that data frames may be larger than one would expect by simply enumerating the bits shown in the tables above. The maximum increase in size of a CAN frame (base format) after bit stuffing is in the case 1111100001110000... which is stuffed as (stuffing bits in bold): 11111**0**00001**1**1110**0**00001....

The stuffing bit itself may be the first of the five consecutive identical bits, so in the worst case there is one stuffing bit per four original bits.

The size is bounded by

$$8n + 44 + \left\lfloor \frac{34 + 8n - 1}{4} \right\rfloor$$

since $8n + 44$ is the size of the frame before stuffing, in the worst case one bit will be added every four original bits after the first one (hence the -1 at the numerator) and, because of the layout of the bits of the header, only 34 out of 44 of them can be subject to bit stuffing.

The bit stuffing rule implies that the maximum time between synchronization edges is 10 bits.

Figure 5.24 shows a complete frame before and after stuffing.

5.2 CAN Bus Error Handling

How CAN Handles Errors

Error handling is built into in the CAN protocol and is of great importance for the performance of a CAN system. The error handling aims at detecting errors in messages appearing on the CAN bus, so that the transmitter can retransmit an erroneous message. Every CAN controller along a bus will try to detect errors within a message. If an error is found, the discovering node will transmit an Error Flag, thus destroying the bus traffic. The other nodes will detect the error caused by the Error Flag (if they haven't already detected the original error) and take appropriate action, that is, discard the current message.

Within the environment where CAN nodes are working, there can be many sources of an error on a CAN bus. Since it is used in critical applications, it is important that it continues to operate at all times. This is achieved at the physical layer by a transceiver which operates when one of the wires is broken. At the data link layer, the protocol mandates that the defective node withdraws from communication. Fault confinement is a critical aspect of the CAN which is achieved by a defective node identifying itself and switching to the bus-off state.

5.2.1 Error Detection Methods

The CAN data link layers are very reliable. All single-bit errors are detected. Multi-bit errors are detected with a high probability; the CAN FD protocol is even a little bit more reliable than the Classical CAN protocol.

The CAN data link layers comprise five error detection mechanisms: three at the message level and two at the bit level. Unlike other communication systems, they do not use acknowledgment messages but instead signals any occurring errors.

CAN protocol implements three detection mechanisms at the message level:

- *Cyclic Redundancy Check:* The CRC safeguards the information in the data and remote frame by adding redundant check bits at the transmission end. At the receiver end, these bits are re-computed and tested against the received bits. If they do not agree, a CRC error has occurred. There is an additional safeguard in the CAN FD protocol: the stuff error counter with parity bit protection.
- *Frame Check Form Error*: This mechanism verifies the structure of the transmitted data and remote frame by checking the bit fields against the fixed format and the frame size. Errors detected by frame checks are designated "format errors."
- *ACK errors*: As mentioned above, all nodes acknowledge received data and remote frames by positive acknowledgment (driving the bus to dominant bit-level during the ACK bit). If the transmitter does not see a dominant bit-level during the ACK bit, this can either mean that there is

Figure 5.25   Form error.

a transmission error which has only been detected by the recipients, that the ACK field has been corrupted, or that there are no receivers.

The CAN protocol also implements two mechanisms for error detection at the bit level:

- **Monitoring Bit error**: The ability of the transmitter to detect errors is based on the monitoring of bus signals. Each node transmits and also observes the bus level and thus detects differences between the bit sent and the bit received. This permits reliable detection of all global errors and errors local to the transmitter.
- **Bit stuffing**: The coding of the individual bits is tested at bit level. The bit representation used by CAN is NRZ (non-return-to-zero) coding, which guarantees maximum efficiency in bit coding. The synchronization edges are generated by means of bit stuffing, that is, after five consecutive equal bits the sender inserts a stuff bit into the bit stream with the complementary value, which is removed by the receivers. The code check is limited to checking the adherence to the stuffing rule.

Detected errors are indicated by means of an Error Frame. It consists of the Error Flag made of six bits of the same bit level and eight bits of recessive value called Error Delimiter.

As it is given latter, if a message fails with any one of these error detection methods, it is not accepted and an error frame is generated from the receiving nodes, causing the transmitting node to resend the message until it is received correctly. However, if a faulty node hangs up a bus by continuously repeating an error, its transmit capability is removed by its controller after an error limit is reached (see Section 5.2.11).

The different errors including Local Errors in EOF and Error Confinement are considered next.

5.2.2 Form Error

Some parts of the CAN message have a fixed format, that is, the standard defines exactly what levels must occur and when. (Those parts are the CRC

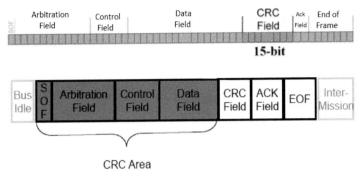

CRC Area

Figure 5.26 CRC error.

Delimiter, ACK Delimiter, End of Frame, and also the Intermission, but there are some extra special error checking rules for that.)

The form error occurs when the "recessive" slots in a CAN message (CRC Delimiter, ACK Delimiter, End of Frame, EOF) field or interframe Space) become "dominant." Any node detecting this will need to transmit out an error frame, thus destroying the current message, and the transmitter will need to retransmit at the next available time.

5.2.3 CRC Error

All messages on the CAN BUS contain a 15-bit field called CRC field. CRC, which is the abbreviation of Cyclic Redundancy Code or Cyclic Redundancy Check, is acting as signature for the transmitted data. It is, except very special cases, unique for each data.

With the Cyclic Redundancy Check the Transmitter calculates a check sum for the CRC bit sequence from the Start of Frame bit until the end of the Data Field. This CRC sequence is transmitted in the CRC Field of the Data or Remote Frame. When the source sends a message on the BUS, all nodes receive the message and immediately start to calculate the CRC of the received data and compare it with that in the received message. If CRCs do not match, a CRC error occurs and an Error Frame is generated. This causes the message to be destroyed and prevents any nodes from using that data. The transmitting node sees the error frame and will then retransmit the original message at the next available time.

The CRC checksum will be used for error detection only. It is not used for error correction. The Hamming Distance of this CRC code is theoretically 6. With this, it is possible to detect up to 5 single bit errors that are randomly scattered about the message or so-called burst errors up to a length of 15 bit.

Figure 5.27   Acknowledge error.

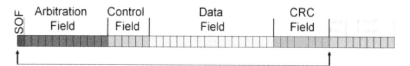

Figure 5.28   Stuff error.

5.2.4 Acknowledge Error

All nodes on the bus that correctly receives a message (regardless of their being "interested" of its contents or not) are expected to send a dominant level in the so-called Acknowledgement Slot in the message.

The Acknowledge errors occurs when the transmitting node checks the ACK Slot bit, which it has sent as a recessive and checks for a dominant.

- If a dominant bit occurs, at least one node received the message correctly.
- If not, an ACK Error occurred, and the transmitter will then transmit out Error Frame to destroy the message it just transmitted and will then retransmit the original message at the next available time.

5.2.5 Stuff Error

The part of the CAN frame between the Start of Frame (SOF) and the CRC delimiter is covered by bit stuffing rules. The CAN bus does not have synchronization information. The nodes synchronize themselves by changes in the bit levels. Therefore, the CAN frame follows the NRZ-5 coding rule. By this rule the bus level cannot have more than 5 bits of the same level. When five consecutive bits of the same level have been transmitted by a node, it will add a sixth bit of the opposite level to the outgoing bit stream. The receivers will remove this extra bit. This is done to avoid excessive DC components on the bus, but it also gives the receivers an extra opportunity to detect errors.

The Stuff error occurs when any node detects 6 consecutive bit of the same polarity between the SOF and end of CRC field. When the error is

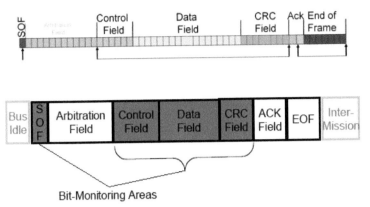

Figure 5.29 Bit error.

detected the node will transmit out an error frame, thus destroying the current message, and then the original message should be repeated by the transmitter at the next available time Keep in mind that error frames intentionally violate the bit stuffing rule.

5.2.6 Bit Error

All nodes perform bit monitoring. A bit error occurs when the transmitter monitors a signal on the bus different from what it sent. In other words, a bit error occurs if a transmitter sends a dominant bit but detects a recessive bit on the bus line or sends a recessive bit but detects a dominant bit on the bus line. An Error Frame is generated and starts with the next bit time.

This error is NOT checked during the ARB or the ACK field that is because during those times, all nodes have equal access to the bus. When the transmitter detects this error, it will transmit out an error frame, thus destroying the current message, and then, the transmitter will need to retransmit that original message at the next available time.

Exceptions:

- During arbitration (standard arbitration procedure)
- In ACK Slot bit (due to valid message acknowledgment)

5.2.7 Local Errors in EOF

If one of the EOF (End of Frame) bits 1 to 6 is detected locally as dominant bits, the node will send an Error Flag to globalize this failure. The CAN

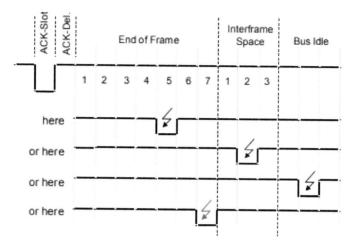

Figure 5.30   Local errors in EOF.

specification reads as follows: The point of time at which a message is taken to be valid is different for the transmitter and the receivers of the message.

- Transmitter: The message is valid for the transmitter, if there is no error until the end of End of Frame. If a message is corrupted, retransmission will follow automatically.
- Receiver: The message is valid for the receiver, if there is no error until the last but one bit of End of Frame. A Receiver, which has sampled a dominant value during the 7th EOF bit, does not regard this as an error. On the other hand, the fixed-form bit contains an illegal bit, and the Receiver may have lost synchronization; therefore, an Overload Frame is transmitted. As the Receiver can inform the sender at the earliest during the next bit time, it's obvious that the Transmitter must wait one additional bit time for his message validation. This is independent of which bit is defined as the validation bit. It does not help to introduce (one or more) additional bits at the end of the message frame.

5.2.8 Message Doubling

The CAN protocol assures with an extreme high probability that no messages are falsified or lost. But it is possible that a message is doubled by a single bit error near the end of End of Frame (EOF). Message doubling can be explained as follows:

If a Transmitter samples a dominant bus level during the last bit of EOF, it must retransmit that message. The dominant bus level can result from:

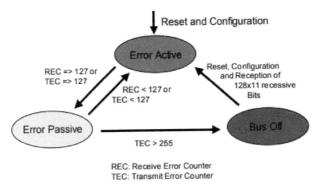

Figure 5.31 Error states of CAN node.

1. a receivers Error Flag reporting a local error in the last but one bit of EOF
2. a disturbance of the last bit of EOF

In case 1, it is likely that there are other receivers which have not been affected by the local error and therefore have already accepted the first message.

In case 2, all receivers have already accepted the first message. After the retransmission, they have got the same message twice.

Therefore, if CAN is used in a disturbed environment:

- Do not use toggle messages
- Do not transmit messages carrying relative data (like angle increments or delta counts)
- Use protected protocols or sequence numbers for data or program segmentation

5.2.9 Error States of CAN Node

Figure 5.31 reflects the error states of CAN node. Three possible states can be identified:

- Error Passive
- Error Active
- Buss Off

To distinguish between temporary and permanent failures, every CAN controller has two Error Counters: The REC (Receive Error Counter) and the TEC (Transmit Error Counter). The counters are incremented upon detected errors respectively are decremented upon correct transmissions or receptions.

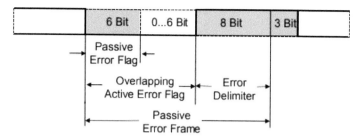

Figure 5.32   Passive error frame.

Depending on the counter values the state of the node is changed: The initial state of a CAN controller is Error Active that means the controller can send active Error Flags. The controller gets in the Error Passive state if there is an accumulation of errors. On CAN controller failure or an extreme accumulation of errors, there is a state transition to Bus Off. The controller is disconnected from the bus by setting it in a state of high resistance. The Bus Off state should only be left by a software reset. After software reset, the CAN controller has to wait for 128 x 11 recessive bits to transmit a frame. This is because other nodes may have pending transmission requests. It is recommended not to start a hardware reset because the wait time rule will not be followed then.

By using these counters, the node is able to recognize if it is the source of the BUS problems and subsequently disconnect itself from the CAN Bus.

This action prevents a single node from commandeering the CAN bus with error frames thus preventing any valid data frames onto the bus.

Using the error counters, a CAN node can not only detect faults but also perform error confinement (see latter).

In order to prevent the bus from being blocked by the Error Frames sent from a faulty or heavily disturbed node, the error indication ability can be restricted for this node. An Error Active node can send a dominant Error Flag while an Error Passive node is only able to send a recessive one. During the transmission of the Error Delimiter a node can detect whether it was the first one in the network reporting the error and thus stopping the data transfer. Error Passive receivers can no longer interrupt the data transfer as a recessive Error Flag does not influence the bus levels. An Error Passive transmitter can still interrupt its own message by sending a passive Error Flag. It is necessary to note here that if one Receiver is in error passive mode no data consistency is guaranteed any more.

INITIATED BY ERROR PASSIVE NODE

Remote or Data Frame	IFS	Suspend Transmission	SOF or Bus Idle

IFS (Interframe Space) = 3 bit
Suspend Transmission = 8 bit

Figure 5.33   Suspend passive transmission.

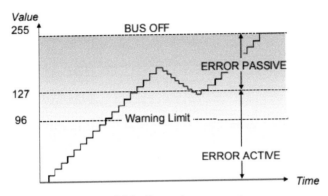

Figure 5.34   Transmit error counter.

To avoid blocking the bus by a disturbed node sending high priority messages a transmission delay was introduced for nodes in Error Passive state. After transmission, an Error Passive node must wait 3 + 8 recessive bits before starting transmission again (additional wait state).

5.2.10 Transmit Error Counter

0–96: The node is in the Error Active state below the warning limit. If it exceeds the counter a value of 96, the CAN controller generates a warning (set error flag, generating interrupt).

97–127: The node is Error Active. If the counter is in this area, it must be assumed that the bus is heavily disturbed.

128–255: In this area, the node is Error Passive. Unfortunately, some of the CAN controllers do not inform the microcontroller about this change of state.

> 255: After reaching this counter value, the node is switched to Bus Off state. In this state, it does not take part in bus activities. The microcontroller is informed by an interrupt on this change of states.

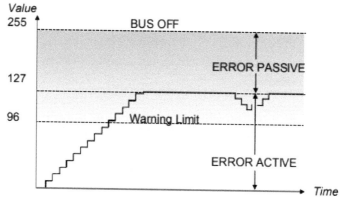

Figure 5.35   Receive error counter.

5.2.11 Receive Error Counter

Theoretically, the CAN Fault Confinement rules could increment REC's value over all limits when a receiver in Error Passive mode detects additional errors without receiving any error-free message. This cannot be implemented in hardware and the REC's value is limited by its actual number of digits. In the Bosch Reference Model, the REC has a resolution of 8 bits. So REC cannot be increased in Passive Error mode.

5.2.12 Recovery from Bus Off

All the error conditions we just discussed generate error frames, and these error frames increment internal transmit and receive Error counters. And in some scenarios, these counters can increment at different rates. Refer to the ISO11898 specification to learn more details.

There are two methods to recover.

- The CAN node can change to configuration then back to Normal Bus mode which clears out the transmit and receive error counters, or
- Detect 128 occurrences of 11 consecutive recessive bits (i.e., long bus idle or 128 valid messages, or a combination of both).

In the following, we discuss the second methods.

Every CAN node will use the state diagram given in Figure 5.31 when connected to the CAN BUS.

Each node starts off in Error active state. This is the state a normal CAN node should be in for normal bus activity. In this state, the CAN node is able to transmit normal data, remote, overload and active error frames. When either the receive error or the transmit error counters increment to greater than 127, the node will enter the Error Passive State. In this state, the CAN node can still transmit out data, remote, and overload frames. But now the node can only transmit passive error frames. Recall the passive error frame (Figure 5.32) consists of all ones, so if the node is still detecting and transmitting error frames, these error frames will be passive and should not cause the other nodes on the network to increment their receive counters.

Once in Error Passive state the Node still have a chance to enter back into error active, which will happen when the error counters drop below 127 and below. The error counters will decrement every time there is a valid message on the bus. If the node continues to increment the Transmit error counter to greater than 255 it will enter bus off. At this point the CAN module disconnects from the bus. In order words, other than sensing the bus levels, it will not transmit or receive any message or error frames. Once in the Bus off state the node can only enter Error Active.

5.2.13 Error Confinement Mechanisms

Fault confinement

The CAN data link layers detect all communication errors with a very high probability. A node detecting an error condition sends an Error Flag and discards the currently transmitted frame. All nodes receiving an Error Flag discard the message, too. In case of local failures, all other nodes recognize the Error Frame sent by the node(s) that detected it and sent by themselves a second time, which results in an eventually overlapping Error Frame. The active Error Frame is made of six dominant bits and an 8-bit recessive delimiter followed by the IMF. This local error globalization method guarantees network-wide data consistency, an important feature in distributed control systems.

If all errors are detected with a very high probability, permanent errors may lead to an unacceptable delay in transmitting messages. In the worst-case, all communication is aborted by means of Error Frames. In order to avoid this, the CAN protocol, as mentioned before, introduces two error counters: one for received messages–Receive Error Counter (REC), and one for transmitted messages–Transmit Error Counter (TEC). They are increased and decreased according to the rules as specified in ISO 11898-1, the standard of the CAN data link layer protocols. Using the error counters, a CAN node can not only detect faults but also perform error confinement.

A node starts out in Error Active mode. When any one of the two Error Counters raises above 127, the node will enter a state known as Error Passive and when the Transmit Error Counter raises above 255, the node will enter the Bus Off state.

- Error Active node will transmit Active Error Flags when it detects errors.
- An Error Passive node will transmit Passive Error Flags when it detects errors.
- A node which is Bus Off will not transmit anything on the bus at all.

The rules for increasing and decreasing the error counters are somewhat complex, but the principle is simple: transmit errors give 8 error points, and receive errors give 1 error point. Correctly transmitted and/or received messages causes the counter(s) to decrease.

Example (slightly simplified): Let's assume that node A on a bus has a bad day. Whenever A tries to transmit a message, it fails (for whatever reason). Each time this happens, it increases its Transmit Error Counter by 8 and transmits an Active Error Flag. Then it will attempt to retransmit the message, and the same thing happens.

When the Transmit Error Counter raises above 127 (i.e., after 16 attempts), node A goes Error Passive. The difference is that it will now transmit Passive Error Flags on the bus. A Passive Error Flag comprises 6 recessive bits, and will not destroy other bus traffic–so the other nodes will not hear A complaining about bus errors. However, A continues to increase its Transmit Error Counter. When it raises above 255, node A finally gives in and goes Bus Off.

What does the other nodes think about node A?–For every active error flag that A transmitted, the other nodes will increase their Receive Error Counters by 1. By the time that A goes Bus Off, the other nodes will have a count in their Receive Error Counters that is well below the limit for Error Passive, that is, 127. This count will decrease by one for every correctly received message. However, node A will stay bus off.

Most CAN controllers will provide status bits (and corresponding interrupts) for two states:

- "Error Warning"–one or both error counters are above 96
- Bus Off, as described above.

Some, but not all, controllers also provide a bit for the Error Passive state. A few controllers also provide direct access to the error counters.

The CAN controller's habit of automatically retransmitting messages when errors have occurred can be annoying at times. There is at least one controller on the market (the SJA1000 from Philips) that allows for full manual control of the error handling.

Fault Confinement Rules

The above discussion can be summarized in the following steps:

1. When a receiver detects an error, the REC will be increased by 1, except when the detected error was a Bit Error during the sending of an Active Error Flag or an Overload Flag.
2. When a receiver detects a dominant bit as the first bit after sending an Error Flag, the REC will be increased by 8.
3. When a transmitter sends an Error Flag, the TEC is increased by 8. Exception 1: If the transmitter is Error Passive and detects an ACK Error because of not detecting a dominant ACK and does not detect a dominant bit while sending its Passive Error Flag. Exception 2: If the transmitter sends an Error Flag because a Stuff Error occurred during arbitration, and should have been recessive, and has been sent as recessive but monitored as dominant.
4. If the transmitter detects a Bit Error while sending an Active Error Flag or an Overload Frame, the TEC is increased by 8.
5. If a receiver detects a Bit Error while sending an Active Error Flag or an Overload Flag, the REC is increased by 8.
6. Any node tolerates up to 7 consecutive dominant bits after sending an Active Error Flag, Passive Error Flag, or Overload Flag. After detecting the fourteenth consecutive dominant bit (in case of an Active Error Flag or an Overload Flag) or after detecting the eighth consecutive dominant bit following a Passive Error Flag, and after each sequence of additional eight consecutive Ďdominant" bits, every transmitter increases its TEC by 8 and every receiver increases its REC by 8.
7. After successful transmission of a frame (getting ACK and no error until EOF is finished), the TEC is decreased by 1 unless it was already 0.
8. After the successful reception of a frame (reception without error up to the ACK Slot and the successful sending of the ACK bit), the REC is decreased by 1, if it was between 1 and 127. If the REC was 0, it stays 0, and if it was greater than 127, then it will be set to a value between 119 and 127.

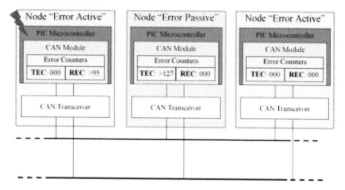

Figure 5.36   Example of error confinement.

9. A node is Error Passive when the TEC equals or exceeds 128, or when the REC equals or exceeds 128. An error condition letting a node become Error Passive causes the node to send an Active Error Flag.
10. A node is Bus Off when the TEC is greater than or equal to 256.
11. An Error Passive node becomes Error Active again when both the TEC and the REC are less than or equal to 127.
12. A node which is Bus Off is permitted to become Error Active (no longer Bus Off) with its error counters both set to 0 after 128 occurrences of 11 consecutive recessive bits have been monitored on the bus

Example of Error Confinement

In this example, we have three nodes that are all in the "Error Active" state which means they are able to interact with the bus normally. By transmitting data/remote/or error active frames.

At some point in the future, it possible for some node to detect error conditions and start transmitting out error frames; either there is noise in the environment, there is a short, or the node has become faulty in some way. PIC microcontrollers with the CAN module come with an internal "error counter warning" interrupt that will trigger when either error counters exceed 95. These features give the CAN node an early warning that the bus might be experiencing an issue. We can see that the node on the left has received this "error counter warning interrupt". Whereas the node in the middle has entered into Error Passive state because it's transmit error counter has become greater than 127. At this point, the node in the middle can still transmit data and remote frames, but is unable to transmit error active frames. Instead, it will need to transmit error passive frames to limit its disturbance of the CAN BUS.

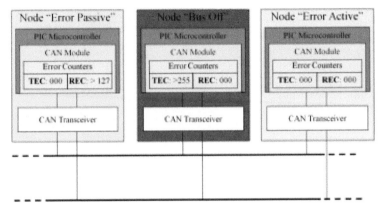

Example of Error Confinement (cont.)

Now we can see that the Node on the left has entered Error Passive, and the node in the middle has entered "Bus Off". The node in the middle now has removed itself from the BUS and will no longer be able to transmit out CAN data or remote frames. Also this "Bus off" node cannot transmit Active or Passive error frames. By allowing this Node to drop off the BUS, the CAN protocol has effectively guaranteed that bandwidth will always be available for critical messages to be transmitted by other nodes on the BUS. Higher Layer Protocol that implements Network Management/Diagnostics should be able to notified the user of such a condition to allow them the chance to service their product. Now, it is possible to ask a question about why the node on the right didn't increment its error counters the entire time through this example. There could be several legitimate reasons why. Maybe this node has been power cycled every time we decide to get a snapshot of the BUS, or ??? Maybe the node is in a "listen Only State" in which it only receives and doesn't transmit anything on the BUS.

5.2.14 Bus Loading

CAN provides a robust, simple, and flexible network solution for manufacturing, automotive and many other applications. The major drawback to CAN is that message latency is non-determinant (due to the existence of Error Frames, Overload Frames and retransmissions), and latency increases with the amount of traffic on the bus. In general, bus utilization should not exceed 30% of the bus capacity to assure that low priority messages do not experience unacceptable delay. Bus utilization is defined as total bit consumption / total bits available, and is calculated as follows:

Table 5.5   Bus loading example

Message	Data (Bytes)	Message Size (Bits)	Rate and Period	Message Bits Consumed
MsgA	0	47	10 trx/s: 100 ms	10 × 47 = 470 bps
MsgB	5	5 × 8 + 47 = 87	2 trx/s: 500 ms	87 × 2 = 174 bps
MsgC	8	8 × 8 + 47 = 111	1 trx/s: 1 s	111 × 1 = 111 bps
...	...	...	...	...
			Total periodic bits consumed	10000 bps

Total bits consumed = 1.1(Total periodic bits consumed) = 11000 bps
Bandwidth consumption = 11000/125000 = 8.8%

Step 1: Choose a time unit = the slowest fixed periodic message on network (usually 1 second).

Step 2: Identify all periodic messages.

Step 3: For each of these messages approximate the total bit size of the message by adding 47 bits to the size of each data field (SOF + Arbitration + RTR + Control + CRC + Acknowledgment + EOF + Interframe Space = 1 + 11 + 1 + 6 + 16 + 2 + 7 + 3 = 47 bits).

Step 4: Calculate the message bits consumed by multiplying the message bit size by the number of transmissions performed in one time unit.

Step 5: Sum all the message bits consumed to estimate the total periodic bits consumed. Multiply this number by a safety factor of 1.1 to account for worst case traffic.

Step 6: Finally, divide the total periodic bits consumed by the total bits available (e.g., 125 kbps or 500 kbps multiplied by the time unit) to arrive at the estimated bandwidth consumption percentage for the network.

5.2.15 Time-triggered Protocols

For real-time control over a network it may be advisable to implement a communication protocol that guarantees that messages meet timing deadlines regardless of the load on the bus. One such protocol that retains the CAN data link layer protocol is "time-triggered CAN," or TTCAN (ISO 11898-4) [9]. A TTCAN message frame incorporates two types of "time windows": exclusive time windows, and arbitrating time windows. Exclusive time windows are assigned to specific messages that are transmitted periodically. Thus, exclusive window messages do not compete for bus access. Arbitrating windows are used for messages that are not time critical. Arbitrating window messages, like normal CAN messages, compete for bus access based on priority through arbitration. Time-triggered CAN requires the existence of a "master node" that periodically broadcasts its time (referred to as global time) in a reference message. For fault tolerance, there must be multiple potential master nodes on the network. If the master node fails (detected by the absence of a reference message), the other potential masters compete for the bus by arbitration, with the highest priority node becoming the new master and broadcasting reference messages. Time-triggered CAN does not re-broadcast corrupted messages, nor does it invoke Error Frames.

A competitive protocol to TTCAN is FlexRay, developed by a consortium of automotive manufacturers and suppliers [10]. The FlexRay communication frame consists of periodically triggered "static" and "dynamic" parts. The static segment is made up of identical length time slots assigned to connected nodes. Each node transmits its messages synchronously in its reserved slot. The static segment also transmits a "synch" frame to provide a global time-base for the network. Unlike CAN, there is no arbitration for the bus. The dynamic segment is essentially a "polling" mechanism wherein each node is given the opportunity to put an event-triggered or asynchronous message on the bus in priority order using a "mini-slotting" timing mechanism. That is, the slot counter increments through each identifier, but whereas in the static segment the counter increments at a periodic rate whether or not there is a message in each slot, in the dynamic segment the counter waits only a brief period for each node to request transmission. If there is no transmission request for a message with a particular slot identifier, the counter continues to increment; if a transmission request occurs, the counter idles until the transmission is complete.

For redundant fault tolerance, FlexRay nodes may be connected to two buses or channels simultaneously.

5.3 CAN FD–The Basic Idea

Because of the bandwidth requirements of the automotive industry, the CAN data link layer protocol needed to be improved. In 2011, Bosch started the CAN FD (flexible data-rate) development in close cooperation with carmakers and other CAN experts. The improved protocol overcomes to CAN limits: You can transmit data faster than with 1 Mbps and the payload (data field) is now up to 64 byte long and not limited to 8 byte anymore. In general, the idea is simple: When just one node is transmitting, the bit rate can be increased,

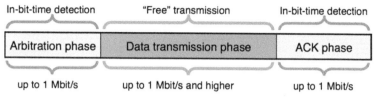

Figure 5.37   CAN FD format.

Classical CAN	Header	Protected payload	Trailer
	SOF Arbitration field Control field	Data field (up to 8 byte) CRC field	ACK field EOF (IMF)

CAN FD8	Header	Protected payload	Trailer	
	SOF Arbitration field Control field	Control field 8-byte data field CRC field	ACK field EOF (IMF)	

CAN FD64	Header	Protected payload	Trailer
	SOF Arbitration field Control field	Control field 64-byte data field CRC field	ACK field EOF (IMF)

Figure 5.38   CAN FD.

because no nodes need to be synchronized. Of course, before the transmission of the ACK slot bit, the nodes need to be re-synchronized.

CAN FD data frames can be transmitted with two different bit rates: In the arbitration phase, the bit rate depends on the network topology and is limited to 1 Mbps; in the data phase, the bit rate is limited by the transceiver characteristics.

Using a ratio of 1:8 for the bit rates in the arbitration and data phase leads to an approximately six-times higher throughput considering that the CAN FD frames use more bits in the header (control field) and in the CRC field.

Faster and longer: Larger payloads improve the protocol efficiency and lead to a higher throughput.

CAN FD–Some protocol details

In order to distinguish between Classical data frames and CAN FD data frames, one of the formerly reserved bits is used. This bit is called FDF (FD

SOF	Arbitration field	Control field	Data field (payload)	CRC field	ACK field	EOF	IMF
1 bit	12 *or* 32* bit	8 *or* 9* bit	0 *to* 64* byte	28 *or* 33 bit**	2 bit	7 bit	3 bit

MSB   LSB

Figure 5.39   CAN FD frame.

FD base frame format (FBFF)

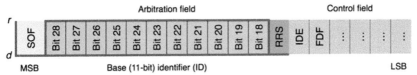

MSB   Base (11-bit) identifier (ID)   LSB

FD extended frame format (FEFF)

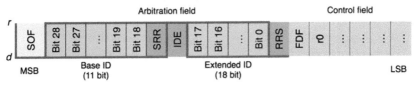

MSB   Base ID   Extended ID   LSB
        (11 bit)   (18 bit)

Figure 5.40   FD base and extended frame format.

frame) bit. If it is of recessive value, the following bit sequence is interpreted as a CAN FD data frame. If it is of dominant value, it is a Classical data or remote frame. In the newly introduced BRS (bit rate switch) bit, the second bit rate is applied, when it is of recessive (r) value. If it is of dominant (d) value, the arbitration phase bit-time setting is used in the data phase, too.

Structure of CAN FD data frames: The fields, as shown in Figure 5.39, are the same as in Classical CAN data frames, but some extensions have been introduced (* stuff-bits are not considered; ** with fixed stuff-bits):

SOF = start-of-frame, CRC = cyclic redundancy check, ACK = acknowledgment, EOF = end-of-frame, IMF = intermission field

Classic and Extended Frame Format: The CAN FD protocol controller has to also support Classical CAN frames. Both CAN protocols (Classical as well as CAN FD) (Figure 5.40) are internationally standardized in ISO 11898-1:2015. CAN FD data frames with 11-bit identifiers use the FBFF (FD base frame format) and those with 29-bit identifiers use the FEFF (FD extended frame format). The CAN FD protocol doesn't support remotely requested data frames.

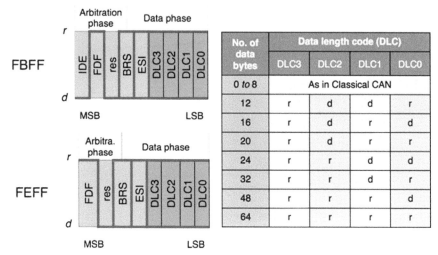

Figure 5.41   FBFF and FEFF.

Two CAN FD frame formats: The IDE bit is recessive in FEFF, the RRS bit is always dominant, and the value of the SRR bit doesn't matter.

RRS = remote request substitution, SRR = substitute remote request, IDE = identifier extension, FDF = flexible data rate format, d = dominant, r = recessive, r0 = reserved

The control field comprises additional bits not provided by the Classical CAN data frames. The FDF (FD format) bit indicates the usage of FD frame formats. At the sample-point of the BRS (bit rate switch) bit, the bit rate switch is performed. This guarantees a maximum of robustness. The following ESI (error state indicator) bit provides information about the error status: a dominant value indicates an error active state.

Extended control field: The DLC (data length code) values not used by the Classical CAN protocol are shown as used by the CAN FD protocol.

IDE (identifier extension), FDF (flexible data rate format), BRS (bit rate switch; recessive, if alternate bit rate), ESI (error state indicator; recessive, if error passive), d = dominant, r = recessive

During the standardization process of the CAN FD protocol, some additional safe guards were introduced in order to improve the communication reliability. This is why the CRC field comprises 17-bit (for frames with payloads up to 16 byte) or 21-bit (for frames larger than 16 byte) polynomials and an 8-bit stuff-bit counter plus a parity bit. The CRC field use fixed-stuff bits (FSB) with an opposite value of the previous bit. All these safe guards

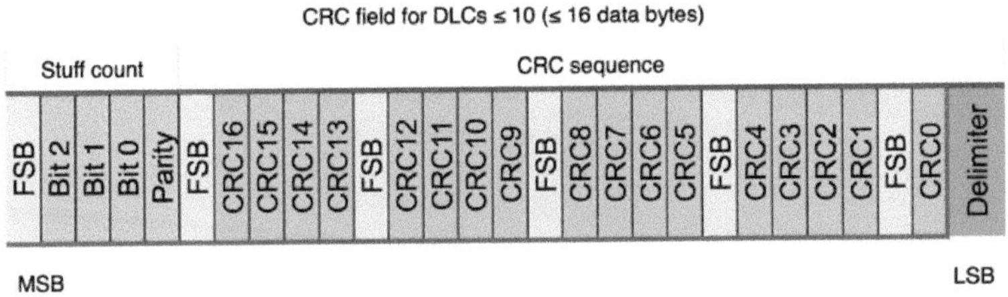

CRC field for DLCs ≤ 10 (≤ 16 data bytes)

Stuff count					CRC sequence																			
FSB	Bit 2	Bit 1	Bit 0	Parity	FSB	CRC16	CRC15	CRC14	CRC13	FSB	CRC12	CRC11	CRC10	CRC9	FSB	CRC8	CRC7	CRC6	CRC5	FSB	CRC4	CRC3	CRC2	CRC1	FSB	CRC0	Delimiter

MSB   LSB

Figure 5.42   CRC fields.

guarantee that all single failures are detected under all conditions. Even the possibility to detect multiple failures has been improved.Medium-term, non-ISO CAN FD controllers might also be on the market–these are not compliant to the ISO 11898-1 standard. They don't implement the above-mentioned additional safe guard features.

Stuff-bit counter: The 3-bit stuff-bit counter is grey-coded and it is protected by a parity bit as well as by the following fixed stuff-bit (FSB)

5.4 ISO 11898 Review and Optional Modes

5.4.1 Bus Monitoring Mode

In an optional bus monitoring mode, the CAN node is able to receive valid data frames and remote frames, but it sends only "recessive" bits on the CAN bus and it cannot start a transmission.

If the optional bus monitoring mode is supported, the MAC sublayer is required to transmit "dominant" bits (ACK bit, overload flag, active error flag), but the bit is rerouted internally so that the MAC sublayer monitors these "dominant" bits, although the CAN bus may remain in recessive state.

The use of the time-triggered communication option requires a single-shot mode. In this mode, the CAN node do not transmit overload and error flags.

The automatic retransmission is disabled.

5.4.2 Time-Triggered Communication (TTC)

In order to guarantee a specific latency time for each transmitted message, an optional scheduling function may be implemented.

The time-triggered communication (TTC) option describes the prerequisites needed for the synchronization of all nodes in the network. With the synchronization of node any message can be transmitted at a specific

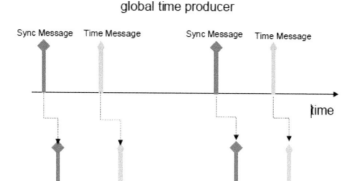

Figure 5.43   Global time producer.

time slot, where it has not to compete for the bus with other messages thus providing predictable latency times by avoiding the loss of arbitration. In order to synchronize the activities of the nodes within the network a common reference point is needed. The SOF bit or the sample point of the last bit of EOF of any message is used as the reference point. The individual presence of a single message at a time is referred to as frame scheduling.

Based on the synchronization of the nodes, the TTC facilitates also the establishment of a global time system in higher-layer protocols. The hardware needed to establish TTC is included between LLC and MAC.

Any node that supports TTC option needs to provide a time base, which is a cyclic up counter of at least 16 bit with either an internal or an external clock. Any message received or transmitted invokes a capture of the time base taken at the SOF recognition of the respective message or at the sample point of the last bit of EOF. After successful message reception, the capture value is provided to the CPU for at least one message and it is readable until the next message is received. It has to be possible to generate at least one programmable event trigger from the above mentioned time base. The trigger should be freely programmable by the CPU in the range of at least 0 to $(2^{16} - 1) \times$ timer clocks.

5.5 CAN in Action

In the following, we will be using the Node with the CAN Module built into the PIC microcontroller as reference to show about the steps a CAN Node

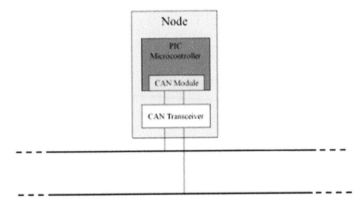

Figure 5.44 CAN node.

will need to go through to transmit out a CAN message and show how CAN works.

1. Transmitting a CAN Message.
2. First, the user software will need to confirm the CAN module is in configuration mode. Then initialize the CAN module bit rate settings so that the module is using the same bit rate as all other nodes on the bus.
3. Then the user should place the CAN module from configuration mode to normal mode.
4. Second, the user software will need load the TXB with the Message ID, DLC, and Data bit, now the user software will "set" the "transmit request bit." This will lock that TX buffer from being written to by the user software...
5. At this point the CAN module takes over.
6. The message is assembled and stuff bits are added where needed Next the CAN module senses the CAN bus looking for the next "quiet time.

When there is quiet time, the module will start transmitting the message while checking for error frames. The CAN transceiver will translate the digital signals from the PICmicro to the transceiver to differential signals to go out onto the BUS Actually the Physical layer bus signal depends on what transceiver is being used (differential was just mentioned because it is common).

If the CAN message is successfully sent onto the CAN bus, meaning there were NO error frames that forced a retransmit, the TX Buffer will unlocks for the next user software writes.

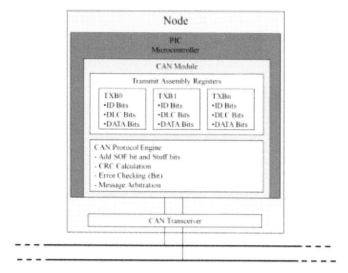

Figure 5.45 CAN message.

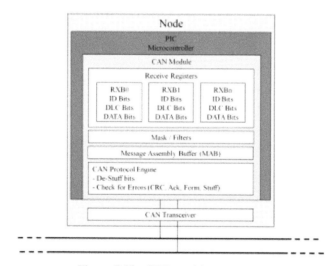

Figure 5.46 CAN receiving message.

1. Receiving a CAN Message

Let's talk about the steps a CAN Node will need to go through to receive CAN message from the CAN bus.

First, the user software will need to confirm whether the CAN module is in configuration mode, and then, the user software will need to initialize the

CAN modules bit rate settings so that the module is using the same bit rate as all other nodes on the bus. The user software should also initialize the mask and filter settings as well. Then, the user should place the CAN module from configuration mode to normal mode.

Next the CAN module will wait until another node on the CAN Bus transmits out a CAN message, it senses a CAN message on the BUS starting with the SOF.

The CAN transceiver will translate the differential signal from the bus into digital signal and pass it along to the CAN module on the Microcontroller. The CAN module will de-stuff the stuff bits and check for message Errors while it is loading the message into the MAB.

If there are no Errors and the entire message has been loaded the message in the MAB, the CAN module will check the received messages ID against the Mask and Filter settings.

If the message ID is a match, then the message will then get stored into the Receiver Buffer, and at that point, the PIC microcontroller (i.e., the User's application) is notified.

If the message DOES Not match the Mask and Filters, the PIC microcontroller would never get notified in other words precious microcontroller cycles would not be wasted on unwanted messages which is why it is important to setup and use the Mask and Filter settings correctly.

DLC = Data Length Code

References

[1] Controller Area Network, Basics Protocols, Chips and Applications; Dr. Konrad Etschberger; ISBN 3 00-007376-0 (www.ixxat.com)
[2] CAN Systems Engineering, From Theory to Practical Applications; Wolfhard Lawrenz, ISBN 0-387-94939-9
[3] Nicolas Navet, Yeqiong Song, Franÿcoise Simonot-Lion, and Cedric Wilwert, "Trends in Automotive Communication Systems," Proceedings of the IEEE, Vol. 93, NO. 6, June 2005 Page(s):1204–1223
[4] Gabriel Leen and Donal Heffernan, "Expanding Automotive Electronic Systems," IEEE Computer Volume 35, Issue 1, Jan. 2002 Page(s):88–93
[5] Jeffrey Cook, Ilya Kolmanovsky, David McNamara, Edward Nelson and K. Venkatesh Prasad, "Control, Computing and Comm
[6] J.M. Eklund, J. Sprinkle, S. Sastry and T.R. Hansen, "Information Technology for Assisted Living at Home: building a wireless infrastructure

for assisted living," IEEE-EMBS 2005 27th Annual International Conference of the Engineering in Medicine and Biology Society, Sept. 2005 page(s): 3931–3934.

[7] CAN in Automation Website, http://www.can-cia.org, viewed September 2006.

[8] Robert Bosch GmbH, "CAN Specification 2.0," 1991.

[9] Holger Zeltwanger, "Time-Triggered Communication on CAN," SAE Paper 2002-01-0437, 2002.

[10] Thomas Fuehrer, Robert Hugel, Florian Hartwich and Harald Weiler, "FlexRay–The Communication System for Future Control Systems in Vehicles," SAE Paper 2003-01-0110, 2003.

[11] "CAN History" (http://www.can-cia.de/index.php?id=161). CAN in Automation.

[12] Building Adapter for Vehicle On-board Diagnostic (http://www.obddiag.net/adapter.html), obddiag.net, accessed 2009-09-09

[13] Comparison of Event-Triggered and Time-Triggered Concepts with Regard to Distributed Control Systems A. Albert, Robert Bosch GmbH Embedded World, 2004, Nürnberg

[14] http://www.semiconductors.bosch.de/en/20/can/2-license.asp

[15] "CAN BUS MESSAGE FRAMES–Overload Frame, Interframe Space" (http://rs232-rs485.blogspot.com/2009/11/can-bus-message-frames-overload.html).

[16] Controller Area Network, Basics Protocols, Chips and Applications; Dr. Konrad Etschberger; ISBN 3–00–007376–0 (www.ixxat.com)

[17] CAN Systems Engineering, From Theory to Practical Applications; Wolfhard Lawrenz, ISBN 0–387–94939–9

[18] MCP2551 Data Sheet, *"High Speed CAN Transceiver"*, DS21667, Microchip Technology, Inc.
AN754, *"Understanding Microchip's CAN Module Bit Timing"*, DS00754, Microchip Technology, Inc.

[19] ISO-11898-2, *"Road Vehicles–Interchange of Digital Information–Part 2: High Speed Medium Access Unit and Medium Dependant Interface"*, International Organization for Standardization.

[20] CAN System Engineering, *"From Theory to Practical Applications"*, Wolfhard Lawrenz, Springer.

External Links

- Bosch specification (http://www.semiconductors.bosch.de/media/pdf/canliteratur/can 2spec.pdf) (old document–slightly ambiguous/unclear in some points, superseded by the standard ISO 11898 (http://www.iso.org/iso/search.h tm?qt=Controller+Area+Network&searchSubmit=Search&sort=rel&t ype=simple&published=true))
- Controller Area Network (CAN) Schedulability Analysis: Refuted, Revisited and Revised (http://www.springerlink.com/content/8n327 20737877071/)
- Pinouts for common CAN-bus connectors (http://www.interfacebus.c om/Can_Bus_Connector_Pinout.html)
- Independent discussion platform CANLIST (http://www.canlist.org/)
- A webpage about CAN in automotive (http://marco.guardigli.it/2010/10 /hacking-your-car.html)
- Controller Area Network (CAN) Schedulability Analysis with FIFO Queues (http://www.cs.york.ac.uk/ftpdir/reports/2011/YCS/462/YC S-2011-462.pdf)

Section III

ZigBee

The section consists of the following three chapters:

Chapter 6: IEEE 802.15.4 Standard and ZigBee
Chapter 7: ZigBee Stack Layer
Chapter 8: ZigBee Pro and ZigBee Security

Summary

This section of the book presents an overview of ZigBee, its applications, characteristics, and security. The following are the key concepts that will be described in the three chapters forming the section:

- ZigBee, a wireless sensor and control network, was developed by the ZigBee Alliance.
- ZigBee applications can be used in home automation, commercial building automation, personal home health care, smart energy, and industrial process monitoring.
- The first ZigBee specification was published in 2004 and supported home control lighting; the ZigBee Alliance no longer supports the 2004 specification.
- In 2006, the ZigBee Alliance published the ZigBee 2006 specification, which was a modification of ZigBee 2004 specification.
- In 2007, ZigBee published ZigBee and ZigBee PRO feature sets.
- ZigBee defines three main types of devices: the coordinator, router, and end device. In addition, devices can act as a trust center or gateway.

- The coordinator is a full-function device (FFD). It performs the critical function of controlling the network, starting a network, and permitting other devices to join or leave the network.
- The trust center performs authentication of devices joining the network, security management, and key distribution.
- ZigBee offers star, tree, and mesh topologies.
- It uses layer architecture for its protocol.
- It uses IEEE 802.15.4 for its physical and MAC layers.
- ZigBee protocol architecture consists of the application, application support sublayer, and network layers
- It offers direct, group, and broadcast addressing.
- A ZigBee end device can have 240 endpoints, where each endpoint may represent different application.
- ZigBee PRO uses a mesh topology, and it is a beaconless network.
- ZigBee PRO offers stochastic addressing, group addressing, asymmetric links, fragmentation, and frequency agility.
- Frequency agility can be used to combat interference.
- ZigBee PRO offers many-to-one routing and multicast routing.
- ZigBee PRO offers standard and high-security modes.
- ZigBee PRO security level is set to 5.
- ZigBee PRO provides link management.
- ZigBee PRO end devices are only powered by battery.
- The ZigBee 2007 specification supports tree and mesh topologies.
- ZigBee PRO selects the best channel at startup and during operation.
- Preventing Spoof Leave Notifications Key.
- Rolling the Network Key.
- Preventing Rejoin Security Attacks.
- Protecting the Network Key.
- Future Scope of ZigBee.
- Advantages of ZigBee Technology.

6

IEEE 802.15.4 and ZigBee

6.1 Introduction

6.1.1 ZigBee General Description

- Over-the air data rates of 250 kbps, 40 kbps, and 20 kbps
- Star or peer-to-peer (mesh) operation
- Allocated 16 bit short or 64 bit extended addresses
- Allocation of guaranteed time slots (GTSs)
- Carrier Sense Multiple Access with Collision Avoidance (CSMA-CA) channel access
- Fully acknowledged protocol for transfer reliability
- Low power consumption
- Energy detection (ED)
- Link quality indication (LQI)
- 16 channels in the 2450 MHz band, 10 channels in the 915 MHz band, and 1 channel in the 868 MHz band
- Aimed to a Low-rate Wireless Personal Area Network (LR-WPAN)

6.1.2 ZigBee: Overview

ZigBee is an open global standard for wireless technology designed by ZigBee Alliance to use low-rate/low-power digital radio signals for personal area networks (PAN). ZigBee operates on the IEEE 802.15.4 specification and is used to create wireless networks that require a low data transfer rate, energy efficiency, and secure networking.

ZigBee is designed for small-scale projects which need wireless connection such as for home automation, medical device data collection, wireless light switches, home energy monitors, traffic management systems, heating and cooling control, and other consumer, industrial equipment that requires short-range low-rate wireless data transfer and other low-power low-bandwidth needs. Such applications use, normally, low-power digital radios for communication. Hence, ZigBee is a low-power, low-data rate and close proximity (i.e., personal area) wireless ad hoc network. The technology defined by the ZigBee specification is intended to be simpler and less expensive than other wireless personal area networks (WPANs), such as Bluetooth or more general wireless networking such as Wi-Fi.

ZigBee employs mesh network topology (actually it is one of the main features of ZigBee), allowing it provides high reliability and a reasonable range.

Its low power consumption limits transmission distances to 10–100 m line-of-sight, depending on power output and environmental characteristics. ZigBee devices can transmit data over long distances by passing data through a mesh network of intermediate devices to reach more distant ones. ZigBee is typically used in low data rate applications that require long battery life and secure networking. ZigBee has a defined rate of 250 kbps, best suited for intermittent data transmissions from a sensor or input device.

One of ZigBee's defining features is the secure communications it is able to provide. This is accomplished through using AES (Advanced Encryption Standard) to encrypt the transactions using 128-bit cryptographic keys (use of AES128 for encryption). This system is based on symmetric keys, which means that both the recipient and originator of a transaction need to share the same key. These keys are either pre-installed, transported by a "trust center" designated within the network or established between the trust center and a device without being transported. Security in a personal area network is most crucial when ZigBee is used in corporate or manufacturing networks.

The ZigBee protocol stack, as mentioned before, is built on top of IEEE 802.15.4, which defines the Media Access Control (MAC) and physical layers for low-rate wireless personal-area network (LR-WPAN). The ZigBee standard offers a stack profile that defines the network, security, and application layers. Developers are responsible for creating their own application profiles or integrating with the public profiles that were developed by the ZigBee Alliance. The ZigBee specification is an open standard that allows manufacturers to develop their own specific applications that require low cost and low power.

The ZigBee Alliance has three network specifications: ZigBee Pro, ZigBee RF4CE, and ZigBee IP. The ZigBee Alliance also publishes application profiles like ZigBee Home Automation, ZigBee Smart Energy, etc., to allow different device classes to interoperate.

Note: ZigBee Alliance consists of more than 270 companies (including Freescale, Ember, Mitsubishi, Philips, Honeywell, and Texas Instruments).

ZigBee Definition: ZigBee is a standards-based wireless technology designed specifically for low-cost, low- power wireless sensor and control networks, particularly Personal Area Networks. Based on an IEEE 802.15 standard, ZigBee operates in the 2.4 GHz band and is intended to be cheaper and simpler to implement than Bluetooth or Wi-Fi.

ZigBee Metaphor: One of the key characteristics of the ZigBee standard is its mesh networking capability. In a large distributed mesh network, a message is relayed from one device to another until it reaches its faraway destination. Similarly, when a group of honey bees, distributed in a large field, want to communicate a message all the way back to their hive, they use message relaying. Each bee performs a specific zigzag dance, which is repeated by the next bee that is slightly closer to the hive. This process is repeated until the message gets to the hive. The name ZigBee was selected as a metaphor for the way devices on the network find and interact with one another.

6.1.3 Evolution of LR-WPAN Standardization

The cellular network was a natural extension of the wired telephony network that became pervasive during the mid-20th century. As the need for mobility and the cost of laying new wires increased, the motivation for a personal connection independent of location to that network also increased. Coverage of large area is provided through (1–2 km) cells that cooperate with their neighbors to create a seemingly seamless network. Examples of standards are GSM, IS-136, IS-95. Cellular standards basically aimed at facilitating voice communications throughout a metropolitan area.

1. IEEE 802.11 Working Group: During the mid-1980s, the IEEE 802.11 working group for WLANs is formed to create a wireless local area network standard. This was formed to cover the growing need of smaller area coverage that is needed for higher user densities and the emergent data traffic.
   Whereas IEEE 802.11 was concerned with features such as Ethernet matching speed, long range (100 m), complexity to handle seamless roaming, message forwarding, and data throughput of 2–11 Mbps, WPANs are focused on a space around a person or object that typically

extends up to 10 m in all directions. The focus of WPANs is low cost, low power, short range, and very small size.

- The IEEE 802.15 Working Group: It is formed to create WPAN standard. This group has currently defined three classes of WPANs that are differentiated by data rate, battery drain, and quality of service (QoS).
- The high data rate WPAN (IEEE 802.15.3): This class is suitable for multi-media applications that require very high QoS.
- Medium rate WPANs (IEEE 802.15.1/Bluetooth): This is handling a variety of tasks ranging from cell phones to PDA communications and has QoS suitable for voice communications.
- The low data rate WPAN (IEEE 802.15.4/LR-WPAN): This class is intended to serve a set of industrial, residential, and medical applications with very low power consumption and cost requirement not considered by the above WPANs and with relaxed needs for data rate and QoS. The low data rate enables the LR-WPAN to consume very little power. This was the base of ZigBee.
- IEEE 802.15.4 defines two layers: Physical Layer and Medium Access Control (MAC) layer. The two layers are the foundation that ZigBee build on to form the ZigBee protocol by adding the Network layer (NWK) and Application layer (APL).

6.1.4 ZigBee History

ZigBee was conceived in 1998, standardized in 2003, and revised in 2006. The ZigBee specification has undergone multiple modifications. The major milestones in its revision history are listed here:

- In 2004, the ZigBee Alliance published its first specification, which supported a home control lighting profile. However, the ZigBee Alliance no longer supports the 2004 specification.
- In February 2006, the ZigBee Alliance published the ZigBee Stack 2006, which contained modifications to ZigBee 2004.
- In October 2007, the ZigBee Alliance published two feature sets called ZigBee and ZigBee PRO. The ZigBee feature set is interoperable with ZigBee PRO. If a network is based on the ZigBee PRO stack, devices from the ZigBee feature set stack can join the network as end devices. Likewise, if a network is based on the ZigBee stack, ZigBee PRO devices can join the network as end devices.

The ZigBee feature set is backward compatible with ZigBee 2006; a ZigBee feature set device can join in a ZigBee 2006 network and vice versa.

6.1.5 ZigBee Application Profiles

The ZigBee Alliance developed the following application profiles:

- **Smart energy:** ZigBee can be used to quickly read electrical, gas, and water meters. The ZigBee smart energy network enables wireless communication between the advanced metering infrastructure (AMI) and the home-area network; that is, the smart energy network will connect home appliances with the utility company for improving energy efficiency and managing peak demand.
- **Commercial building automation:** In a commercial building, ZigBee can be an integral tool in building maintenance. ZigBee wireless can be used to monitor smoke-detector operation and fire-door position. Suppose that a high-rise building contains 50 floors, with each floor having 50 rooms, and each room is equipped with a smoke detector. For safety reasons, each smoke detector must be tested every month. This requires checking 2,500 rooms! Instead of requiring that someone manually test the 2,500 smoke detectors, ZigBee allows a central station to remotely monitor each smoke detector. A ZigBee device may also be used to turn on and off a light without using any wire.
- **Home automation:** ZigBee home automation profile defines devices that are used for residential and commercial applications. ZigBee can be used to remotely control lighting, heating, cooling, and door-locking mechanisms. It can also remotely monitor smoke detectors and home security systems.
- **Personal, home, and hospital care (PHHC):** This profile is used for monitoring the personal health of a patient at home without limiting a patient's mobility. For example, it can remotely monitor blood pressure and heart rate.
- **Telecom applications:** Embedding a ZigBee device into a mobile phone or PDA creates a new device called a ZigBee mobile device. A ZigBee mobile device can be used to communicate with other ZigBee devices. Users of ZigBee mobile devices can send and receive messages and share ring tones, contacts and images. More important, the ZigBee mobile device can even communicate with ZigBee devices that use different application profiles. For example, a ZigBee mobile device

can be used to alert emergency services when a PHHC-enabled device detects a critical problem with the patient's health.

- **Remote control for consumer electronics (ZigBee RF4CE):** Currently, most remote controllers are using infrared (IR) technology, which requires line of sight; ZigBee RF4CE is a protocol that uses radio frequency (RF) to replace IR technology for remote controllers used in consumer electronics.
- **Industrial process monitoring and control:** ZigBee offers solutions for wireless sensor and control. Therefore, it can be used for monitoring and controlling industrial processes without using wire. For example, in inventory tracking, each piece of equipment can be tagged with a wireless sensor and can then be located by a ZigBee node. This process is called radio frequency identification (RFID). ZigBee can monitor machine condition and the performance of operating equipment within a plant. ZigBee can record and transmit such critical information as temperature, pressure, flow, tank level, humidity, and vibration.

In conclusion and concerning ZigBee applications, ZigBee standard may be used into a market that is usually not covered by another wireless technologies (Figure 6.1). ZigBee provides cut down in data rates, while most wireless standards are striving to go faster and ZigBee fits on 8-bit microcontrollers, while other wireless protocols add more and more features. ZigBee looks to control a light or send temperature data to a thermostat, while other wireless technologies aim to deliver streaming high-definition media or to provide the last mile to the Internet. ZigBee is designed to run for years, while other such

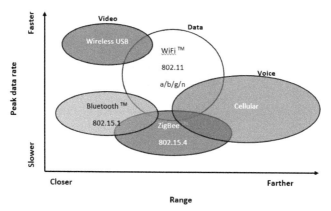

Figure 6.1   Wireless technologies compared.

technologies may work for hours or perhaps days with batteries and ZigBee products can typically provide decades or more of use, while other wireless technologies provide 12 to 24 months of shelf life for a product. Moreover, the slogan for ZigBee is also indicates the Wireless Control That Simply Work.

Low-power wireless mesh networking is provided by ZigBee and it can support up to thousands of devices in a network, different from Bluetooth, UWB, and Wireless USB, which are some other personal area network standards. Technologically controlled home construction and its automation, health care sector, smart power and energy, communication department, and other services such as retail sector are the diverse application areas to which ZigBee Alliance has wide advantages based on these characteristics. Home patient monitoring is one of the applications of ZigBee and by wearable devices, for example, a patient can measure the blood pressure and heart rate. The patient wears a ZigBee device that incorporates with a sensor that collects information related to health on a periodic basis. The data are now wirelessly transmitted to a local server, and initial analysis is performed there. The vital information is finally sent to the patient's nurse or physician via the Internet for further analysis. The ZigBee standard was introduced and developed by ZigBee Alliance.

6.2 ZigBee Wireless Sensor and Control Network

6.2.1 ZigBee Network Characteristics

Several standards currently exist for wireless networks, including Bluetooth, Wi-Fi, and WiMax. ZigBee is a new standard for wireless sensor and control networks. It has the following characteristics:

- Low battery consumption. A ZigBee end device should operate for months or even years without needing its battery replaced.
- Low cost.
- Low data rate. The maximum data rate for a ZigBee device is 250 kbps.
- Easy to implement.
- Supports up to 65,000 nodes connected in a network.
- ZigBee can automatically establish its network.
- ZigBee uses small packets compared with Wi-Fi and Bluetooth.

Table 6.1 shows a comparison of ZigBee characteristics with those of Wi-Fi and Bluetooth.

Table 6.1 shows that Bluetooth, while similar in functionality to ZigBee, does not offer the range of topologies, and its standby current is nearly 70

Table 6.1   ZigBee, Bluetooth, and Wi-Fi characteristics

	Wi-Fi IEEE 802.11	Bluetooth IEEE 802.15.1	ZigBee IEEE 802.15.4
Application	Wireless LAN	Cable replacement	Control and monitor
Frequency bands	2.4 GHz	2.4 GHz	2.4 GHz, 868 MHz, 915 MHz
Battery life (days)	0.1–5	1–7	100–7,000
Nodes per network	30	7	65,000
Bandwidth	2–100 Mbps	1 Mbps	20–250 kbps
Range (m)	1–100	1–10	1–75 and more
Topology	Tree	Tree	Star, tree, cluster tree, and mesh
Standby current	20 * 10–3 amps	200 * 10–6 amps	3 * 10–6 amps
Memory	100 KB	100 KB	32–60 KB

times more than ZigBee. Of the three wireless networks under comparison, ZigBee is the only one that offers mesh topology. In addition, a ZigBee end device can be in sleep mode and still keep its association with its network. ZigBee is considered a more sophisticated network when compared to either Bluetooth or Wi-Fi.

6.2.1.1 ZigBee versus Bluetooth

ZigBee looks rather like Bluetooth but is simpler, has a lower data rate and spends most of its time snoozing. This characteristic means that a node on a ZigBee network should be able to run for 6 months to 2 years on just two AA batteries. The operational range of ZigBee is 10–75 m compared to 10 m for Bluetooth (without a power amplifier). ZigBee sits below Bluetooth in terms of data rate (see Figure 6.1). The data rate of ZigBee is 250 kbps at 2.4 GHz, 40 kbps at 915 MHz and 20 kbps at 868 MHz whereas that of Bluetooth is 1 Mbps. ZigBee uses a basic master-slave configuration suited to static star networks of many infrequently used devices that talk via small data packets. It allows up to 254 nodes. Bluetooth's protocol is more complex since it is geared towards handling voice, images and file transfers in ad hoc networks. Bluetooth devices can support scatternets of multiple smaller non-synchronized networks (piconets). It only allows up to 8 slave nodes in a basic master-slave piconet set-up. When ZigBee node is powered down, it

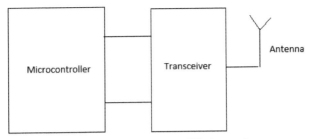

Figure 6.2 Architecture of ZigBee node.

can wake up and get a packet in around 15 ms, whereas a Bluetooth device would take around 3 s to wake up and respond.

6.2.2 ZigBee Device Types and Operating Modes

A ZigBee network consists of ZigBee nodes (devices). The node architecture is shown in Figure 6.2. A node consists of a microcontroller, a transceiver, and an antenna. A ZigBee node uses stack profiles, which are developed by software. A node can be used for a wide variety of applications–for example, lighting control, smoke-detector, and home-security monitoring. Therefore, a node can support multiple subunits, and each subunit has an application object that describes the subunit function.

The nodes can be classified from two points of views: physical and logical.

a. ZigBee Devices–Physical Classification

A node can operate as either a full-function device (FFD) or reduced-function device (RFD).

i. **Fully functional devices (FFD)**: Full Function Devices can perform all existed operations within the standard, involving routing techniques, coordination tasks and sensing task. The FFD can support as router, coordinator or an end device (It can be either RFD or FFD based on its targeted application). A normal FFD in a ZigBee network will be powered from an AC-fed main supply, as it must always be active and hearing to the network.

ii. **Reduced functional devices (RFD)**: This is a device that has a reduced level of functionality. It has the minimum computing ability. The primary objective of RFD is to find an existed network for transferring the data, examine if there is any pending data, and forward request to the network coordinator for the data. The RFD's usually go to the sleep mode when they are not transferring any data and this decreases the battery consumption. The RFDs do not send packets and must be related with an FFD. In other words RFDs can only talk to FFDs as they contain no routing functionality. RFD is typically an end node (device), that is, sensors and actuators which perform specific tasks, that is, monitoring lighting conditions, recording temperature data, and having control over the external devices.

RFDs are often known as child devices as they need other parent devices with which to communicate.

From the given definitions, an FFD can perform all the tasks that are defined by the ZigBee standard, and it operates in the full set of the IEEE 802.15.4 MAC layer. An RFD performs only a limited number of tasks.

b. ZigBee Devices: Logical Devices

ZigBee Device Object defines three types of logical devices, each having a specific role as illustrated (Figure 6.3). The three types are:

- **Coordinator (C) or Coordinator Device (CD):** A coordinator device (CD) is an FFD and responsible for establishing, executing and managing the overall ZigBee network. CD is responsible for configuring the security level of the network and configuring the address of the

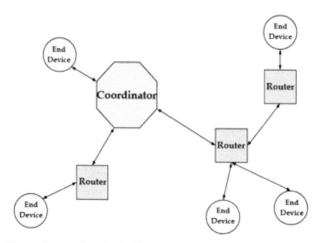

Figure 6.3   Types of network nodes in ZigBee architecture (ZigBee stack) (https://mmbnet works.atlassian.net/wiki/spaces/SKB/pages/39518242/Zigbee+Network+Concepts).

"Trust Center" (The default value of this address is the ZigBee coordinator's own address; otherwise, the ZigBee coordinator may designate an alternate Trust Center). The ZigBee coordinator also maintains a list of currently associated devices and facilitates support of orphan scan and rejoin-processing to enable previously associated devices to rejoin the network. Each network has exactly one coordinator. The coordinator performs, in general, the following functions:

- o Starts the Network: Always first coordinator need to be installed for establishing ZigBee network service, it starts a new PAN (Personal Area Network), and once started other ZigBee components as router (R) and End devices (E) can join the network (PAN).
- o Selects the channel to be used by the network and also PAN ID
- o Assigns how addresses are allocated to nodes or routers
- o Permits other devices to join or leave the network
- o Holds a list of neighbors and routers
- o Transfers application packets
- o It will not go to sleep mode.
- o It is mains powered (AC) and support child devices.
- o Stores information about the network, including acting as the Trust Center and repository for security keys (see "Trust Center").

- **End device (E or ED):** A ZigBee End Device (ZED or ED) can be an RFD. An RFD operates within a limited set of the IEEE 802.15.4 MAC layer, enabling it to consume less power. The end device (child) can be connected to a router or coordinator (parent). These devices do not have to stay awake the entire time, while the devices relating to the other two categories have to. In other words, it operates at low duty-cycle power, meaning it consumes power only while transmitting information. ZigBee architecture is designed so that an end device transmission time is short. The end device performs the following functions:

- o Joins or leaves a network
- o Transfers application packets

ZigBee End Device (ZED) requires the least amount of memory, and, therefore, can be less expensive to manufacture than a ZigBee Router (ZR) or ZigBee Coordinator (ZC).
- **Router (R):** A router is an FFD. ZigBee router is an intermediary device in a network which sends the data from the source node to the destination node. These devices forward the data as well as sense the data from their

surrounding atmosphere. Routers need the Coordinator's (in general the Trust Center's) permission to join the network if security is enabled on the network and can also double-up as end devices. A router is used in tree and mesh topologies to expand network coverage. The function of a router is to find the best route to the destination over which to transfer a message. A router performs all functions similar to a coordinator except the establishing of a network. This means that at certain occasions, routers can allow other routers and end devices to join the network and will maintain a list of currently associated devices and facilitate support of orphan scan and rejoin processing to enable previously associated devices to rejoin the network. Since routers link multiple sections of a network, they cannot be put to sleep. The Router is mains powered (AC) and support child devices.

Besides the three logical types of nodes, the network also has the following:

- **ZigBee Trust Center (ZTC): It is a device trusted by devices within a network to distribute keys for the purpose of network.** It is also responsible for end-to-end application configuration management and on device authentication. The Trust center could be the coordinator device or a device designated as the Trust center by the coordinators of the ZigBee Network.

  All members of the network recognize exactly one Trust Center, and there is exactly one Trust Center in each secure network. It is configured to operate in **either standard or high security mode** and may be used to help establish end-to-end application keys either by sending out link keys directly (i.e., key-escrow capability) or by sending out master keys. These keys are generated at random.

  Role of Trust Center in Key Generation and Distribution:

  For purpose of trust management, a device accepts an initial master or active network key originating from its Trust Center via unsecured communication channel. For purposes of network management, a device accepts an initial active network key and updated network keys only from its Trust Center. For purposes of configuration, a device accepts master keys or link keys intended for establishing end-to-end security between two devices only from its Trust Center. Aside from the initial master key or network key, additional link, master, and network keys are generally only accepted if they originate from a device's Trust Center via secured key transport.

– **Standard Security Mode**: It is designed for residential applications. In this mode, the Trust Center maintains a list of devices, master keys, link keys and network keys with all the devices in the network; however, it maintains a standard network key and controls policies of network admittance. In this mode, each device that joins the network securely shall either have a global link key or a unique link key depending upon the application in use. It is required that the trust center have prior knowledge of the value of the link key and the type (global or unique) in order to securely join the device to the network.

– **High security mode**: It is designed for high-security commercial applications. In this mode, the Trust Center maintains a list of devices, master keys, link keys and network keys that it needs to control and enforce the policies of network key updates and network admittance. It also mandates the implementation of key establishment using SKKE and entity authentication.

Standard and high-security modes will be discussed in detail in Chapter 8.

• **ZigBee gateway:** The ZigBee gateway is used to connect the ZigBee network to another network, such as a LAN, by performing protocol conversion.

6.2.2.1 Operating modes

The current ZigBee protocols support beacon-enabled and non-beacon-enabled networks. In non-beacon-enabled networks, an unslotted CSMA/CA channel access mechanism is used. In this type of network, ZigBee Routers typically have their receivers continuously active, requiring a more robust power supply. However, this allows for heterogeneous networks in which some devices receive continuously while others only transmit when an external stimulus is detected. The typical example of a heterogeneous network is a wireless light switch: The ZigBee node at the lamp may constantly receive, since it is connected to the AC mains supply, while a battery-powered light switch would remain asleep until the switch is thrown. The switch then wakes up, sends a command to the lamp, receives an acknowledgment, and returns to sleep. In such a network, the lamp node will be at least a ZigBee Router, if not the ZigBee Coordinator; the switch node is typically a ZigBee End Device.

In beacon-enabled networks, the ZigBee Routers transmit periodic beacons to confirm their presence to other network nodes. Nodes may sleep between beacons, thus lowering their duty cycle and extending their battery life. Beacon intervals depend on data rate; they may range from 15.36 ms to 251.65824 s at 250 kbps, from 24 ms to 393.216 s at 40 kbps, and from 48 ms to 786.432 s at 20 kbps. However, low duty cycle operation with long beacon intervals requires precise timing, which can conflict with the need for low product cost.

In general, the ZigBee protocols minimize the time the radio is on, to reduce power use. In beaconing networks, nodes only need to be active while a beacon is being transmitted. In non-beacon-enabled networks, power consumption is decidedly asymmetrical: Some devices are always active while others spend most of their time sleeping.

Except for the Smart Energy Profile 2.0, ZigBee devices are required to conform to the IEEE 802.15.4 2003 Low-rate Wireless Personal Area Network (LR-WPAN) standard. The standard specifies the lower protocol layers–the physical layer (PHY) and the Media Access Control portion of the data link layer (DLL). The basic channel access mode is "carrier sense, multiple access/collision avoidance" (CSMA/CA). That is, the nodes talk in the same way that humans converse; they briefly check to see that no one is talking before he or she start, with three notable exceptions. Beacons are sent on a fixed timing schedule and do not use CSMA. Message acknowledgments also do not use CSMA. Finally, devices in beacon-enabled networks that have low latency real-time requirements may also use Guaranteed Time Slots (GTS), which, by definition, do not use CSMA.

6.2.3 ZigBee Topologies

Network topology is the design of the elements such as links or nodes of a communication network. It can be used to illustrate various types of telecommunication networks.

ZigBee uses the IEEE 802.15.4 2003 specification for its physical layer and MAC layer. IEEE 802.15.4 offers star, tree, cluster tree, and mesh topologies (called also peer-to-peer topology); however, ZigBee supports only star, tree, and mesh topologies. It uses an association hierarchy; a device joining the network can either be a router or an end device, and routers can accept more devices. While building ZigBee network, the choice of topology must take into consideration which nodes are line-powered or battery-powered,

expected battery lifetime, amount of network traffic required, latency requirements, the cost of the solution, etc. The details of the ZigBee topologies are
given in Section 6.3 while discussing IEEE 802.15.4 topologies.

6.2.3.1 Forming the ZigBee network

The following steps describe the method of forming ZigBee network:

- Coordinator searches for suitable RF channel which is usable and not
  interfering with Wireless LAN frequencies in use. This is because
  WLAN also operates in the same 2.4 GHz bands. This is done on all
  the 16 channels. It is also referred as energy scan.
- Coordinator starts the network by assigning a PAN ID to the network.
  Assignment is done in two ways: Manual (pre-configured) and dynamic
  (obtained by checking other PAN IDs of networks already in the operation nearby so that PAN ID does not conflict with other networks). Here
  Coordinator also assigns network address to itself, that is, 0x0000.
- The coordinator at this stage completes its configuration and is ready
  to accept network joining request queries from routers and end devices
  who wish to join the PAN.
- In addition to above, Coordinator(C) sends broadcast beacon request
  frame on remaining quiet channel (see Section 6.4). This is also referred
  as beacon scan or PAN scan. By this, Coordinator receives PAN ID of
  routers(R) and end devices(E) present nearby. It also comes to know
  whether R/E allow join or not.
- At this stage, Routers and End devices can join the network by sending
  association request to Coordinator (C). C will respond with association
  response.

6.2.3.2 Joining the ZigBee Network

This section discusses how a router or end device joins ZigBee network.
There are two ways to join a ZigBee network: MAC association and network
re-join.

- First one is implemented by device underlying MAC layer and second
  one is implemented by network layer, despite the name may also be used
  to join a network for the first time.
- MAC association can be performed between Coordinator (C) and Router
  (R)/End device (E) or Router (R) and End device (E) or Router (R) and
  another Router (R).

- Let us assume that Coordinator(C) has already established the PAN network. Hence next step for R or E is to find out whether C is allowing joining or not. So they do PAN scan or send beacon request frame.
- After they come to know that they can join the network, they will send association request frame and will join the network as soon as they receive the association response.
- As mentioned above whether or not C or R allows a new device to join depends on two main factors:
  – Permit joining attribute
  – Number of end device children it already has.

6.2.4 End Device Addressing

When a device joins a ZigBee network, the ZigBee coordinator or router assigns a logical 16-bit address to the device. Also, each device has a 64-bit IEEE address, and no two devices can have the same IEEE address in the entire world. Short addresses (16 bits) can be used by devices in a network. The advantage of using the 16-bit address is to extend the life of the battery. A 16-bit address reduces the size of the frame. A smaller frame size results in shorter transmission time. Less transmission time means greater battery life. The disadvantage of using the 16-bit address is that two nodes in different networks can have the same address.

The methods of assigning addresses for routers and end devices are given in Chapter 8.

6.2.5 Depth of a Network, Number of Children, and Network Address Allocation

- **Depth of a network:** The depth of a network is determined by the number of routers (hops) from the coordinator to the farthest device, where farthest is defined by number of hops. In a star topology, the depth of a network is one.
- **Number of children:** The number of end devices (children) that are connected to a router or coordinator. The coordinator sets the maximum number of children connected to a router.
- **Address allocation:** In a tree topology, each coordinator holds information about the network, such as the maximum number of children (the number of end devices connected to each router), maximum number of routers, and uses this information to assign an address to each router.

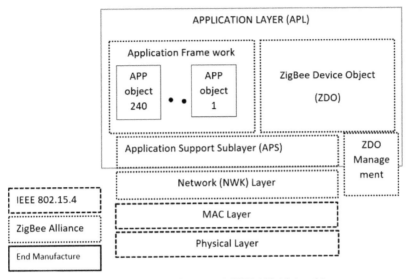

Figure 6.4 Structure of ZigBee and IEEE 802.15.4 architecture.

The routers, then, assign the addresses to their respective end devices (children). In a mesh topology, however, each router assigns a random address to its respective end devices. Address allocation is given in Chapter 8.

6.3 ZigBee Protocol Stack Overview

The ZigBee protocol stack consists of four layers–Physical layer, Medium access control (MAC) layer, Network layer (NWK) and Application layer (APL)–as shown in Figures 6.4 and 6.5. Each layer provides a set of services exposed to the upper layer via a service access point (SAP). Each SAP supports a number of service primitives to achieve the required functionality. The two lower layers, the physical and MAC layers, are governed by IEEE 802.15.4 standard and the network and application layers are governed by the ZigBee standard. The application layer framework consists of APS (application support sublayer) and the ZDO (ZigBee device objects). Application objects defined by the manufacturers use the framework and share APS and security with the ZDO. IEEE 802.15.4 has two PHY layers that operate in two separate frequency ranges, 868/915 MHz and 2.4 GHz. The lower frequency PHY layer covers both the 868 MHz European band and the 915 MHz band,

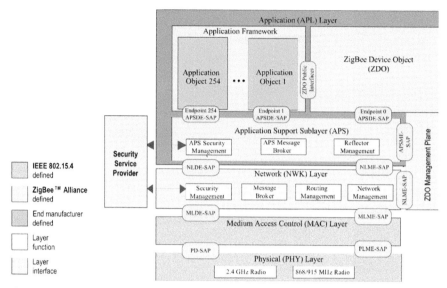

Figure 6.5   Outline of the ZigBee stack architecture (https://research.kudelskisecurity.com/ 2017/11/01/zigbee-security-basics-part-1/).

used in countries such as USA and Australia. The higher frequency PHY layer is used virtually worldwide. The IEEE 802.15.4 MAC controls access to the radio channel using a CSMA-CA mechanism. Its responsibilities may also include transmitting beacon frames, synchronization, and providing a reliable transmission mechanism.

The structure of ZigBee and IEEE 802.15.4 architecture is repeated in more detail in Figure 6.4, in order to explain the different locks.

Figures 6.4 and 6.5 show the ZigBee stack as three sections:

- Foundation layers (this is PHY and MAC layers defined by IEEE 802. 15. 4)
- Application and Interface Section (or ZigBee layers). The layers are built on the 802.15.4 to form ZigBee. This consists of the network layer, the ZigBee device object (ZDO), the application sublayer, and security management
- Manufacturer application: Manufacturers of ZigBee devices can use the ZigBee application profile or develop their own application profile.

The physical layer and the MAC layer will be discussed in Section 6.4.4 while discussing 802.15.4. The network and application layers are discussed next.

6.3.1 Application Layer (APL)

The Application layer (APL), consists of ZigBee device objects (ZDOs), Application support sublayer (APS), and the Application Framework (manufacturer application objects). The ZigBee application layer consists of the APS sublayer, ZDO and the defined by the manufacturer application objects.

6.3.1.1 ZigBee Device Objects (ZDO)

ZigBee Device Objects (ZDO) are applications which employ network and application support layer primitives to implement ZigBee End Devices, ZigBee Routers, and ZigBee Coordinators. It provides an interface between the application objects, the device profile, and the APS. The ZDO is located between the application framework and the application support sublayer. The ZDO is responsible for the following:

- Initializing the application support sublayer (APS), the network layer (NWK), and the security service provider.
- Assembles configuration information from the end applications to determine and implement device and service discovery, security management (key loading, key establishment, key transport and authentication), network management (network discovery, leaving/joining a network, resetting a network connection and creating a network), binding, node, and group management.

ZigBee Device Object (ZDO) manages the security policies and the security configuration of a device. Objects listed as mandatory in the specifications document reside on all ZigBee devices.

The ZDO presents public interface to the application objects in the application framework layer for control of device and network functions by the application objects. The ZDO interfaces with the lower portions of the ZigBee protocol stack, on endpoint 0, through the ASPDE-SAP for data, and through the ASPME-SAP and NLME-SAP for control messages. The public interface provides address management to the device, discovery, binding, and security functions within the application framework layer for the ZigBee protocol stack

Endpoints and ZDO

The endpoints are what most people associate with ZigBee. Each endpoint houses what's called an application object which is basically a device profile with whatever extra functionality you decide to add. When the device is started, all the endpoints will register themselves with the application framework and provide descriptions of their device profile and their capabilities. Endpoint 0 is a special endpoint and always contains the ZigBee Device Object (ZDO). This object implements the ZigBee Device Profile which has multiple functions, one of them being the network manager.

The user application can manage the network by making requests and handling callbacks to this object, which is why it's important to know about it. In general, the ZigBee endpoints are going to be the main interface between the user application and the stack.

When mentioning that the network management deviates slightly from the stack's layer hierarchy, it means that the ZDO's network manager object bypasses the APS and interfaces directly to the networking layer. This is clear from Figure 6.5 where the ZDO wraps around the APS and the NWK layers.

6.3.1.2 APS (application support sublayer)

The APS sublayer provides an interface between a NHLE (next higher layer entity) and the network layer.

The *APS* provides an interface between the network layer (NWK) and application layer (APL) through a general set of services for use by both the ZigBee device object and the defined by manufacturer application object. It provides services for the establishment and maintenance of security relationships.

The APS sublayer provides two services accessed through two SAPs (Service Access Points). These are the APS data service (**APS data entity**), accessed through the APSDE-SAP (APS sublayer data entity SAP), and the APS management service (**APS management entity**), accessed through the APSME-SAP (APS Management Entity SAP). APSDE: provides data transmission services between application entities and APMSE: provides security services, binding of devices and group management). APSME-ASP maintains a database of managed objects known as the AIB (APS Information Base). The APS layer allows frame security to be based on link keys or the network key. The APS layer is responsible for the processing steps needed to securely transmit outgoing frames, securely receive incoming frames, and securely establish and manage cryptographic keys. Upper layers control the management of cryptographic keys by issuing primitives to the APS layer. Figure 6.6 shows the APS sublayer model.

- APSDE: The APSDE provides data service to the network layer and both ZDO and application objects to enable the transport of application PDUs between two or more devices. The devices themselves must be located on the same network.

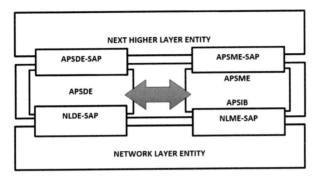

Figure 6.6 APS sublayer model.

- APSME: The APSME provides a management services to allow an application to interact with the stack. The APSME shall provide the ability to match two devices together based on their services and their needs. This service is called binding service, and the APSME shall be able to construct and maintain a table to store this information. This database is referred to as the APS sublayer information base (AIB).

In summary, the Application Sublayer interfaces the ZigBee application layer to the ZigBee networking layer, and it provides a common set of data transport services to all the endpoints. There are also a couple of other services that the APS provides and we'll get into that when we discuss the app layer in more detail.

6.3.1.2.1 ZigBee APS frame format
Figure 6.7 represents the APS Frame Format.

6.3.1.2.2 APS Ack frames
APS Ack frames do not have group address field, have extended header, and no payload as shown in Figure 6.8.

6.3.1.2.3 APS command frames
APS Command Frames (Figure 6.9) are used for key establishment and switching, removing a device from the network.

6.3.1.3 Application framework
The ZigBee application framework is the environment in which application objects are hosted on ZigBee devices. Up to 240 distinct application objects

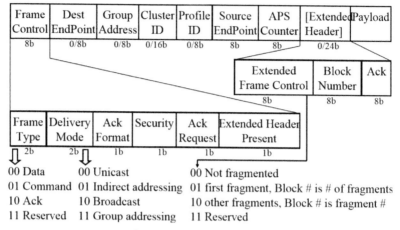

Figure 6.7   APS frame format.

Frame Control	Dest EndPoint	Cluster ID	Profile ID	Source EndPoint	APS Counter	Extended Header
8b	8b	0/16b	0/16b	8b	8b	24b

Figure 6.8   APS Ack frame format.

Frame Control	Group Address	APS Counter	APS Command ID	APS Command Payload
8b	0/16b	8b	8b	

Figure 6.9   APS command frame.

can be defined. These are usually manufacturer defined application objects. It defines application profiles (agreements for messages, message formats, and processing actions that enable developers to create an interoperable, distributed application employing application entities that reside on separate devices) and clusters. Each object is identified by an endpoint address from 1 to 240. Two additional endpoints are defined for APSDE-SAP usage. The endpoint 0 is reserved for the data interface to the ZDO, and endpoint 255 is reserved for the data interface function to broadcast data to all application objects. Endpoints 241-245 are reserved for future use.

- Application profiles: Application profiles are agreements for messages, message formats, and processing actions that enable developers to create and interoperable, distributed application employing applications entities that reside on separate devices. These applications profiles enable applications to send commands, request data and process commands and requests.

- Clusters: Clusters are identified by a cluster identifier, which is associated with data flowing out of, or into, the device. Cluster identifiers are unique within the scope of a particular application profile.

In summary, the Application Framework is just a glorified multiplexer and container for all of the endpoints. All of the endpoints register themselves with the AF, and when a data frame comes into the application layer, the AF will check its destination endpoint and forward it there.

6.3.2 Network Layer (NWK)

The network layer ensures correct operation of the IEEE 802.15.4-2003 MAC sublayer and provides a suitable service interface to the application layer. Its capabilities and structure are those typically associated to such network layers, including routing. It interfaces with application layer via **data entity** (NLDE: generates network level PDU, provides topology-specific routing and security) and the **management entity** (NLME: configures a new device, starts a network, performs joining, rejoining and leaving a network functionality, provides addressing capabilities, neighbor discovery, route discovery, reception control and routing). The NWK layer is responsible for the processing steps needed to securely transmit outgoing frames and securely receive incoming frames. The NWK layer's frame-protection mechanism uses the Advanced Encryption Standard (AES) and CCM* (Enhanced counter with CBC-MAC mode of operation) for authentication and confidentiality.

In general, the Network Layer has the following components:

- Network Layer Data Entity (NLDE): Makes NPDU from NSDU
- Network Layer Management Entity (NLME): Configure new device, neighbor discovery, route discovery, joining/leaving a network, ...
- Network Layer Information Base (NIB): Capabilities (RFD/FFD), Security level, protocol version, route discovery time, max retries for route discovery, neighbor table, ...

The routing protocol used by the network layer is AODV. In AODV, to find the destination device, AODV broadcasts out a route request to all of its neighbors. The neighbors then broadcast the request to their neighbors and onward until the destination is reached. Once the destination is reached, it sends its route reply via unicast transmission following the lowest cost path back to the source. Once the source receives the reply, it will update its routing table for the destination address of the next hop in the path and the path cost.

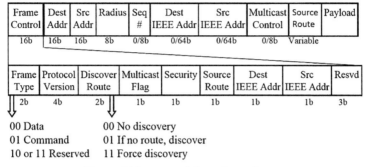

00 Data   00 No discovery
01 Command   01 If no route, discover
10 or 11 Reserved   11 Force discovery

Figure 6.10   Network layer frame format.

Chapter 7 deals with this layer.

6.3.2.1 ZigBee network layer frame format
This is given in Figure 6.10.
   In Figure 6.10:

Radius: Defines the hop limit

Multicast Flag: If "1", Multicast control present.

Source Rout: The Header contains route

Destination/Source IEEE Address: This is the IEEE address given to the node

Broadcast Frame Address:

   ○ FFFF → means All devices
   ○ FFFD → Means Devices with Receiver ON
   ○ FFFC → Means Coordinator/Routers
   ○ FFFB → Means Low-Power Routers

6.3.3 Security Services

As one of its defining features, ZigBee provides facilities for carrying out secure communications, protecting establishment and transport of cryptographic keys, cyphering frames (frame protection), and controlling and managing devices. It builds on the basic security framework defined in IEEE 802.15.4. This part of the architecture relies on the correct management of symmetric keys and the correct implementation of methods and security policies.

   ZigBee security is the subject of Chapter 8. In this section, as part of describing the ZigBee stack, brief idea about security services offered by ZigBee is given.

6.3.3.1 Basic security model

The basic mechanism to ensure confidentiality is the adequate protection of all keying material. Trust must be assumed in the initial installation of the keys, as well as in the processing of security information. For an implementation to globally work, its general conformance to specified behaviors is assumed.

Keys are the cornerstone of the security architecture; as such their protection is of paramount importance, and keys are never supposed to be transported through an insecure channel. A momentary exception to this rule occurs during the initial phase of the addition to the network of a previously unconfigured device. The ZigBee network model must take particular care of security considerations, as ad hoc network may be physically accessible to external devices. Also the state of the working environment cannot be predicted.

Within the protocol stack, different network layers are not cryptographically separated, so access policies are needed, and conventional design assumed. The open trust model within a device allows for key sharing, which notably decreases potential cost. Nevertheless, the layer which creates a frame is responsible for its security. If malicious devices may exist, every network layer payload must be ciphered, so unauthorized traffic can be immediately cut off. The exception, again, is the transmission of the network key, which confers a unified security layer to the grid, to a new connecting device.

6.3.3.2 Security architecture

The security architecture includes security mechanisms at two layers of the stack: the network and the APS layers. The two layers are responsible for secure transport of their frames. APS, also, provides services for the establishment and maintenance of security relationships. The ZDO manages the security policies and the security configuration of a device.

ZigBee uses 128-bit keys to implement its security mechanisms. A key can be associated either to a network, being usable by both ZigBee layers and the MAC sublayer, or to a link, acquired through pre-installation, agreement or transport. Establishment of link keys is based on a master key which controls link key correspondence. Ultimately, at least, the initial master key must be obtained through a secure medium (transport or pre-installation), as the security of the whole network depends on it. Link and master keys are only visible to the application layer. Different services use different one-way variations of the link key to avoid leaks and security risks.

Key distribution is one of the most important security functions of the network. A secure network will designate one special device which other devices trust for the distribution of security keys: the trust center. Ideally, devices will have the center trust address and initial master key preloaded; if a momentary vulnerability is allowed, it will be sent as described above. Typical applications without special security needs will use a network key provided by the trust center (through the initially insecure channel) to communicate.

Thus, the **trust center** maintains both the network key and provides point-to-point security. Devices will only accept communications originating from a key supplied by the trust center, except for the initial master key. The security architecture is distributed among the network layers as follows:

- The MAC sublayer is capable of single-hop reliable communications. As a rule, the security level it is to use is specified by the upper layers.
- The network layer manages routing, processing received messages and being capable of broadcasting requests. Outgoing frames will use the adequate link key according to the routing if it is available; otherwise, the network key will be used to protect the payload from external devices.
- The application layer offers key establishment and transport services to both ZDO and applications.

The security levels infrastructure is based on CCM*, which adds encryption- and integrity-only features to CCM.

6.3.3.3 Network layer security

When a frame originating at the network layer needs to be secured, or when a frame originates at a higher layer and the NWK secure all frame attribute in the NIB is TRUE, ZigBee shall use the frame-protection mechanism, unless the security enable parameter of the NLDE-DATA. Request primitive is FALS, explicitly prohibiting security. The Network layer's frame-protection mechanism shall make use of the AES (Advanced Encryption Standard) and use CCM as specified. The security level applied to a network frame shall be determined by the NWK security level attributed in NIB. Upper layers manage network layer security by setting up active and alternate network keys and by determining which security level to use.

6.3.3.4 APL layer security

When a frame originating at the APL LAYER needs to be secured, the APS sublayer shall handle security. The APS layer allows frame security to be based on link keys or the network key. Figure 6.11 shows an example of the

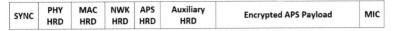

SYNC	PHY HRD	MAC HRD	NWK HRD	APS HRD	Auxiliary HRD	Encrypted APS Payload	MIC

Figure 6.11 APL layer security framework.

security fields they may be included in an APL frame. The APS layer is also responsible for providing applications and the ZDO within key establishment, key transport, and device management services.

6.3.3.5 Trust center role

For security purposes, ZigBee defines the role of Trust Center. The trust center is the device trusted by devices within a network to distribute keys for the purposes of network and end-to-end application configuration management. All members of the network shall recognize exactly one Trust center, and there shall be exactly one trust center in each secure network. In high security, commercial applications a device can be pre-loaded with the Trust center address and initial master key. Alternatively, if the application can tolerate a moment of vulnerability, the master key can be sent via an in-band unsecured key transport. If not pre-loaded, a device's Trust center defaults to the ZigBee coordinator or a device is designated by the ZigBee Coordinator. In low security, residential applications a device securely communicates with its Trust Center using the current networks key, which can be preconfigured or sent via an in-band unsecured key transport. For purposes of trust management, a device accepts an initial master or active network key originating from its trust center via unsecured key transport. For purposes of network management, a device accepts initial active network key and updated network keys only from its Trust Center. For purposes of configuration, a device accepts master keys or link keys intended for establishing end-to-end security between two devices only form its Trust Center. Aside from the initial master key or network key, additional link, master, and network keys are generally only accepted if the originate form a device's Trust Center via secured key transport.

6.3.4 ZigBee Address Assignment

ZigBee allocate 16-bit address for each node. Two schemes are in use for allocating the address:

- Distributed Scheme
- Stochastic Scheme

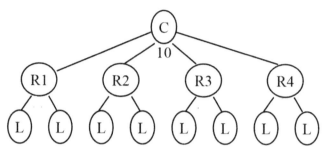

Figure 6.12   Network of Example 6.1.

a. Distributed Scheme

In this scheme, which is good for tree topology, each child is allocated a sub-range of addresses. The scheme limits the maximum depth L, the maximum number of children per parent C, and the maximum number of routers R. The address of any child, the nth child, is a function of the parent address and the limits (L, C, and R) as follows:

Address of the nth child = Address of parent + (n-1) S(d) + 1, where

$$S(d) = \begin{cases} 1 + C(L-d) & \text{if } R = 1 \\[2mm] \dfrac{CR^{L-d-1} - 1 - C + R}{R-1} & \text{if } R > 1 \end{cases}$$

Example 6.1: Distributed Scheme

Consider a network given in Figure 6.12 with the following parameters:

- Maximum depth L = 2
- Maximum number of Routers R = 4
- Maximum number of Children per parent: C = 3
- Coordinator: d = 0. Skip

Assume that the address of the coordinator is 10 (decimal), calculate the addresses of the different devices (R1 to R4) and all end devices.

Let us start by calculating S(0), and S(1):

$$S(0) = \frac{C\,R^{L-d-1} - 1 - C + R}{R-1} = \frac{3 \times 4^{2-0-1} - 1 - 3 + 4}{4-1} = 4$$

$$S(1) = \frac{3x\,4^{2-1-1} - 1 - 3 + 4}{4-1} = 1$$

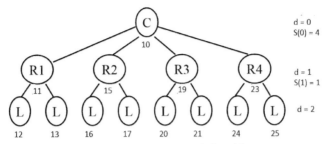

Figure 6.13 The network with the addresses.

Considering the equation:

$$\text{Address of the } n^{\text{th}} \text{ child} = \text{Address of parent} + (n-1) * S(d) + 1$$

The address will be as follows:

Address of R1 $= 10 + 1 = 11$
Address of R2 $= 10 + 1 + 1 \times 4 = 15$
Address of R3 $= 10 + 1 + 2 \times 4 = 19$
Address for R3 $= 10 + 1 + 3 \times 4 = 23$
Similarly for the next level, we use $S(1) = 1$
Children of R1 are assigned addresses 12 and 13
Children of R2 are assigned addresses 16 and 17
Children of R3 are assigned addresses 20 and 21
Children of R4 are assigned addresses 24 and 25

Figure 6.13 shows the network with the allocated addresses.

b. Stochastic Address Assignment

In this scheme, the parent draws a 16-bit random number between 1 and $2^{16} - 1$ and assign it to the new child. To avoid any address conflict, the parent advertises the assigned number to the network. If another node has that address, an address conflict message is returned, and the parent draws another number and repeat (see Section 8.3.1).

6.4 IEEE 802.15.4 Standard

IEEE 802.15.4 standard describes Wireless Personal Area Network (WPAN). WPANs are used to convey information over relatively short distance. Unlike WLAN (Wireless local area networks), connections effected via WPANs involve little or no infrastructure. This feature allows small, power-efficient, inexpensive solutions to be implemented for a wide range of devices. The IEEE 802.15.4 standards' scope is to define the physical layer and medium access control sublayer specification for low data rate wireless connectivity with fixed portable, and moving devices with no battery or very limited battery consumption requirements typically operating in the POS (personal operating space) of 10 m. It is foreseen that, depending on the application, a longer range at a lower data rate may be an acceptable trade-off. It is the intent of this project to work toward a level of coexistence with other wireless device in conjunction with coexistence task Groups, such as 802.15.2 and 802.11/ESTI-BRAN/MMAC 5GSG.

Overview

A LR-WPAN is a simple, low-cost communication network that allows wireless connectivity in application with limited power and relaxed throughput requirements. The main objectives of an LR-WPAN are ease of installation, reliable data transfer, short-range operation, extremely low cost, and a reasonable battery life, while maintaining a simple and flexible protocol. Some of the characteristics include:

- Over-the-air (OTA) data rates of 250 kbps, 40 kbps, and 20 kbps
- Star or peer-to-peer operation (also called Mesh operation)
- Allocated 16-bit short or 64-bit extended addresses
- Allocation of GTS (Guaranteed Time Slots)
- CSMA-CA (Carrier Sense Multiple Access with Collision Avoidance channel access)
- Fully acknowledged protocol transfer reliability
- Low power consumption
- ED (Energy Detection)
- LQI (Link Quality Indication)
- 16 channels in the 2450 MHz band, 10 channels in the 915 MHz band, and 1 channel in the 868 MHz band

6.4.1 IEEE 802.15.4 Standard

The IEEE 802.15.4 standard has undergone a number of releases. In addition to this there are a number of variants of the IEEE 802.15.4 standard to cater for different forms of physical layer, etc. These are summarized below in Table 6.2.

Table 6.2   IEEE 802.15.4 standard summary

IEEE 802.15.4 Standard Summary	
IEEE 802.15.4 Version	Details and comments
IEEE 802.15.4–2003	This was the initial release of the IEEE 802.15.4 standard. It provided for two different PHYs–one for the lower frequency bands of 868 and 915 MHz, and the other for 2.4 GHz.
IEEE 802.15.4–2006	This 2006 release of the IEEE 802.15.4 standard provided for an increase in the data rate achievable on the lower frequency bands. This release of the standard updated the PHY for 868 and 915 MHz. It also defined four new modulation schemes that could be used–three for the lower frequency bands, and one for 2.4 GHz.
IEEE 802.15.4a	This version of the IEEE 802.15.4 standard defined two new PHYs. One used UWB technology and the other provided for using chirp spread spectrum at 2.4 GHz.
IEEE 802.15.4c	Updates for 2.4 GHz, 868 MHz and 915 MHz, UWB and the China 779-787 MHz band.
IEEE 802.15.4d	2.4 GHz, 868 MHz, 915 MHz and Japanese 950–956 MHz band.
IEEE 802.15.4e	This release defines MAC enhancements to IEEE 802.15.4 in support of the ISA SP100.11a application.
IEEE 802.15.4f	This will define new PHYs for UWB, 2.4 GHz band and also 433 MHz
IEEE 802.15.4g	This will define new PHYs for smart neighborhood networks. These may include applications such as smart grid applications for the energy industry. It may include the 902–928 MHz band.

Although new versions of the standard are available for use by any of the higher layer standards, ZigBee still uses the initial 2003 release of the IEEE 802.15.4 standard.

6.4.2 Device Types

Two different device types can participate in the LR-WPAN network: FFD (fully functioning device) and RFD (reduced-function-device). The details of the two types are given in Section 6.2.2.

6.4.3 Network Topology

IEEE 802.15.4 offers star, tree, cluster tree, and mesh topologies; however, ZigBee supports only star, tree, and mesh topologies. Some references state that 802.15.4 may operate in either of two topologies: the star topology or

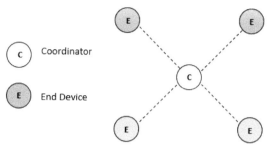

Figure 6.14   Star topology.

peer-to-peer topology (which is also called the mesh topology). Next we introduce all these topologies starting with those supported by 802.15.4.

6.4.3.1 Star topology

Star topology is defined by the underlying 802.15.4 specification which Zig-Bee builds on. The star topology is the simplest and most limited one in the 802.15.4 and accordingly in ZigBee. It consists of a coordinator positioned in the center and several end devices (nodes), as shown in Figure 6.14.

After an FFD is activated for the first time, it may establish its own network and become the PAN coordinator. In the star topology communication is established between devices and a single central controller (Figure 6.14), called the PAN coordinator. PAN coordinator may also have specific application, but it can be used to initiate, terminate, or route communication around the network. Any packet exchange between end devices must go through the coordinator. The PAN coordinator is the primary controller of the PAN. All devices operating on a network of any of the stare or the peer-to-peer topologies shall have unique 64-bit extended addresses. This address can be used for direct communication within the PAN, or it can be exchanged for a short address allocated by the PAN coordinator when the device associates.

All-star networks operate independently from all other star networks currently in operation. This is achieved by choosing PAN identifier, which is not currently used by any other network within the radio sphere of influence. Once the PAN identifier is chosen, the PAN coordinator can allow other device to join its network. Both FFDs and RFDs may join the network.

The PAN coordinator may be main powered, while the devices will be more likely be battery powered. Applications that benefit from the star topology include home automation, personal computers (PC) peripherals, toys and games, and personal health care.

The peer-to-peer topology also has a PAN coordinator. But it differs for the star topology in that a device can communicate with any other device as long as they are in range of one another.

The disadvantage of this topology is the operation of the network depends on the coordinator of the network, and because all packets between devices must go through coordinator, the coordinator may become bottlenecked (Failure of the coordinator can shut-down the whole network). Also, the star hub can become bottlenecked with network bandwidth. On the other hand, having a star network topology will limit the overall distances that can be covered because of the limited number of hops.

6.4.3.2 Mesh topology, also, peer-to-peer topology

A key component of the ZigBee protocol is the ability to support mesh networking. In a mesh network, nodes are interconnected with other nodes so that multiple pathways connect each node. Mesh topology, also referred to as a peer-to-peer network, and consists of one coordinator, several routers, and end devices, as shown in Figure 6.15.

Connections between nodes are dynamically updated and optimized through sophisticated, built-in mesh routing table. It is possible to expand the network range by adding more devices into the network. If during the

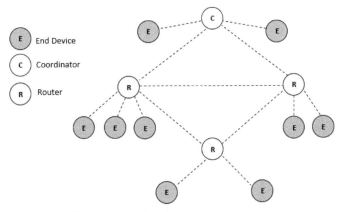

Figure 6.15 Mesh (peer-to-peer) topology.

transmission one of the paths fail, the node will find the alternate path to reach to the destination therefore eliminating dead zones. Using this mesh topology, it is easier for user to add or remove the device because they can communicate with any destination device in the network.

Mesh networks are decentralized in nature; each node is capable of self-discovery on the network. Also, as nodes leave the network, the mesh topology allows the nodes to reconfigure routing paths based on the new network structure. The characteristics of mesh topology and ad hoc routing provide greater stability in changing conditions or failure at single nodes.

Applications such as industrial control and monitoring, wireless sensor networks, asset and inventory tracking would benefit from such a topology.

The following are the characteristics of a mesh topology:

1. A mesh topology is a multihop network; packets pass through multiple hops to reach their destination.
2. The range of a network can be increased by adding more devices to the network.
3. It can eliminate dead zones.
4. A mesh topology is self-healing, meaning during transmission, if a path fails, the node will find an alternate path to the destination.
5. A peer-to-peer network can be ad hoc, self-organizing and self-healing.
6. Devices can be close to each other so that they use less power.
7. Adding or removing a device is easy.
8. Any source device can communicate with any destination device in the network.
9. Compared with star topology, mesh topology requires greater overhead.
10. Mesh routing uses a more complex routing protocol than a star topology.
11. It also allows multiple hops to route messages from any device to any other device in the network. It can provide reliability by multipath routing.

6.4.3.3 Tree topology

In this topology, the network consists of a central node (root tree), which is a coordinator, several routers, and end devices, as shown in Figure 6.16. The function of the router is to extend the network coverage. The end nodes that are connected to the coordinator or the routers are called children. Only routers and the coordinator can have children. Each end device is only able to communicate with its parent (router or coordinator). The coordinator and routers can have children and, therefore, are the only devices that can be

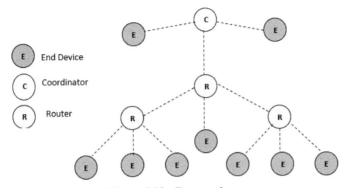

Figure 6.16   Tree topology.

parents. An end device cannot have children and, therefore, may not be a parent. An end device can interact with another end device only through its parent node and there is no direct link among end devices. A special case of tree topology is called a cluster tree topology.

Tree networks may employ beacon-oriented communication as per the IEEE 802.15.4 standard.

The disadvantages of tree topology are:

- If one of the parents becomes disabled, the children of the disable parent cannot communicate with other devices in the network.
- Even if two nodes are geographically close to each other, they cannot communicate directly.

6.4.3.4 Cluster tree topology

The cluster-tree network is a special case of the peer-to-peer network in which most devices of the FFDs. An RFD may connect to a cluster tree network as a leaf node at the end of a branch, because it may only associate with one FFD at a time. Any of the FFDs may act as a coordinator and provide synchronization services to other devices or other coordinators. Only one of these coordinators can be the overall PAN coordinator, which may have greater computational resources than any other device in the PAN. The coordinator forms the first cluster by establishing itself as the CLH (cluster head) with a CID (Cluster Identifier) of zero, choosing an unused PAN identifier, and broadcasting beacon frames to neighboring devices. A candidate device receiving a beacon frame may request to join the network

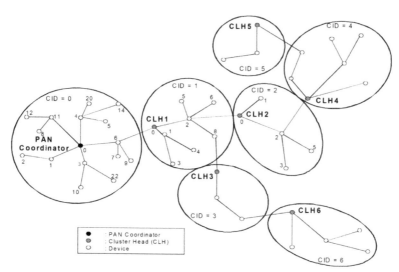

Figure 6.17   Formation of cluster tree network.

Note: It is also possible to consider that the cluster tree topology is a special case of tree topology in which a parent with its children is called a cluster, as shown in Figure 6.18. Each cluster is identified by a cluster ID. ZigBee does not support cluster tree topology, but IEEE 802.15.4 does support it.

at the CLH if the PAN coordinator permits the device to join, it will add the nee device as a child device in its neighbor list. Then the newly joined device will add the CLH parent in its neighbor list and begin transmitting periodic beacons. Other candidate devices may then join the network at that device. If the original candidate device is not able to join the network at the CLH, it will search for another parent device. The simplest form of a cluster tree network is a single cluster network, but larger networks are possible by forming a mesh of multiple neighboring clusters. Once predetermined application or network requirements are met, the PAN coordinator may instruct a device to become the CLH of a new cluster adjacent to the first one. Other devices gradually connect and form a multi cluster network structure.

The formation of the cluster tree is shown in Figure 6.17.

6.4.4 802.14.5 Architecture

The LR-WPAN architecture is defined in terms of a number of blocks in order to simplify the standard. These blocks are called layers. Each layer responsible for one part of the standard offers services to the higher layers.

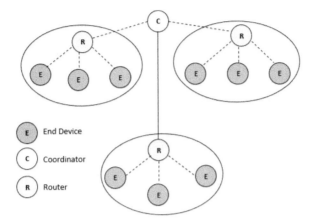

Figure 6.18 Cluster tree topology.

The interfaces between the layers serve to define the logical links that are describes in this standard. An LR-WPAN device comprises of a PHY, which contains the RF (radio frequency) transceiver along with its low level control mechanism, and a MAC sublayer that provides access to the physical channel for all types of transfer. Figure 6.19 shows these blocks in graphical representation. The upper layer consists of a network layer, which provides network configuration, manipulation, and message routing, and an application layer, which provides the intended function of the device. The definition of these upper layers is outside the scope of this standard. An IEEE 802.2 Type 1 LLC (logical link control) can access the MAC sublayer through the SSCS (service specific convergence sublayer). The LR-WPAN architecture can be implemented either as embedded device or as devices requiring the support of an external device such as a PC.

As shown in Figures 6.4 and 6.5, the IEEE 802 15.4 physical layer together with Medium Access Control (MAC) layer act as foundation layers for ZigBee Architecture.

6.4.4.1 PHY layer
The physical and electrical characteristics are defined by the Physical Layer. This layer is responsible for data transmission and reception. Mapping bits of information permits them to travel through the air by modulation, and spreading techniques are the basic task of physical layer.

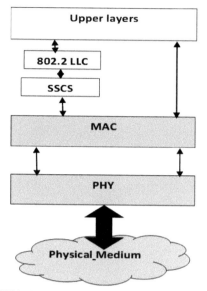

SSCS: Service Specific Convergence Sublayer
Figure 6.19   WPAN architecture.

The PHY provides, in general, two services: the PHY data service and the PHY management service interfacing to the PLME (physical layer management entity). The PHY data service enables the transmission and reception of the PPDU (PHY protocol data units) across the physical radio channel. The features of the PHY are activation and deactivation of the radio transceiver, ED (Receiver Energy Division), LQI, channel selection, CCA (clear channel assessment), and transmitting as well as receiving packets across the physical medium. The standard offers two PHY options based on the frequency band. Both are based on direct sequence spread spectrum (DSSS). The radio shall operate at one of the following license-free band:

- 868–868.6 MHz (Europe)
- 902–928 MHz (North America)
- 2400–2483.5 MHz (Worldwide)

The data rate is 250 kbps at 2.4 GHz, 40 kbps at 915 MHz and 20 kbps at 868 MHz.

The receiver ED measurement is intended for use by a network layer as part of channel selection algorithm. It is an estimate of the received signal over power within the band width of an IEEE 802.15.4 channel. No attempt is made to identify or decode signals on the channel. The ED time should be equal to 8 symbols periods. Upon reception of a packet, the PHY the PSDU length, PSDU itself and link quality in the PD-DATA, indication primitive. The LQI measurement is a characterization of strength and/ or quality of a received packet. The measurement may be implemented using receiver ED, a signal-to-noise estimation or a combination of these methods.

The use of LQI result is up to the network or application layers. CCA is performed according to at least one of the following three methods.

- Energy above threshold. CCA shall report a busy medium upon detecting any energy above the ED threshold.
- Carrier sense only. CCA shall report a busy medium only upon the detection of a signal with the modulation and spreading characteristics of IEEE 802.15.4. This signal may be above or below the ED threshold.
- Carrier sense with energy above threshold. CCA shall report a busy medium only upon the detection of a signal with the modulation and spreading characteristics of the IEEE 802.15.4 with energy above the ED threshold.

During operation, the physical layer receives the MAC protocol data units and processes them through channel coding, interleaving, baseband modulation, multi-antenna encoding, precoding, resource and antenna mapping. The choice of an appropriate modulation and coding scheme, as well as multi-antenna transmission mode, is critical to achieve the desired reliability and system throughput in mobile wireless data communications. Typical mobile radio channels tend to be dispersive and time-variant and exhibit severe Doppler effects, multipath delay variation, intra-cell and inter-cell interference, and fading. A good and robust design of the physical layer ensures that the system can normally operate and overcome the above deleterious effects and can provide the maximum throughput and lowest latency under various operating conditions.

PHY Constants

Constant	Description	Value
aMaxPHYPacketSize	The maximum PSUD size (in octets) the PHY shall be able to receive	127
aTurnaroundTime	RX-to-TX or TX-to-RX maximum turnaround time	12 symbol periods

PHY PIB (PIB: PAN Information Base) Attributes

Attribute	Identifier	Type	Range	Description
phyCurrent Channel	0 x 00	Integer	0-26	The RF channel to use for all following transmissions and receptions
phyChannels Supported	0 x 01	Bitmap	See description tion	The 5 most significant bits (MSB) ($b_{27} \ldots b_{31}$) of *phyChannelsSupported* shall be reversed and set to 0, and the 27 LSBs ($b_0 \ldots b_{26}$) shall indicate the status (1 = available, 0 = unavailable) for each of the 27 valid channels (b_k shall indicate the status of channel k).
phyTransmit Power	0 x 02	Bitmap	0x00– 0xbf	The 2 MSBs represent the tolerance on transmit power: 00 = ± 1 dB 01 = ± 3 dB 10 = ± 6 dB The 6 LSBs represent a signed integer in twos complement format, corresponding to the nominal transmit power of the device in decibels relative to 1 mW. The lowest value of *phyTransmitPower shall be interpreted as less than or equal to -32 dBm.*
phyCCAMode	0 x 03	Integer	1–3	The CCA mode

6.4.4.1.1 Function of physical layer in ZigBee architecture

Physical Layer is responsible for the following functions:

- Activation and deactivation of transmission and reception.
- Channel selection and its assessment.
- Sending and receiving of packets.
- Energy detection within the channel.

Table 6.3 Physical layer frequency band

Frequency Band	Country	Data Rate	Channel Numbers
868.3 MHz	European countries	20 kbps	0
902–928 MHz	United States	40 kbps	1–10
2.405 GHz	Worldwide	250 kbps	11–26

Table 6.4 IEEE 802.15.4 RF channel details

Frequency Band (MHz)	Channels Available	Throughput Available (kbps)	Region Use Allowable
868–868.6	1	20	Europe
902–928	10 (2003 rel)30 (2006 rel)	30	USA
2 400	16	250	Global

Table 6.5 Summary of the frequency bands and data rates of ZigBee PHY version

ZigBee PHY Version	Frequency Band (MHz)	Chip Rate (kchip/s)	Modulation Type	Bit Rate (kbps)	Symbol Rate (ksymbol/s)	Symbols
868 MHz band	868–868.6	300	BPSK	20	20	Binary
915 MHz band	902–928	600	BPSK	40	40	Binary
2450 MHz band	2400–2483.5	2000	OQPSK	250	62.5	16-ary orthogonal

6.4.4.1.2 Frequency band, data rate, and channel numbering

The physical layer performs modulation on outgoing signals and demodulation on incoming signals. It transmits information and receives information from a source. Table 6.3 shows the physical layer frequency band, data rate, and channel numbers.

The higher data rate at 2.4 GHz is attributed to a higher-order modulation scheme. Lower frequency provides longer range due to lower propagation losses. Low rate can be translated into better sensitivity and larger coverage area. Higher rate means higher throughput, lower latency or lower duty cycle. This information is summarized in Table 6.5.

6.4.4.1.3 IEEE 802.15.4 frequencies and frequency bands

The IEEE 802.15.4 frequency bands align with the license free radio bands that are available around the globe. Of the bands available, the 2.4 GHz (2 400 MHz) band is the most widely used in view of the fact that it is available globally and this brings many economies of scale.

With new allocations arising as a result of issues such as the digital dividend and other countries adopting and using IEEE 802.15.4, other frequencies and bands are being considered. These include: 314-316 MHz, 430-434 MHz, and 779-787 MHz frequency bands in China and the 950 MHz-956 MHz band in Japan. Other frequencies are also being considered for UWB variants of IEEE 802.15.4.

6.4.4.1.4 ZigBee frequency bands and data rates

Concerning ZigBee, it has two bands of operation 868/915 MHz and 2450 MHz; 868/915 band provides about 20–40 kbps, and 2450 MHz band provides about 250 kbps data rates. Tables 6.5–6.7 cover in detail the ZigBee frequency bands 868, 915, 2450 MHz range and provides data rates supported by each. Figure 6.20 shows channelization and spectrum spacing.

Channelization and Spectrum Spacing

A total of 27 channels, numbered from 0 to 26, are available across the three frequency bands. 16 channels are available in the 2450 MHz band, 10 in the 915 MHz band, and 1 in the 868 MHz band. The center frequency of these channels is defined as follows:

$F_c = 868.3$
$F_c = 906 + 2 (k-1)$ k from 1 to 10, and
$F_c = 2405 + 5 (k-11)$ k from 11 to 26

This is given in Tables 6.6 and 6.7.

Following are 11–26 ZigBee channels in the 2400 MHz band.

Table 6.6   ZigBee channels, 915 MHz ISM band

ZigBee Channels (1–10)	Center Frequency (MHz)
1	906
2	908
3	910
4	912
5	914
6	916
7	918
8	920
9	922
10	924

Table 6.7 ZigBee channels, 2.4 GHz ISM band

ZigBee Channels (11–26)	Center Frequency (MHz)
11	2405
12	2410
13	2415
14	2420
15	2425
16	2430
17	2435
18	2440
19	2445
20	2450
21	2455
22	2460
23	2465
24	2470
25	2475
26	2480

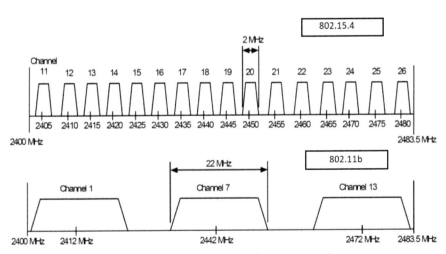

Figure 6.20 Channelization and spectrum spacing.

6.4.4.1.5 IEEE 802.15.4 modulation formats

There were two different modulation schemes defined for IEEE 802.15.4 in the original standard released in 2003. Both these air interface or radio interface configurations are based on direct sequence spread spectrum, DSSS techniques. The one for the lower frequency bands provides a lower data rate in view if the smaller channel width, whereas the format used at 2.4 GHz enables data to be transferred at rates up to 250 kbps.

The 2006 release of the 802.15.4 standard upgraded the number of areas of the air interface and the modulation schemes. There were four different physical layers that were defined. Three used the DSS approach using either binary or offset quadrature phase shift keying, BPSK and OQPSK. An optional physical layer approach was defined using amplitude sift keying, ASK.

In case of 2.4 GHz every 4 bits are mapped into one symbol and encoded into one of 16 32-bit quasi orthogonal PN codes. The coding is shown in Table 6.8. O-QPSK modulation is used with the 2.4 GHz.

Table 6.8   Coding in case of 2.4 GHz

Data Symbol (Decimal)	Data Symbol (Binary) (b_0, b_1, b_2, b_4)	Chip values $c_0 c_1 - c_{30} c_{31}$
0	0 0 0 0	1 1 0 1 1 0 0 1 1 1 0 0 0 0 1 1 0 1 0 1 0 0 1 0 0 0 1 0 1 1 1 0
1	1 0 0 0	1 1 1 0 1 1 0 1 1 0 0 1 1 1 0 0 0 0 1 1 0 1 0 1 0 0 1 0 0 0 1 0
2	0 1 0 0	0 0 1 0 1 1 1 0 1 1 0 1 1 0 0 1 1 1 0 0 0 0 1 1 0 1 0 1 0 0 1 0
3	1 1 0 0	0 0 1 0 0 0 1 0 1 1 1 0 1 1 0 1 1 0 0 1 1 1 0 0 0 0 1 1 0 1 0 1
4	0 0 1 0	0 1 0 1 0 0 1 0 0 0 1 0 1 1 1 0 1 1 0 1 1 0 0 1 1 1 0 0 0 0 1 1
5	1 0 1 0	0 0 1 1 0 1 0 1 0 0 1 0 0 0 1 0 1 1 1 0 1 1 0 1 1 0 0 1 1 1 0 0
6	0 1 1 0	1 1 0 0 0 0 1 1 0 1 0 1 0 0 1 0 0 0 1 0 1 1 1 0 1 1 0 1 1 0 0 1
7	1 1 1 0	1 0 0 1 1 1 0 0 0 0 1 1 0 1 0 1 0 0 1 0 0 0 1 0 1 1 1 0 1 1 0 1
8	0 0 0 1	1 0 0 0 1 1 0 0 1 0 0 1 0 1 1 0 0 0 0 0 1 1 1 0 1 1 1 1 0 1 1
9	1 0 0 1	1 0 1 1 1 0 0 0 1 1 0 0 1 0 0 1 0 1 1 0 0 0 0 0 0 1 1 1 0 1 1 1
10	0 1 0 1	0 1 1 1 1 0 1 1 1 0 0 0 1 1 0 0 1 0 0 1 0 1 1 0 0 0 0 0 0 1 1 1
11	1 1 0 1	0 1 1 1 0 1 1 1 1 0 1 1 1 0 0 0 1 1 0 0 1 0 0 1 0 1 1 0 0 0 0 0
12	0 0 1 1	0 0 0 0 0 1 1 1 0 1 1 1 1 0 1 1 1 0 0 0 1 1 0 0 1 0 0 1 0 1 1 0
13	1 0 1 1	0 1 1 0 0 0 0 0 0 1 1 1 0 1 1 1 1 0 1 1 1 0 0 0 1 1 0 0 1 0 0 1
14	0 1 1 1	1 0 0 1 0 1 1 0 0 0 0 0 0 1 1 1 0 1 1 1 1 0 1 1 1 0 0 0 1 1 0 0
15	1 1 1 1	1 1 0 0 1 0 0 1 0 1 1 0 0 0 0 0 0 1 1 1 0 1 1 1 1 0 1 1 1 0 0 0

Table 6.9 Generic packet format

Octets:4	1	1		Variable
Preamble	SFD	Frame length (7 bits)	reserved(1 bit)	PSDU
SHR		PHR		PHY payload

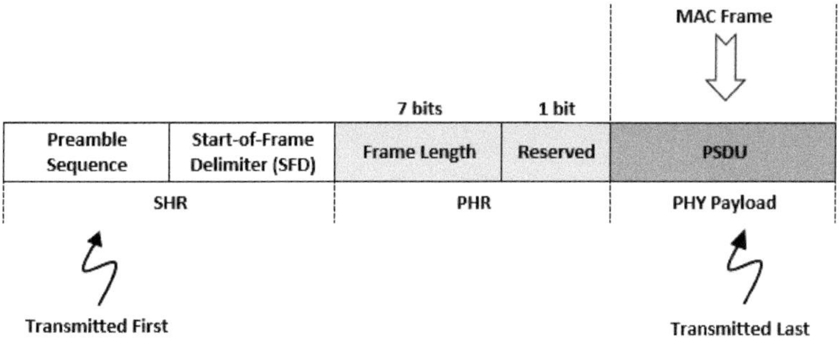

Figure 6.21 PPDU format.

Pulse Shaping

The half-sine pulse shape used to represent each baseband chip is:

$$p(t) = \begin{cases} \sin\left(\pi\dfrac{t}{2T_c}\right), & 0 \le t \le 2\,T_c \\ 0, & otherwise \end{cases}$$

6.4.4.1.6 ZigBee physical layer frame format-PPDU

ZigBee PHYSICAL layer (PHY) frame format consists of subfields: preamble, SFD, frame length, and PSDU field descriptions.

The PHY Protocol Data Unit (PPDU) format is shown in Figure 6.21 and Table 6.9. The PPDU consists of three components: the Synchronization header (SHR), the PHY header (PHR), and the PHY payload.

The SHR enables the receiver to synchronize and lock into the bit stream. The PHR contains frame length information. The PHY payload is provided by upper layers and includes data or commands that need to be transmitted to another device.

The SHR consists of a preamble and a start-of-frame delimiter (SFD), see Table 6.10.

Preamble field

The preamble field is used by the receiver to obtain chip and symbol synchronization. The bits in the preamble field in all PHYs, except for the ASK

Table 6.10   Preamble field lengths and durations

PHY Option	Length		Duration (μs)
868 MHz BPSK	4 octets	32 symbol	1600
915 MHz BPSK	4 octets	32 symbol	800
868 MHz ASK	5 octets	2 symbol	160
915 MHz ASK	3.75 octets	6 symbol	120
868 MHz O-QPSK	4 octets	8 symbol	320
915 MHz O-QPSK	4 octets	8 symbol	128
2.4 GHz O-QPSK	4 octets	8 symbol	128

Table 6.11   SFD field format (except for ASK PHYs)

Bits	0	1	2	3	4	5	6	7
Values	1	1	1	0	0	1	0	1

PHYs, are binary zeros (There are 32 zeros in the preamble field and used for synchronization). The preamble in an 868 MHz ASK PHY is generated by repeating certain sequence twice. The duration of this preamble is 160 μs. In a 915 MHz ASK PHY, a specific sequence is repeated six times and takes 120 μs. The lengths and durations of the preambles in all PHY options are listed in Table 6.10.

Start-of-Frame Delimiter (SFD)

The SFD field indicates the end of the SHR and start of the PHR. The SFD is an 8-bit field shown in Table 6.11. The lengths of SFD fields are provided in Table 6.12.

Table 6.12   SFD field lengths

PHY Option	Length	
868 MHz BPSK	1 octets	8 symbol
915 MHz BPSK	1 octets	8 symbol
868 MHz ASK	2.5 octets	1 symbol
915 MHz ASK	0.625 octets	1 symbol
868 MHz O-QPSK	1 octets	2 symbol
915 MHz O-QPSK	1 octets	2 symbol
2.4 GHz O-QPSK	1 octets	2 symbol

Table 6.13 Frame length values

Frame Length Values	PHY Payload
0 to 4	Reserved
5	Acknowledgment MPDU
6 to 8	Reserved
9 to *aMaxPHYPacketSize*	Any other MPDU

Frame Length

The next field in a PHY packet is the frame length, which specifies the total number of octets in the PHY payload (PSDU). The PSDU length can be any value from 0 to 127 octets. But practically, based on IEEE 802.15.4-2006, the PSDU length is either 5 octets for a MAC acknowledgment frame or 9–127 for any other MPDU. The frame length values of 0–4 and 6–8 are reserved for potential future applications (Table 6.13).

PHY Service Data Unit (PSDU)

The last field is the PHY Service Data Unit (PSDU). The content of the PSDU is provided by the MAC as a MAC frame. In IEEE 802.15.4, the first bit that will be transmitted is the leas significant bit (LSB) of the SHR. The most significant bit (MSB) of the last octet of the PHY payload is transmitted last.

6.4.4.2 Receiver energy detection (ED)

The receiver energy detection (ED) measurement is intended for use by a network layer as part of channel selection algorithm. It is an estimate of the received signal power within the bandwidth of an IEEE 802.15.4 channel. No attempt is made to identify or decode signals on the channel. The ED time should be equal to 8 symbol periods. The ED result shall be reported as an 8-bit integer ranging from 0x00 to 0xff. The minimum ED value (0) shall indicate received power less than 10dB above the specified receiver sensitivity. The range of received power spanned by the ED values shall be at least 40dB. Within this range, the mapping from the received power in decibels to ED values shall be linear with an accuracy of $+-6$ dB.

6.4.4.3 Link quality indication (LQI)

Upon reception of a packet, the PHY sends the PSDU length, PSDU itself and link quality (LQ) in the PD-DATA.indication primitive. The LQI measurement is a characterization of the strength and/or quality of a received packet. The measurement may be implemented using receiver ED, a signal-to-noise estimation or a combination of these methods. The use of LQI result is up to the network or application layers. The LQI result should be reported as an integer ranging from 0x00 to 0xff. The minimum and maximum LQI values should be associated with the lowest and highest quality IEEE 802.15.4 signals detectable by the receiver and LQ values should be uniformly distributed between these two limits.

6.4.4.4 Clear channel assessment (CCA)

The clear channel assessment (CCA) is performed according to at least one of the following three methods:

- Energy above threshold. CCA shall report a busy medium upon detecting any energy above the ED threshold.
- Carrier sense only. CCA shall report a busy medium only upon the detection of a signal with the modulation and spreading characteristics of IEEE 802.15.4. This signal may be above or below the ED threshold.
- Carrier sense with energy above threshold. CCA shall report a busy medium only upon the detection of a signal with the modulation and spreading characteristics of IEEE 802.15.4 with energy above the ED threshold.

7

ZigBee Stack Layers

This chapter concentrates on the ZigBee Stack Layers. It starts by introducing the details of the MAC layer followed by the Routing Layer. The information given are enough to help any engineer developing ZigBee network.

7.1 ZigBee Mac Layer

The MAC sublayer (Figure 7.1) provides two services: the MAC data service and the MAC management service. Interfacing to the MAC sublayer management entity (MLME) service access point (SAP) (MLME-SAP). The MAC data service enables the transmission and reception of MAC protocol data units across the PHY data service. The features of the MAC sublayer are beacon management, channel access, GTS management, frame validation, acknowledged frame delivery, association, and disassociation.

In addition, the MAC sublayer provides means for implementing application appropriate security mechanisms. The slotted CSMA-CA shall be used if superframe structure is used in PAN. If beacons are not being used in the PAN or a beacon cannot be located in a beacon-enabled network, an unslotted CSMA-CA algorithm is used.

- Slotted CSMA-CA: the back off period boundaries of every device in the PAN are aligned with the superframe slot boundaries of the PAN coordinator. Each time a device wishes to transmit data frames during the CAP, it shall locate the 11 boundary of the next back off period.
- Unslotted CSMA-CA: The back off period of one device is not required to be synchronized to the back off period of another device.

After introducing the MAC superframe, brief discussion on CSMA-CA will be given.

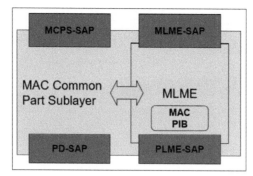

MLME: MAC Layer Management Entity   PLME: Physical Layer Management Entity
MCPS: MAC Common Part Sublayer

Figure 7.1   MAC sublayer.

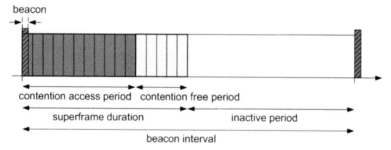

Figure 7.2   Superframe structure.

7.1.1 Superframe Structure

The LR-WPAN standard allows the optional use of a superframe structure. The format of the superframe is defined by the coordinator. The superframe is bounded by network beacons and is divided into 16 equally sized slots. The beacon frame is sent in the first slot of each superframe. If a coordinator does not want to use the superframe structure, it may turn off the beacon transmissions. The beacons are used to synchronize the attached devices, to identify the PAN and to describe the structure of superframes.

The superframe can have an active and an inactive portion (see Figure 7.2). During the inactive portion, the coordinator shall not interact with its PAN and may enter a low-power mode.

The active portion consists of contention access period (CAP) and contention free period (CFP). Any device wishing to communicate during the CAP shall compete with other devices using a slotted CS MACA

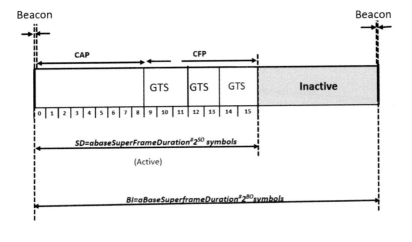

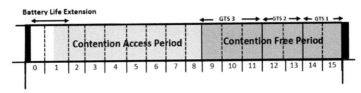

Figure 7.3 Superframe format: Details of the active period of the superframe.

mechanism. On the other hand, the CFP contains guaranteed time slots (GTSs) (Figure 7.3). The GTSs always appear at the end of the active superframe starting at a slot boundary immediately following the CAP. The PAN coordinator may allocate up to seven of these GTSs and a GTS can occupy more than one slot period. The duration of different portions of the superframe are described by the values of "*macBeaconOrder*" and "*macSuperFrameOrder*". *macBeaconOrder* describes the interval at which the coordinator shall transmit its beacon frames. The beacon interval (BI) is related to the *macBeaconOrder*, BO, as follows:

$$BI = aBaseSuperFrameDuration * 2^{BO}, \quad 0 \leq BO \leq 14.$$

The superframe is ignored if BO = 15. The value of *macSuperFrameOrder* describes the length of the active portion of the superframe. The superframe duration, SD, is related to *macSuperFrameOrder*, SO, as follows:

$$SD = aBaseSuperFrameDuration * 2^{SO}, \quad 0 \leq SO \leq 14.$$

If SO = 15, the superframe should not remain active after the beacon. The active portion of each superframe is divided into a "*aNumSuperFrameSlots*" equally spaced slots of duration 2^{SO} * *aBaseSlotDuration* and is composed of three parts: a beacon, a CAP and CFP. The beacon is transmitted at the start of slot 0 without the use of CSMA. The CAP starts immediately after the beacon. The CAP shall be at least *aMinCAPLength* symbols unless additional space is needed to temporarily accommodate the increase in the beacon frame length to perform GTS maintenance. All frames except acknowledgment frames or any data frame that immediately follows the acknowledgment of a data request command that are transmitted in the CAP shall use slotted CSMA-CA to access the channel. A transmission in the CAP shall be complete one IFS period before the end of the CAP. If this is not possible, it defers its transmission until the CAP of the following superframe.

An example of superframe structure is shown in Figure 7.3. The CFP, if present, shall start on a slot boundary immediately following the CAP and extends to the end of the active portion of the superframe. The length of the CFP is determined by the total length of all of the combined GTSs. No transmissions within the CFP shall use a CSMA-CA mechanism. A device transmitting in the CFP shall ensure that its transmissions are complete one IFS period before the end of its GTS. IFS time is the amount of time necessary to process the received packet by the PHY. Transmitted frames shall be followed by an IFS period. The length of IFS depends on the size of the frame that has just been transmitted. Frames of up to *aMaxSIFS-FrameSize* in length shall be followed by a SIFS whereas frames of greater length shall be followed by a LIFS. The PANs that do not wish to use the superframe in a nonbeacon-enabled shall set both *macBeaconOrder* and *macSuperFrameOrder* to 15. In this kind of network, a coordinator shall not transmit any beacons, all transmissions except the acknowledgment frame shall use unslotted CSMA-CA to access channel, and GTSs shall not be permitted.

7.1.1.1 Updating superframe configuration

In a beacon-enabled network, the NWK layer can request the MLME to start a superframe structure. The NWK layer provides the necessary parameters, including but not limited to the length of the active period and how often the beacons must be transmitted.

One of the parameters in the superframe configuration is the BLE option (see Figure 7.4). This option allows the beaconing coordinator to turn off its receiver for a period of time equal to *macBattLifeExtPeriods* after transmitting its beacon frame to conserve battery energy. This period is in addition to the required IFS period after transmitting any frame.

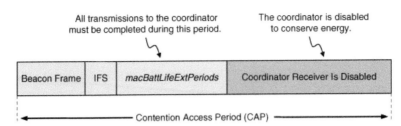

Figure 7.4 If the BLE option is selected, the coordinator receiver is active only for a limited time after the IFS period.

If the BLE option is set to false, the beaconing coordinator must keep its receiver active for the entire CAP.

QoS by Superframe
- A coordinator may reserve some time slots as Guaranteed Time Slots (GTS)
- GTSs form contention-free period (CFP), which is always behind contention access period (CAP)

7.1.2 CSMA-CA Algorithm

As mentioned above, if superframe structure is used in the PAN, then slotted CSMA-CA (Carrier Sense Multiple Access/Collision Avoidance) shall be used. If beacons are not being used in the PAN or a beacon cannot be located in a beacon-enabled network, unslotted CSMA-CA algorithm is used. In both cases, the algorithm is implemented using units of time called backoff periods, which is equal to "*aUnitBackoffPeriod*" symbols. In slotted CSMA-CA channel access mechanism, the backoff period boundaries of every device in the PAN are aligned with the superframe slot boundaries of the PAN coordinator. In slotted CSMA-CA, each time a device wishes to transmit data frames during the CAP, it shall locate the boundary of the next backoff period. In unslotted CSMA-CA, the backoff periods of one device do not need to be synchronized to the backoff periods of another device. Each device has 3 variables: NB, CW and BE. NB is the number of times the CSMA-CA algorithm was required to backoff while attempting the current transmission. It is initialized to 0 before every new transmission. CW is the contention window length, which defines the number of backoff periods that need to be clear of activity before the transmission can start. It is initialized to 2 before each transmission attempt and reset to 2 each time the channel is assessed to be busy. CW is only used for slotted CSMA-CA. BE is the backoff exponent, which is related to how many backoff periods a device shall wait before attempting to assess the channel. Although the receiver of the device is enabled during the channel assessment portion of this algorithm, the device shall discard any frames received during this time.

In slotted CSMA-CA, NB, CW and BE are initialized and the boundary of the next backoff period is located.

The CSMA-CA algorithm works as follows:

1. ***Initialization***: If a device wishes to transmit a frame using CSMA-CA, it first initializes the local variables **BE:=macMinBE** for the *backoff exponent* and **NB:=0** for the number of successive backoffs before the current transmission.

2. ***Backoff***: Before a station attempts to send a frame, it waits for a random integer number between **0** and **2^BE-1** complete *backoff periods* of length ***aUnitBackoffPeriod***. If slotted CSMA-CA is used, transmissions are synchronized with the beacon, and therefore the backoff starts at the beginning of the next backoff period; if unslotted CSMA-CA is used, the backoff starts immediately. The first backoff period of each superframe starts with the transmission of the beacon. If the backoff has not been completed at the end of the CAP, it resumes at the start of the next superframe.

3. ***Clear Channel Assessment***: After completing its backoff, the station performs a *clear channel assessment (CCA)*. If, after eight symbol periods, the channel is assessed to be busy, both **BE** and **NB** are incremented by one, up to a maximum of ***aMaxBE*** for **BE** and *macMaxCSMABackoffs*+1 for **NB**. If **NB** exceeds ***macMaxCSMABackoffs***, the protocol terminates with a *channel access failure*; if not, the protocol returns to the backoff step. If the channel is assessed to be free, the data frame can be transmitted. In slotted CSMA-CA, two CCAs, each starting at the beginning of a backoff period, have to be performed.

4. ***Starting the transmission***: In slotted CSMA-CA, a transmission can only start at a backoff period boundary, only if all steps (two CCAs, frame transmission, and acknowledgment) can be completed at least one interframe space (IFS) period before the end of the CAP. If the MAC sublayer cannot proceed, it shall wait until the start of the CAP in the next superframe and repeat the evaluation.

5. ***Channel is Busy***: If the channel is assessed to be busy (step 4), the MAC sublayer shall increment both NB and BE by one, ensuring that BE shall

be no more than *aMaxBE*. In slotted CSMA-CA, CW can also be reset to 2. If the value of NB is less than or equal to *macMaxCSMABackoffs*, the CSMA-CA shall return to step 2, else the CSMA-CA shall terminate with a Channel Access Failure status.

6. **Channel Assessed Idle**: If the channel is assessed to be idle (step 5), in a slotted CSMA-CA, the MAC sublayer shall ensure that contention window is expired before starting transmission. For this, the MAC sublayer first decrements CW by one. If CW is not equal to 0, go to step 3 else start transmission on the boundary of the next backoff period. In the unslotted CSMA-CA, the MAC sublayer start transmission immediately if the channel is assessed to be idle.

7.1.3 Data Transfer Model

Three types of data transfer transactions exist: from a coordinator to a device, from a device to a coordinator and between two peer devices. The mechanism for each of these transfers depend on whether the network supports the transmission of beacons. When a device wishes to transfer data in a nonbeacon-enabled network, it simply transmits its data frame, using the unslotted CSMA-CA, to the coordinator. There is also an optional acknowledgment at the end as shown in Figure 7.5a. When a device wishes to transfer data to a coordinator in a beacon-enabled network, it first listens for the network beacon. When the beacon is found, it synchronizes to the

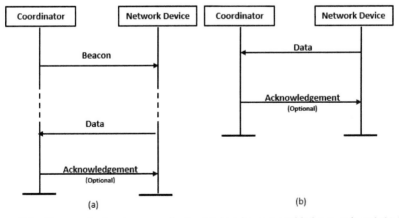

Figure 7.5 Communication to a coordinator (a) in a beacon-enabled network and (b) in a nonbeacon-enabled network.

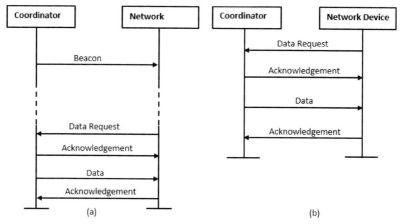

Figure 7.6   Communication from a coordinator to device (a) in a beacon-enabled network and (b) in a nonbeacon-enabled network.

superframe structure. At the right time, it transmits its data frame, using slotted CSMA-CA, to the coordinator. There is an optional acknowledgment at the end as shown in Figure 7.5b. The applications transfers are completely controlled by the devices on a PAN rather than by the coordinator. This provides the energy-conservation feature of the ZigBee network. When a coordinator wishes to transfer data to a device in a beacon-enabled network, it indicates in the network beacon that the data message is pending. The device periodically listens to the network beacon, and if a message is pending, transmits a MAC command requesting this data, using slotted CSMA-CA. The coordinator optionally acknowledges the successful transmission of this packet. The pending data frame is then sent using slotted CSMA-CA. The device acknowledged the successful reception of the data by transmitting an acknowledgment frame. Upon receiving the acknowledgment, the message is removed from the list of pending messages in the beacon as shown in Figure 7.6a. When a coordinator wishes to transfer data to a device in a nonbeacon-enabled network, it stores the data for the appropriate device to make contact and request data. A device may make contact by transmitting a MAC command requesting the data, using unslotted CSMA-CA, to its coordinator at an application-defined rate. The coordinator acknowledges this packet. If data are pending, the coordinator transmits the data frame using unslotted CSMA-CA. If data are not pending, the coordinator transmits a data frame with a zero-length payload to indicate that no data were pending. The device acknowledges this packet as shown in Figure 7.6b. In a peer-to-peer network, every device can communicate with any other device in its transmission radius. There are two options for this. In the first case, the node will listen constantly and transmit its data using unslotted CSMA-CA. In the second case, the nodes synchronize with each other so that they can save power.

7.1.4 Network Formation: Starting a PAN

A PAN shall be started when the user app decides to form a network instead of joining an existing one, it will instruct the ZDO to call the network formation function. Only a router that is coordinator-capable can form a network and this is indicated in the application layer's information base. It is just a fancy term for the app layer's configuration table.

When the network formation function is called, a list of allowed channels needs to be supplied to which may be limited to a subset of the total available channels (16 channels at 2.4 GHz). It is usually based on certain requirements like avoiding channels that overlap with an existing 802.11 network. To avoid any interference it is recommended, in all cases, to locate any routine node a few feet away from an 802.11 access point.

The network formation function will call the MAC's "energy detection scan" and active scan services and perform scans on the supplied channel list. When the scans are finished, the MAC's scan confirm function will return the energy readings and network scan descriptors to the function via the MAC's scan confirmation. From there, the network formation function will need to decide on the channel to join. The usual criterion is to choose a channel with the lowest energy reading (lowest amount of traffic) and the fewest networks. If you have access to the source code, you can also modify the channel decision function to add additional criteria.

Once the channel is decided on, the newly crowned coordinator will decide on a PAN ID and set the channel in the radio. The final step is for the NWK layer to call the MAC start service which configures the MAC layer. After that, confirmations go back all the way up to the user app (Figure 7.7).

The Energy Detection Scan and Assigning a PAN ID are given next in detail.

a. Energy Detection Scan

A PAN starts by an Energy Detection Scan in which the FFD/coordinator searches for suitable RF channel which is not usable and not interfering with Wireless LAN frequencies in use. The reason of this step is because WLAN also operates in the same 2.4 GHz bands.

NWK Formation Sequence Diagram

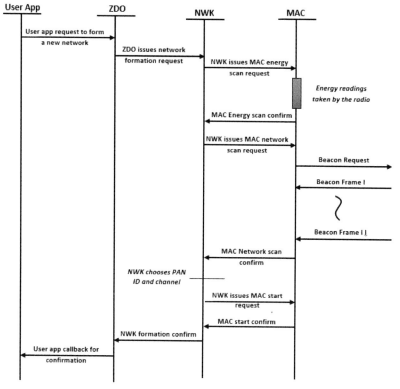

Figure 7.7   Network formation sequence diagram.

The active scan allows the FFD to locate any coordinator transmitting beacon frames within its POS (personal operating space). An active channel scan is requested over a specified set of logical channels. For each logical channel, the device shall first switch to the channel and send a beacon request command. The device shall then enable its receiver for at most *aBaseSuperframeDuration* $* (2n + 1)$ symbols, where n is between 0 and 14. During this time, the device shall reject all nonbeacon frames and record the information contained in all unique beacons in a PAN descriptor structure. If the coordinator of a beacon-enabled PAN receives the beacon request command, it shall ignore the command and continue transmitting its beacons as usual. If the coordinator of a nonbeacon-enabled PAN receives this command, it shall transmit a single beacon frame using unslotted CSMACA.

The active scan on a particular channel terminates when the number of PAN descriptors stored equals this implementation-specified maximum or *aBaseSuperframeDuration* $* (2n + 1)$ symbols, where n is between 0 and 14, have elapsed. The entire scan shall terminate when the number of PAN descriptors stored equals the implementation-specified maximum or every channel in the set of available channels has been scanned.

b. Assigning a PAN ID

In this step, the coordinator starts the network by assigning a PAN ID to the network. Assignment the PAN ID is done by two ways: Manual (pre-configured) and dynamic (Obtained by checking other PAN IDs of networks already in operation nearby so that PAN ID does not conflict with other networks) or from the list of PAN descriptors returned from the active channel scan (step a above). An ED scan allows the FFD obtain a measure of the peak energy in each requested channel. During the ED scan, the MAC sublayer shall discard all frames received over the PHY data service. An ED scan is performed over a set of logical channels. For each logical channel, repeatedly perform an ED measurement for *aBaseSuperframeDuration* $* (2n + 1)$ where n is the value of the *scanDuration*. The maximum ED measurement obtained during this period shall be noted before moving onto the next channel in the channel list. The ED scan shall terminate when either the number of channel ED measurements stored equals the implementation-specified maximum or energy has been measured on each of the specified logical channels.

Case of conflict

In some instances, a situation could occur in which two PANs exist in the same POS with the same PAN identifier. If this conflict happens, the coordinator and its devices shall perform PAN identifier conflict resolution procedure. The PAN coordinator shall conclude that a PAN identifier conflict is present if either a beacon frame is received by the PAN coordinator with the PAN coordinator subfield set to 1, that is, transmitted by the PAN coordinator, and the PAN identifier is equal to *macPANId* or a PAN ID conflict notification command is received by the PAN coordinator from a device on its PAN. A device shall conclude that a PAN identifier conflict is present if a beacon frame is received by the device with the PAN coordinator subfield set to 1, the PAN identifier equal to *macPANId*, an address that is not equal to both *macCoordShortAddress* and *macCoordExtendedAddress*. On the detection of the PAN identifier conflict by a device, it shall generate the PAN ID conflict notification command and send it to the PAN coordinator. If the PAN ID conflict notification command is received correctly, the PAN coordinator shall send an ack and resolve the conflict. On the detection of the PAN identifier conflict by a coordinator, the coordinator shall first perform an active scan and then select a new PAN identifier based on the information from the scan. The coordinator shall then broadcast the coordinator realignment command containing the new PAN identifier with the source PAN identifier field equal to the value in macPANId. Once the coordinator realignment field has been sent, the coordinator shall set macPANId to the new PAN identifier.

c. Broadcasting Beacon Request

The coordinator at this stage completes its configuration and is ready to accept network joining request queries from routers and end devices who wish to join the PAN.

In addition to above, Coordinator(C) sends broadcast beacon request frame on remaining quiet channel. This is also referred as beacon scan or PAN scan. By this Coordinator receives PAN ID of routers(R) and end devices (E) present nearby. It also comes to know whether R/E allow join or not.

d. Routers and End Device Joining the network

At this stage Routers and End devices can join the network. There are two ways to join a ZigBee network: MAC association and network re-join.

- First one is implemented by device underlying MAC layer and second one is implemented by network layer, despite the name may also be used to join a network for the first time.
- MAC association can be performed between Coordinator (C) and Router (R)/End device (E) or Router (R) and End device (E) or Router (R) and other Router (R).
    - Let us assume that Coordinator(C) has already established the PAN network. Hence next step for R or E is to find out whether C is allowing joining or not. So they do PAN scan or send beacon request frame.
    - After they come to know that they can join the network, they will send association request frame and will join the network as soon as they receive the association response.
- As mentioned above whether or not C or R allow a new device to join depends on two main factors:
    - Permit joining attribute
    - Number of end device children it already has.

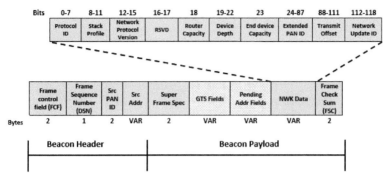

Figure 7.8 Beacon NWK payload format.

7.1.5 Network Joining

After forming the network, Routers and End devices can join the network.

7.1.5.1 Network discovery

As the name implies, the ZigBee network discovery service is used to discover the existing networks on the current channel. It is mostly just used when the device is started to find out if there are any suitable networks to join, although it can also be called at any time via the user app.

When a network discovery is requested by the ZDO (or user app), the discovery function will call the MAC's active scan service which, in turn, will broadcast a beacon request. When other devices see the beacon request, they will respond with an 802.15.4 beacon frame. Considering Figure 7.8 an 802.15.4 beacon frame contains MAC information about the responding device as well as a beacon payload for generic data. Within that payload, the responding device will include ZigBee network information such as the protocol ID and version, amount of routers and end devices allowed to join, the device profile that is being used, and other somewhat useful information.

When the beacons from the scan request are received, the device will add both the MAC and NWK info to its scan descriptor list and its neighbor table. After all of the beacons have been collected, a network discovery confirmation will be sent to the ZDO along with the list containing all the scan descriptors. The ZDO or the user app would then need to decide which network to join based on certain join criteria. It's here that the user can specify if they only want their device to join certain networks or even if there is a specific device they'd like to join to.

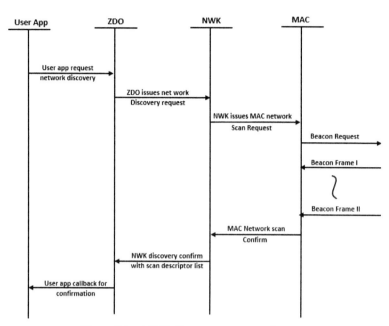

Figure 7.9   NWK discovery sequence diagram.

7.1.5.2 Network join

Joining a device or allowing a device to join is probably one of the most complicated processes in ZigBee. There are actually two sides to the network join function: the child side which sends the request and the parent side which processes the request and sends the response. To get a thorough understanding of the join process, each sequence will be treated separately and also we refer the reader to Section 7.1.6 which gives details of the 802.15.4 association sequence.

7.1.5.2.1 Network join–child

The first part of the join process for the child is to do a network discovery. This is usually done when the device is first started and is not associated with any network as mentioned previously. Once the network discovery is finished and the potential parent has been decided on according to the join criteria, then it is time for the network join process to start.

When the potential parent has been chosen, a network join request is called by the ZDO. The network join request will call the MAC's association service and issue an association request to the potential parent. From there, the procedure follows the MAC's association sequence until the association response is received from the potential parent.

When this response is received, it will get passed up to the network layer via the MAC's association response. If the join was successful, the device will update it as NWK and MAC information tables to include the new network address, PAN ID, and also update the neighbor table to specify its parent. Once the administrative work is taken care of, the network join confirmation is sent up to the ZDO where it can inform the application about the join status. If the join status was unsuccessful, then the ZDO/user app will choose another potential parent from the neighbor table and retry the join procedure until it eventually joins a network or runs out of potential parents.

One of the last things that occur after a successful join is that the device will broadcast a device announcement informing everyone on the network that it has joined the network as well as it is 16-bit network address and 64-bit IEEE address. This is important because if the device was previously joined to the network with a different network address, the other devices will be able to find out from it is IEEE address and can clear all references to the old network address. Also, the address info will be added to everyone's address map which tracks all the devices on the network. Figure 7.10 shows the sequence.

7.1.5.2.2 Network join–parent

The parent side of the join process is slightly easier. When a MAC association request arrives at the potential parent, it sends an indication to the network layer that a device is trying to join. The potential parent will then search its neighbor table to see if the 64-bit IEEE address already exists. If it does, then that means that the device was already previously joined and the parent will just issue the same network address to it. If not, and the parent is allowing devices to join it, then it will simply add the device to its neighbor table specifying that it's a child device and generate a new network address for it. This all gets packaged up and sent out as a MAC association response. Again the rest goes according to the MAC's association service.

7.1.6 Association and Disassociation

Association refers to the relationship of a station to an access point; **reassociation** allows an **association** to be moved to a different access point; **disassociation** allows an **association** to be terminated. Every station must

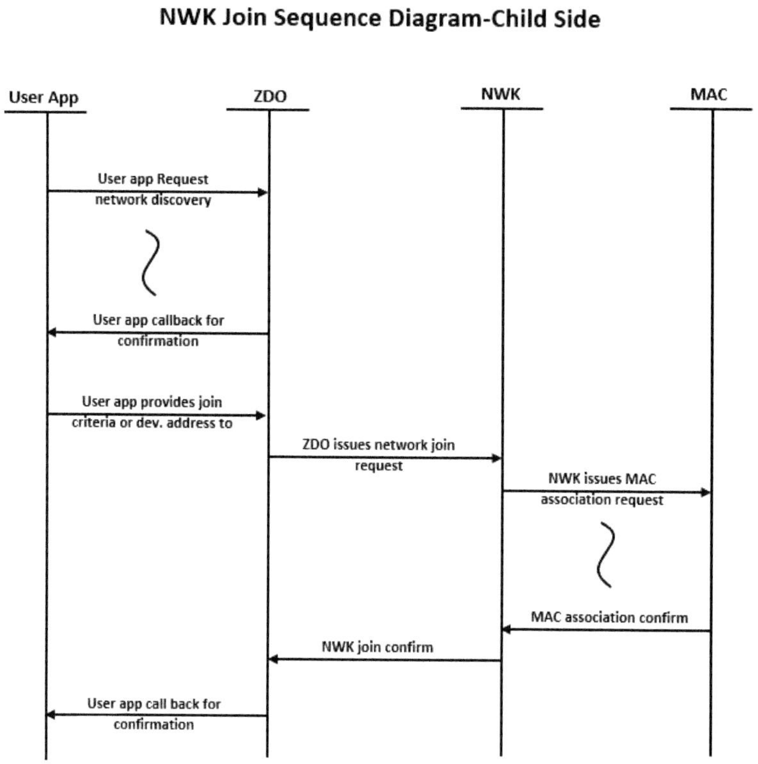

NWK Join Sequence Diagram-Child Side

Figure 7.10 NWK join sequence diagram–child side.

provide authentication, deauthentication, privacy, and **MAC** service data unit (MSDU) delivery.

Device Association

An FFD may indicate its presence on a PAN to other devices by transmitting beacon frames. This allows other devices to perform device discovery. An FFD that is not a PAN coordinator shall begin transmitting beacon frames only when it has successfully associated with a PAN. Association of a device starts after having completed either an active channel scan or a passive channel scan. The passive scan, like an active scan, allows a device to locate any coordinator transmitting beacon frames within its POS whereas there beacon request command is not required for passive scan. The results of the channel scan are then used to choose a suitable PAN. A device shall attempt to associate only with a PAN that is currently allowing association.

NWK Join Sequence Diagram-Parent Side

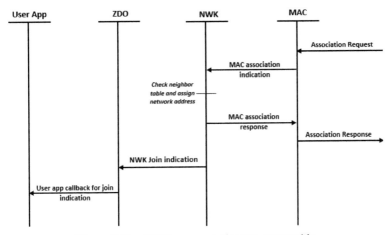

Figure 7.11 NWK sequence diagram–parent side.

How to choose a suitable pan with which to associate from the list of pan descriptors returned from the channel scan is up to application. Following the selection of a PAN with which to associate, the next higher layers request that MLME configures the *phyCurrentChannel* to the appropriate logical channel on which to associate, *macPANId* to the identifier of the PAN with which to associate and *macCoordExtendedAddress* or *macCoordShortAddress* to the address of the coordinator with which it associates. An unassociated device shall initiate the association procedure by sending an associate request command to the coordinator of an existing PAN. If the association request command is received correctly, the coordinator shall send an acknowledgment. This acknowledgment, however, does not mean that the device has associated. The coordinator needs time to determine whether the current sources available on a PAN are sufficient to allow another device to associate. This decision should be made within *aResponseWaitTime* symbols. If already associated, remove all information. If sufficient resources are available, the coordinator shall allocate a short address to the device and generate an association response command containing the new address and a status indicating the successful association. If there are not enough resources, the coordinator shall generate an association response command containing a status indicating failure. This response is sent to the device using indirect transmission (pending, request,...). On the other side, the device, after getting the acknowledgment frame, waits for the response for *aResponseWaitTime*

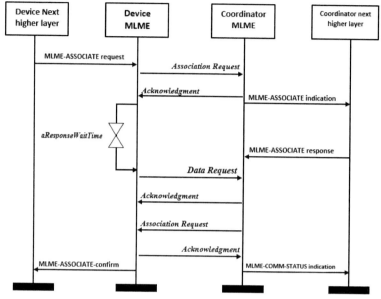

Figure 7.12   Device association.

symbols. It either checks the beacons in the beacon-enabled network or extract the association response command from the coordinator after *aResponseWaitTime* symbols. On reception of association response command, the device shall send an acknowledgment. If the association is successful, store the addressed of the coordinator with which it has associated. The association procedure is shown in Figure 7.12.

Device Disassociation

When a coordinator wants one of its associated devices to leave the PAN, it shall send the disassociation notification command to the device using indirect transmission. Upon reception of the packet, the device should send the acknowledgment frame. Even if the *ack* is not received, the coordinator shall consider the device disassociated. If an associated device wants to leave the PAN, it shall send a disassociation notification command to the coordinator. Upon reception, the coordinator sends ack. Even if the *ack* is not received, the device shall consider itself disassociated. An associated device shall disassociate itself by removing all references to the PAN. A coordinator shall disassociate a device by removing all references to that device (see Figure 7.13).

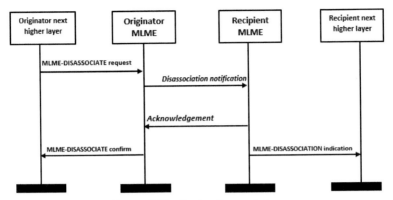

Figure 7.13   Device disassociation.

7.1.6.1 Orphan notification

As mentioned before, a device must be associated with a network to be able to communicate with other devices in the network. A device that was previously associated with a network but has lost its association is considered an *orphaned device*. A device that leaves a network using the disassociation procedure is not considered an orphaned device. If the NWK layer of a device faces repeated communications failures, it may conclude that the device has been orphaned.

For example, if the device transmits a frame that requires an acknowledgment and does not receive the acknowledgment after waiting for *macAck-WaitDuration* symbols, the device may repeat the data transmission up to *macMaxFrameRetires*. If still, after *macMaxFrameRetires* attempts, no acknowledgment is received, the device counts this as a *single* communication failure. The application developer decides on the number of communication failures that will be tolerated before declaring the device an orphan.

The NWK layer of an orphaned device can instruct its MLME to perform either one of the following procedures:

• Reset the MAC and then perform the association procedure.
• Perform the orphaned device realignment procedure.

The reset and association procedures were discussed in the preceding sections. The realignment procedure (shown in Figure 7.14) starts with a device sending an orphan notification command to the coordinator. The MLME of the coordinator notifies the NWK layer of the presence of an orphaned device using the *MLME-ORPHAN.indication* primitive.

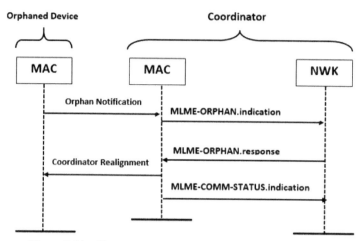

Figure 7.14   The sequence chart for orphan device notification.

The NWK layer of the coordinator verifies the address of the orphan device, and in its response (using the *MLME-ORPHAN.response* primitive) the NWK layer confirms whether the device was previously associated with this coordinator.

If the device was previously associated with this coordinator, the MLME of the coordinator sends the realignment command to the orphaned device. The realignment command is used to deliver network settings. When the command is successfully transmitted to the orphaned device, the MLME uses the *MLME-COMM-STATUS.indication* primitive to report its success to the NWK layer.

If the device was not associated with this coordinator, the MLME of the coordinator does not perform any action. The orphaned device waits for *macResponseWaitTime* symbols (a MAC attribute) and if it does not receive any realignment command, the orphaned device assumes it is not associated with any coordinator in its range.

7.1.7 Synchronization

For PANs supporting beacons, synchronization is performed by receiving and decoding beacon frames. For PANs not supporting beacons, the synchronization is performed by polling the coordinator for data. In a beacon enabled network, devices shall be permitted to acquire synchronization only with beacons containing the PAN Identifier specified in *macPANId*. If tracking is specified in the *MLMESYNC.request* primitive, the device shall attempt to acquire the beacon and keep track of it by regular and timely activation of its receiver. It shall enable its receiver at a time prior to the next expected beacon frame transmission, that is, just before the known start of the next superframe. If tracking is not specified, the device shall attempt to acquire the beacon only once.

To acquire beacon synchronization, a device shall enable its receiver and search for at most *aBaseSuperframeDuration* $*$ $(2n + 1)$ symbols, where n is the *macBeaconOrder*. If a beacon frame containing the current PAN identifier of the device is not received, the MLME shall repeat the search. Once the number of missed beacons reached *aMaxLostBeacons*, the MLME notifies the next upper layer by issuing MLME-SYNC-LOSS.indication with a reason BEACON-LOSS. The MLME shall timestamp each received beacon frame at the same symbol boundary within each frame, the location of which is implementation specific. In a nonbeacon-enabled network, the devices shall be able to poll the coordinator for data at the discretion of the next higher layer. On receipt of MLME-POLL.request primitive, the MLME follow the procedure for extracting pending data from the coordinator. Another problem with synchronization is orphaned device. If the next higher layer receives repeated communication failures following its requests to transmit data, it may conclude that it has been orphaned. A single communications failure occurs when a device transaction fails to reach the coordinator, that is, an acknowledgment is not received after *aMaxFrameRetries* attempts at sending data. If the next higher layer concluded that the device has been orphaned, it may either reset the MAC sublayer and perform the association procedure or perform the orphaned device realignment procedure. If the decision is for orphaned device alignment, orphan scan is performed. During the orphan scan, the MAC sublayer shall discard all frames received over the PHY data service that are not coordinator realignment MAC command frames. For each logical channel over a specified set of logical channels, the device sends an orphan notification command. The device shall then enable its receiver for at most *aResponseWaitTime* symbols. If the device successfully receives a coordinator realignment command within this time, the device shall disable its receiver.

If a coordinator receives the orphan notification command, it searches its device list for the device sending the command. If the coordinator finds a record of the device, it shall send a coordinator realignment command to the orphaned device. Otherwise, it shall ignore the packet. The orphan scan terminates when the device receives a coordinator realignment command or the specified set of logical channels has been scanned.

7.1.8 Transmission, Reception, and Acknowledgment

In order to transmit a data or a beacon or a MAC command frame, the MAC sublayer shall copy the value of masDSN into the sequence number field of the MHR of the outgoing frame and then increment it by one. The source address field shall contain the address of the device sending the frame. If the device has been allocated a short address, it shall use that address in preference to its 64-bit extended address. If the source address field is not present, the originator of the frame shall be assumed to be a PAN coordinator and the destination address shall contain the address of the recipient. The destination address shall contain the intended recipient of the frame, which may be either a 16-bit short address or a 64-bit extended address. If the destination address field is not present, the recipient of the frame shall be assumed to be the PAN coordinator. The destination and source address may be in different PANs, which is identified by the PAN identifier fields. In beacon-enabled PAN, the transmitting device shall attempt to find the beacon before transmitting. If it cannot find the beacon, it shall use unslotted CSMA-CA. Once the beacon is found, it transmits in the appropriate portion of the superframe. Transmission in the CAP shall use slotted CSMA-CA and those in GTS shall not use CSMA-CA. In a nonbeacon-enabled network, the frames are transmitted using unslotted CSMA-CA. Upon reception of packets, the MAC sublayer shall discard all its received frames that do not contain a correct value in their FCS field in the MFR. Receiver is important in terms of energy consumption. Each device may choose whether the MAC sublayer is to enable its receiver during idle periods. During these idle periods, the MAC sublayer shall still service transceiver task requests from the next higher layer. On completion of each transceiver task, the MAC sublayer shall request that the PHY enables or disables its receiver, depending on whether *macRxOnWhenIdle* is set to TRUE or FALSE, respectively. If beacon is enabled, the value of *macRxOnWhenIdle* shall be considered only during idle periods of the CAP. Another energy conserving feature of the standard is the indirect transmission feature. The transactions start by the devices themselves rather than the coordinator. In other words, either the coordinator needs to indicate in its beacon when messages are pending for devices or the devices themselves need to poll the coordinator to determine whether they have any messages pending. A device on a beacon-enabled PAN can determine whether any frames are pending for it by examining the contents of the received beacon frame. If the address of the device is contained in the address list field of the beacon frame, the MLME of the device shall send a data request command to the coordinator during the CAP. Upon reception of this command, the coordinator shall send an ack. It indicates whether any data are pending for that device in the *ack* frame. On receipt of the *ack*, the device shall enable its receiver for at most *aMaxFrameResponseTime* CAP symbols in a beacon-enabled PAN or symbols in a nonbeacon-enabled PAN to receive the corresponding frame from the coordinator. If there is data pending, the coordinator should send the frame else send a frame containing zero length payload, indicating that no data are present. The data frame is transmitted without using CSMA-CA if the MAC sublayer can commence transmission of the data frame between *aTurnaroundTime* and *aTurnaround-Time + aUnitBackoffPeriod* symbols and there is time remaining in the CAP for the message, appropriate IFS and acknowledgment and using CSMA-CA otherwise. A frame transmitted with the acknowledgment request field set to 1 shall be acknowledged by the recipient. If the intended recipient correctly receives the frame, it shall generate and send an acknowledgment frame containing the same DSN from the data or MAC command frame that is being acknowledged. The transmission of the *ack* shall commence between *aTurnaroundTime* and *aTurnaroundTime + aUnitBackoffPeriod* symbols after the reception of the last symbol of the data or MAC command frame.

7.1.9 GTS Allocation and Management

A GTS allows a device to operate on the channel within a portion of the superframe that is dedicated exclusively to that device. A device shall attempt to allocate and use a GTS only if it is currently tracking the beacons. A GTS shall be allocated only by the PAN coordinator and it shall be used only for communications between the PAN coordinator and a device. A single GTS can extend over one or more superframe slots. The PAN coordinator may allocate up to seven GTSs at the same time, provided there is sufficient capacity in the superframe. A GTS shall be allocated before use, with the PAN coordinator deciding whether to allocate a GTS based on the requirements of the GTS request and the current available capacity in the superframe. GTS shall be allocated on a first-come-first-serve basis and all GTSs shall be placed contiguously at the end of the superframe and after the CAP. Each GTS shall be deallocated when the GTS is no longer required, and a GTS can be deallocated at any time at the discretion of the PAN coordinator or by the device that originally requested the GTSs. A device that has been allocated GTS may also operate in the CAP. The management of the GTSs shall be undertaken by the PAN coordinator only. For each GTS, the PAN coordinator shall be able to store its starting slot, length, direction and associated device address. The GTS direction is specified as either transmit or receive. Each device may request one transmit GTS and/or one receive GTS. For each allocated GTS, the device shall be able to store its starting slot, length and direction. If a device has been allocated a receive GTS, it shall enable its receiver for the entirety of the GTS. In the same way, a PAN coordinator shall enable its receiver for the entirety of the GTS if a device has been allocated a transmit GTS. A device is instructed to request the allocation of a new GTS through the GTS request command, with GTS characteristics (direction, length,...) set according to the requirements of the intended application. On receipt of this command, the PAN coordinator shall send an acknowledgment frame. Following the *ack* transmission, the PAN coordinator shall first check if there is available capacity in the current superframe based on the remaining length of the CAP and the desired length of the requested GTS. The superframe shall have available capacity if the maximum number of GTSs has not been reached and allocating a GTS of the desired length would not reduce the length of the CAP to less than *aMinCAPLength*. The PAN coordinator shall make its decision within *aGTSDescPersistenceTime* superframes. On receipt of the ack from the coordinator, the device shall continue to track the beacons and wait for at most *aGTSDescPersistenceTime* superframes. If no GTS descriptor in the superframe, notify the next upper layer of failure. When the coordinator determines whether capacity is available for the requested GTS, it shall generate a GTS descriptor with the requested specifications and the short address of the requested device. It indicates the length and the start of the GTS in the superframe and notifies the next upper layer of the new GTS allocation. If there was not sufficient capacity to allocate the requested GTS, the start slot shall be set to 0 and the length to the largest GTS length that can currently be supported.

This GTS descriptor shall remain in the beacon frame for *aGTSPersistenceTime* superframes. On receipt of the beacon frame, the device shall process the descriptor and notify the next upper layer of the success. In the same way, a device is instructed to request the deallocation of an existing GTS through the GTS request command using the characteristics of the GTS it wishes to deallocate. From this point on, the GTS to be deallocated shall not be used by the device and then an *ack* from the PAN coordinator to the device. The PAN coordinator then deallocates the request of the GTS characteristics in the packet matches those in its allocation. The PAN coordinator shall also ensure that any gaps occurring in the CFP, appearing due to the deallocation of a GTS, are removed to maximize the length of the CAP. The MLME of the PAN coordinator shall also attempt to detect when a device has stopped using a GTS using the following rules: For a transmit frame GTS, the MLME of the PAN coordinator shall assume that the device is no longer using the GTS if a data frame is not received for at least $2 * n$ superframes. For receive GTSs, the MLME of the PAN coordinator shall assume that the device is no longer using its GTS if an acknowledgment frame is not received for at least $2 * n$ superframes. The value of n is equal to 28-macBeaconOrder if $0 \leq macBeaconOrder \leq 8$ and 1 if $9 \leq macBeaconOrder \leq 14$.

7.1.10 MAC Layer Frame

As mentioned in the ZigBee protocol stack that **ZigBee MAC layer frame** composed of MAC header, MAC payload and FCS. Each MAC frame will contain Frame control field (16 bit), which carry frame type, addressing fields and other control flags. This MAC control field contain frame type field, which is the main differentiating factor in identifying one MAC frame with the other. It is 3 bit in length.

The MAC frames are divided into following four major categories, which is used by ZigBee devices to establish connection to the PAN by exchanging system information:

1. Beacon
2. Data
3. Acknowledgment
4. MAC command

In the following, we discuss the different types of frames and the different fields.

General MAC Frame Format

The general MAC frame is shown in Figure 7.15 adopted in ZigBee technology at MAC layer. It consists of three sections: the MAC header (MHR), the MAC payload, and the MAC footer (MFR). The size of each field is shown in

Generic MAC Frame format:

Byte: 2	1	0/2	0/2/8	0/2	0/2/8	Var	2
Frame control	Sequence number	Destination PAN ID	Destination Address	Source PAN ID	Source Address	Frame (Data) Payload	FCS
			Addressing fields				
		MAC Header (MHR)				MAC Payload	MFR

Figure 7.15 MAC frame.

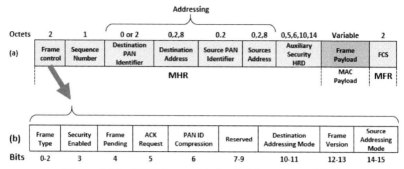

Figure 7.16   MAC frame with detailed frame control field.

Table 7.1   Frame control field

Bits: 0–2	3	4	5	6	7–9	10–11	12–13	14–15
Frame type	Security enabled	Frame pend-ing	Ack. request	Intra-PAN	Reserved	Dest. address-ing mode	Reserved	Source address-ing mode

octets. This part is also referred as MPDU or MAC Protocol Data Unit. This gets embedded into PPDU (physical PDU) frame of ZigBee.

Generic MAC layer frame has frame control field of 2 bytes (octets). Figure 8xx shows the details of 16 bits of the frame control field. The contents carry useful information such as frame type, source and destination addressing modes. Frame type specifies whether the frame is beacon frame, data frame, ACK of data, MAC command frame etc. Figure 7.16b and Table 7.1 outline the "Frame-type subfield" with 3 bits.

Next the definitions of the different subfields are given:

Frame-type Subfield: The first field is the frame control (Figure 7.16b) that defines the frame type (beacon, data, acknowledgment, and MAC command). Table 7.2 gives the possible types.

Security-enabled Subfield: The second part of the frame control field is the "Security-enabled" subfield. If this subfield is set to 1, this frame has security protection and the auxiliary header will be part of the MAC frame. Otherwise, the size of the auxiliary header is zero.

Frame Pending Subfield: The "frame pending" subfield is used as part of the indirect data transmission method and, if it is set to 1, it means that there is data pending at the transmitting device for the recipient device. If

Table 7.2 Frame type subfield

Frame type value (b2,b2,b0)	Description
000	Beacon frame
001	Data frame
010	Acknowledgment
011	MAC command
100–111	Reserved

the acknowledgment request subfield is set to 1, the recipient device must send an acknowledgment frame back.

PAN ID Subfield: When communicating within the same PAN, the PAN identifier will be the same for both source and destination devices; therefore, it is unnecessary to repeat them both in one frame. The PAN ID compression subfield helps avoid the unnecessary repeat of the PAN identifier. If the PAN ID compression field is set to 1, only the destination PAN identifier will be included in the frame, and the source PAN identifier is assumed to be the same as the destination.

Destination and Source Subfields: The destination and source addressing mode subfields determine the addressing mode (16-bit short address or 64-bit extended address). The length of the destination and address fields in the MAC frame depends on the addressing mode.

Frame Version: The next subfield is the frame version. The IEEE 802.15.4 standard may be updated over time, and the frame version subfield determines what version of the IEEE 802.15.4 is used to construct the frame.

Sequence Number Subfield: The next field in the MAC frame, the sequence number, can contain either a beacon sequence number (BSN) ore a data sequence number (DSN). The sequence numbering helps distinguishing between various sequences. For example, if two received frames have the same sequence number, it means that the same frame was retransmitted. If the first frame was detected successfully, the second frame (with the same sequence number) can be ignored.

The values of BSN and DSN are stored as MAC PIB attributes (*macBSN* and *macDSN*, correspondingly). The BSN is used only in the beacon frames. The DSN is used in any type of frame other than a beacon frame. A device initializes the *macDSN* value (or *macBSN* value if it is a beacon frame) to a random number and increment it once after each transmission.

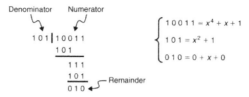

Figure 7.17   A simple modulo 2 division example.

Auxiliary Security Header: The auxiliary security header is an optional field in the MAC frame and contains information such as security level and the type of security keys used to protect the MAC frame.

FCS Subfield: The last field in the MAC frame is always the Frame Check Sequence (FCS) field, which the receiver uses to check for any possible error in the received frame. The details of the FCS are provided in the following section.

Calculation of the Frame Check Sequence: The IEEE 802.15.4 uses 16-bit FCS based on the International Telecommunication Union (ITU) Cyclic Redundancy Check (CRC) to detect possible errors in the data packet. The basic concept of CRC is as follows. In the transmitting device, all the bits in the MHR and the MAC payload are treated as coefficients of a polynomial. This polynomial is then divided by another polynomial, which is known by both the receiver and transmitter. The remainder of this division is called FCS and is added to the end of the frame as the MAC footer (MFR). The recipient device will perform the same division and expects to get the same remainder. If the remainder calculated by the recipient device is not the same as the remainder provided by the transmitting device in MFR, the recipient device will conclude that the frame is received with errors.

A very simple example of polynomial division is shown in Figure 7.17. The division is based on modulo 2 binary arithmetic. The numerator is binary number 10011, which is equivalent to $x^4 + x + 1$ polynomial:

$$10011 \rightarrow 1 \times x^4 + 0 \times x^3 + 0 \times x^2 + 1 \times x^1 + 1 \times x^0 = x^4 + x + 1$$

If 10011 is divided by 101, the remainder will be 010.

The FCS in IEEE 802.15.4 is generated by going through the following steps:

- Define the polynomial M(x) to represent the MHR and MAC payload sequence of bits.

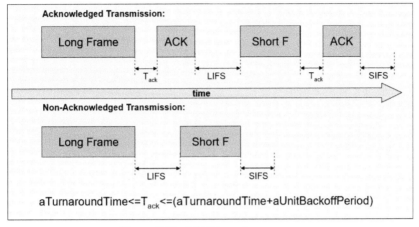

Figure 7.18 MAC inter-frame space.

- Multiply M(x) by x^{16} to create M(x)$\times x^{16}$.
- Divide the M(x)$\times x^{16}$ polynomial by the following polynomial:

$$G_{16}(x) = x^{16} + x^{12} + x^5 + 1$$

The remainder of this division is placed in the MFR as the frame check sequence.

MAC Inter-frame Space IFS: The MAC Inter-frame spaces (IFS) are shown in Figure 7.18.

7.1.10.1 Data frame

The data frame format is shown in Figure 7.19. As discussed above, the data payload is passed to the MAC sublayer and is referred to as the MSDU (MAC Service Data Unit). The MSDU is prefixed with an MHR and appended with an MFR. The MHR contains the frame control, sequence number, and addressing information fields. The MFR is composed of a 16 bit FCS. The MHR, MSDU and MFR together form the MAC data frame. The MPDU is passed to the PHY as the PHY data frame payload. The PSDU is prefixed with an SHR, containing the preamble sequence and SFD fields, and a PHR containing the Length of the PSDU in octets. The preamble sequence and data SFD enables the receiver to achieve symbol synchronization. The SHR, PHR and PSDU together form the PHY data packet.

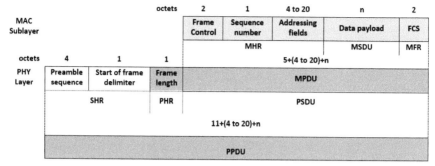

Figure 7.19   Data frame structure.

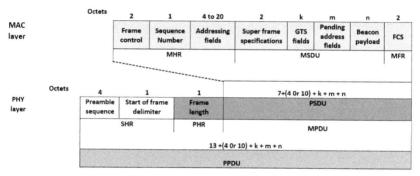

Figure 7.20   Beacon frame format.

7.1.10.2 Beacon frame

Depending on the parameters of the MLME-START request primitive, the FFD may either operate in a beaconless mode or may begin beacon transmissions either as the PAN coordinator or as a device on a previously established PAN. Figure 7.20 shows the structure of a beacon frame, which originates from the MAC sublayer. The entire MAC frame is used as a payload in a PHY packet. The content of the PHY payload is referred to as the PHY Service Data Unit (PSDU). A coordinator can transmit network beacons in a beacon-enabled network. The MSDU (MAC service data unit) contains the superframe specification, pending address specification, address list, and beacon payload fields.

The MSDU is prefixed with a MHR (MAC header) and appended with a MFR (MAC footer). The MHR contains the MAC frame control fields, BSN (beacon sequence Number), and addressing information fields. The MFR contains a 16 nit FCS (frame Check sequence). The MHR, MSDU and MFR together form the MAC beacon frame. The MPDU passed to the PHY as the PHY beacon packet payload. The PSDU is prefixed with the SHR (synchronization header), containing the preamble sequence and SDF (start of frame delimiter), fields, and a PHR containing the length of the PSDU in octets. The preamble sequence enables the receiver to achieve symbol synchronization. The SHR, PHR and PSDU together form the PHY beacon packet.

The role of the subfields of the MAC layer format and the physical layer PHY format is as follows:

In the PHY packet, the preamble field is used by the receiver for synchronization. The start-of-frame delimiter (SDF) indicates the end of SHR and start of PHR. The frame length specifies the total number of octets in the PHY payload (PSDU).

The MAC frame consists of three sections: the MAC header (MHR), the MAC payload, and the MAC footer (MFR). The frame control field in the MHR contains information defining the frame type, addressing fields, and other control flags. The sequence number specifies the beacon sequence number (BSN). The addressing field provides the source and destination addresses. The auxiliary security header is optional and contains information required for security processing.

The MAC payload is provided by the NWK layer. The *superframe* is a frame bounded by two beacon frames. The superframe is optionally used in a beacon-enabled network and helps define GTSs. The GTS field in the MAC payload determines whether a GTS is used to receive or transmit.

The beacon frame is not only used to synchronize the devices in a network but is also used by the coordinator to let a specific device in a network know there is data pending for that device in the coordinator. The device, at its discretion, will contact the coordinator and request that it transmit the data to the device. This is called *indirect transmission*. The pending address field in the MAC payload contains the address of the devices that have data pending in the coordinator. Every time a device receives a beacon, it will check the pending address field to see if there is data pending for it.

The beacon payload field is an optional field that can be used by the NWK layer and is transmitted along with the beacon frame. The receiver uses the Frame Check Sequence (FCS) field to check for any possible error in the received frame.

7.1.10.2.1 Beacon generation

An FFD that is not the PAN coordinator shall begin transmitting beacon frames only when it has successfully associated with a PAN. This primitive

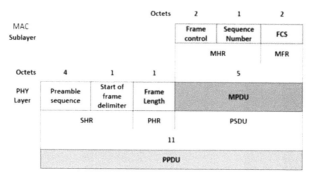

Figure 7.21   Acknowledgment frame structure.

also includes *macBeaconOrder* and *macSuperFrameOrder* parameters that determine the duration of the beacon interval and the duration of the active and inactive portions. The time of the transmission of the most recent beacon shall be recorded in *macBeaconTxTime* and shall be computed so that its value is taken at the same symbol boundary in each beacon frame, the location of which is implementation specific.

7.1.10.3 Acknowledgment frame

The MAC acknowledgment frame, shown in Figure 7.21, is the simplest MAC frame format and does not carry any MAC payload. The frame originates from the MAC sublayer. The MAC acknowledgment frame is constructed form an MHR and an MFR. The MFR contains the MAC frame control and data sequence number fields. The MFR is composed of a 16-bit FCS. The MHR and MFR together form the MAC acknowledgment frame. The MPDU is passed to the PHY as the PHY acknowledge frame payload. The PSDU is prefixed with the SHR, containing the preamble sequence and SFD fields, and the PHR containing the length of the PSDU in octets. The SHR, PHR and PSDU together form the PHY acknowledgment packet.

The acknowledgment frame is sent by one device to another to confirm successful reception of a packet.

7.1.10.4 MAC command frame

The MAC commands such as requesting association or disassociation with a network are transmitted using the MAC command frame.

Figures 7.22 and 7.23 show the structure of the MAC command frame, which originates from the MAC sublayer. The MSDU contains the command type (e.g., association request or data request) filed and command specific

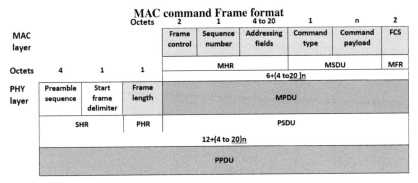

Figure 7.22 MAC command frame structure.

MAC command Frame format

Octets:2	1	section 7.2.2.4.1 IEEE 802.15.4	1	Variable	2
Frame control	sequence number	Addressing fields	Command Frame Identifier	Command Payload	FCS
MHR			MAC Payload		MFR

Figure 7.23 MAC command frame format.

data, called the command payload. The MSDU is prefixed with an MHR and appended with an MFR. The MHR contains the MAC frame control, data sequence number, and addressing information fields. The MFR contains 16 bit FCS. The MHR, MSDU and MFR together form the MAC command frame. The MPDU is passed to the PHY as the PHY command frame payload. The PSDU is prefixed with the SHR, containing the preamble sequence and SFD fields, and the PHR containing the length of the PSDU in octets. The SHR, PHR and PSDU together form the PHY command packet.

ZigBee protocol supports different command frames for difference use case as shown in the following table "MAC command frames". Command frame identifier defines the frame type as given in Table 7.3.

7.1.11 Channel Access Mechanism

The LR-WPAN uses two types of channel access mechanism, depending on the network configuration. Non beacon enabled networks use an unslotted CSMA-CA channel access mechanism. Each time a device wishes to transmit data frames or MAC commands, it shall wait for a random period. If the channel is found to be idle, following the random back off, the device shall transmit its data. If the channel is found to be busy, following the random

Table 7.3   MAC command frames

Command Frame ID	Command Name	RFD	
		Tx	Rx
0x01	Association request (Tx)	X	
0x02	Association response (Rx)		X
0x03	Disassociation notification (Tx, Rx)	X	X
0x04	Data request (Tx)	X	
0x05	PAN ID conflict notification (Tx)	X	
0x06	Orphan notification (Tx)	X	
0x07	Beacon request (Tx)	X	
0x08	coordinator realignment (Rx)		X
0x09	GTS request		
0x0A–0xFF	Reserved		

back off, the device shall wait for another random period before trying to access the channel again. Acknowledgment frames shall be sent without using a CSMA-CA mechanism. Beacon-enabled networks use a slotted CSMA-CA channel access mechanism, where the back-off slots are aligned with start of the beacon transmission. Each time a device wishes to transmit data frames during the CAP, it shall locate the boundary of the next back off slot and then wait for a random number of back off slots. If the channel is busy, following this random back off, the device shall wait for another random number back off slots before trying to access the channel again. If the channel is idle, the device can begin transmitting on the next on the next available back off slot boundary. Acknowledgment and beacon frames shall be sent without using a CSMA-CA mechanism.

7.1.12 GTS Allocation and Management

The GTS allows a device to operate on the channel within a portion of the superframe. It is dedicated exclusively to that device. If it is currently tracking the beacons, a device shall attempt to allocate and use a GTS only. The GTS shall be allocated only by the PAN coordinator and it shall be used only for communication between the PAN coordinator and a device. Single GTS can extend over one or more super frame slots. The PAN coordinator may allocate up to seven GTSs at the time, provide there is sufficient capacity in the super frame. The PAN coordinator must store this information, and these parameters are included in the GTS request command. In radical state, PAN coordinator deallocates one or more GTS to keep minimum CAP length.

- Staring slot
- Length
- Direction; direction of data flow, from device to coordinator, from coordinator to device
- Associated device address

Each device can transmit GTS or receive GTS and device can allocate and use GTS when tracking beacon. If synchronization with all PAN coordination is broken, all GTS allocations are lost, too.

7.1.13 Related MAC Standards

In this chapter, we see the trends of Current ZigBee research. Especially there are several important standards and working groups. First IEEE 802.15.4 is researched for mesh network. Next there are many task groups: 802.15.4e research WPAN enhancements, 802.15.4f task group made for RFID, 802.15.4g is working for smart utility, etc.

The IEE 802.15 task group 5 is chartered to determine the necessary mechanism that must be present in the PHY and MAC layer of WPANs to enable mesh networking.

IEE 802.15.15.5: The mesh network is a PAN that employs one of two connection arrangements, full mesh topology or partial mesh topology. In full mesh topology, each node is connected directly to each of the other, but some of the nodes are connected only to those other nodes with which they exchange the most data. Mesh networks have the capability to provide:

- Extension of network coverage without increasing transmit power or receive sensitivity
- Enhanced reliability via route redundancy
- Easier network configuration
- Better device battery life due to fewer retransmissions

IEEE 802.15.4e WPAN Enhancement: The IEEE 802. 15.4e is chartered to define MAC enhancement to the existing standard 802.15.4-2006. The intent of this amendment is to enhance and add functionality to the 802.15.4-2006 MAC to better support the industrial markets and permit compatibility with no modification being proposed within the Chinese WPAN. This group has defined the application spaces that it will address along with the MAC behavior changes and addition that are required to enable those applications spaces. The application spaces are Factory Automation, Process Automation, Asset Tracking, General sensor Control, Home medical Health and Monitor, telecom Application Neighborhood Area Networks, Audio.

IEEE 802.15.4F for RFID: The IEEE 802.15.4f is chartered to define new wireless physical layer and enhancements to the 802.15.4-2006 mac layer which are required to support new PHY for active RFID system. An RFID tag is a device which is typically attached to an asset or person with a unique identification and the ability to produce its own radio signal not derived from an external radio signal. Tag applications include wireless sensor telemetry, control, and location determination. To generate a radio signal, active RFID tags must employ some source of power. Traditionally, this has been accomplished by integrated batteries, although designs exist for such devices that harvest ambient energy from the surrounding environment.

IEEE 802.15.4g for Smart Utility: The role of IEEE 802.15. Smart Utility Networks is to create a PHY amendment to 802.15.4 to provide a global standard that facilitates very large scale process control applications such as the utility smart-grid network capable of supporting large, geographically diverse networks with minimal infrastructure, with potentially millions of fixed endpoints. There are other task groups. TG6 (Task Group 6) is researching for body area network, TG7 was made for visible light communication and 1 GHz has been chartered to explore the feasibility of terahertz for wireless communication.

7.2 ZigBee Routing (Network) Layer

Nowadays wireless sensor networks are widely employed to improve business process models of industries, web service applications, defense sector, Internet of Things (IoT), space exploration, medical sector, etc. Mobile ad hoc networks (MANET) and wireless sensor networks (WSN) are the two major categories of wireless ad hoc network. In both types, there is no fixed topology and there is no centralized management system. The two networks are self-configured dynamic networks in which nodes are free to move and will therefore alter its linkage to any other devices at any point of time. Routing among the nodes is one among the most important function in data communication networks. The routing techniques (protocols) that are in use with fixed topology networks or those having centralized management cannot be used with ad hoc networks. MANET and WSN are, accordingly, in the need for different types of protocols that are designed especially for routing wireless ad hoc networks.

MANET routing Protocols: Concerning MANET, the routing protocols in use can be classified in many ways. One of them is directly related to the capabilities of the nodes forming the network: MANET composed of high capabilities mobile devices for data communication and supports different protocol like unicast, multicast and hybrid routing. MANET routing protocols can be further classified as proactive and reactive routing protocols.

Proactive Routing: The topology information will be periodically updated by the proactive routing protocol, so this always has an optimized as well as updated routing scheme. OLSR and DSDV are the typical routing protocols which are proactive in nature. In short, the data transmission request is received through an application, routing protocol will be reactive and involves in the route-finding scheme. Similarly, when it leads to delay in process to find out a path, it will not produce the control packet overhead, since no data will be there for transmission.

Reactive Routing: Reactive routing protocols include AODV, DSR, as well as TORA. MANET routing protocols gives the optimal routing path where if there is proactive or reactive routing to the provided starting and reaching destination pair. But also, the necessary specified capacity of routing may be very high to save completely with provided routing paths in a device which is limited for resources. In addition, it should exchange the control packets to furnish and to find out the proper routing as well as other transmissions of the same packets, the interferences of these packets may cause severe low rat and narrow bandwidth channels.

Before discussing the ZigBee routing protocols, a forwarding technique that are commonly used with many protocols will be introduced: Broadcasting and the neighbor table.

7.2.1 Broadcasting

Broadcasting plays an important role in ZigBee and is used for many functions. Two of the most prominent are route discovery and group transmissions. Route discovery is the process of locating a path to a destination address whose route is unknown. ZigBee uses a modified form of AODV (Ad hoc On-demand Distance Vector) which is based on flooding the network with rout request. The flooding part occurs by broadcasting route requests and have them propagate through the network until the destination is reached.

Group transmissions are a method of transmitting data to all devices within a certain group. A broadcast is used to transmit the data and the frame will be discarded by any members that do not belong to the group. Along with those functions, there are numerous other smaller functions that utilize broadcasts in both the ZDO (ZigBee Device Object or endpoint 0 on all ZigBee devices) and the ZCL (ZigBee Cluster Library).

Broadcasting: To understand broadcasts, we at first remind the reader with the different device types that mentioned in Section 6.2.2. There are three types of ZigBee devices: the coordinator, routers, and end devices. The coordinator is just a router that starts the network. It always has a network address of 0 and mainly performs the function of scanning the network and selecting the channel and ID for the network. A router is a device that has the capability to forward frames and usually, is able to accept child devices. An unfortunate attribute of routers and coordinators in ZigBee is that they are unable to sleep which is a standard complaint among users that are investigating using ZigBee for wireless sensor networking. This limitation means that ZigBee routers usually need to be attached to a MAINs power supply.

An end device has no resources to forward frames and can only join and communicate with a parent router. The simplified communication capabilities allow most of the MAC, NWK, and APS management functions to be stripped out and should result in a very small memory footprint. Sleepy end devices are able to be duty-cycled where they sleep most of the time and awaken periodically to poll its parent for any buffered messages. It uses 802.15.4 indirect transmission for the polling. Duty cycling the end device allows it to consume very little power, thus increasing the battery life which is one of the most important factors in wireless sensor networking.

As a fact it is possible to target broadcasts based on the device type. There are four broadcast addresses that can be used depending on the receiving devices. The addresses are shown in the table below.

Address	Audience
0xFFFF	All devices
0xFFFD	All devices with receiver on permanently
0xFFFC	Routers and coordinators
0xFFFB	Low-power routers

Transmitting a network broadcast frame in ZigBee actually sets off a chain of events. If a new broadcast is received, either from another device

Table 7.4 Broadcast transaction table (BTT)

Field	Description
Src Address	16-bit network address of broadcast initiator
Sequence Number	Network layer frame sequence number
Expiration Time	Amount of time before this entry expires

or from a higher layer, a broadcast transaction record is created. If the frame was received from another device, a copy of the frame is also made and sent up to the next layer for processing.

The broadcast transaction record is used to track the source address and sequence number of the broadcast. These two pieces of information are used to uniquely identify a broadcast frame. This is important because once the broadcast frame is forwarded, all neighbors within the transmission range will re-send the broadcast frame and each device will get multiple copies of it. As long as the device has the broadcast transaction record, the device will know that it has already received and processed the frame so the device can discard the copies.

Broadcast Transaction Record Entry: The record that was created actually goes into a table called the Broadcast Transaction Table, or BTT (Table 7.4).

The BTT implements what is called a passive acknowledgment system, and is used to ensure that all known neighbors have received the broadcast sent by the device. As mentioned before, when a broadcast is transmitted, all devices that receive it will broadcast a copy. Each time a copy of the broadcast arrives, the address of the sender will be added to the BTT to mark that it has relayed the broadcast. After a broadcast timeout, if all neighbors haven't relayed the broadcast, meaning they aren't present in the BTT, then the original sender will need to do a broadcast retry. This happens until the max retries (usually 3) or all the neighbors show up in the BTT.

The problem with the BTT is that it is not very deterministic. If there are a lot of neighbors, that is, the network is dense, then the broadcast transaction table has the potential to become large, and accordingly, it takes a large part of the space of the RAM (the memory). Thus there is an option in the ZigBee specification to forgo the broadcast transaction table. The tradeoff is that the device will need to broadcast the frame the maximum amount of retries for any broadcast. At first glance, this would be desirable because you can get rid of the BTT which has an unknown number of entries. However, this also means that each broadcast will be retried three times (the default retry number), taking a toll on all devices on the network. Each received frame, whether a duplicate or not, requires RAM since it needs to get to the network layer before it can be checked and discarded. If many devices on the network have no broadcast table, then each broadcast would generate a huge amount of traffic, possibly triggering some devices to run out of memory. This let the ZigBee designer (or user) to be careful while using the broadcast transmissions

7.2.2 ZigBee Routing Protocols

This chapter is mainly concentrating on ZigBee, which comes under the category of wireless sensor networks, accordingly this section is concentrating on ZigBee routing protocols. ZigBee routing algorithm can be thought of an hierarchical routing strategy with table-driven optimizations applied where possible.

ZigBee routing techniques mainly falls into two categories:

1. The reactive routing protocol Ad hoc On-demand Distance Vector (AODV), or its modification AODVjr, (AODV junior).
2. ZigBee hierarchical algorithm which also called tree routing protocol or Cluster-Tree algorithm.

Many other routing algorithms are known, but all of them have their roots in these basic categories. Some of the ZigBee routing protocols are:

1. Ad hoc on-demand distance vector (AODV)
2. AODVjr (AODV junior)
3. Dynamic source routing (DSR)
4. Cluster tree algorithm
5. Tree hierarchical routing
6. Many-to-one routing or multipath routing
7. Neighbor table-based routing techniques, etc.

Next subsections discuss in detail the two main algorithms, AVOD and Cluster tree, and the other algorithms are briefly given.

7.2.2.1 AODV: Ad hoc on-demand distance vector

The AODV (Ad Hoc On-demand Distance Vector) routing is a pure on demand route acquisition algorithm: nodes that do not lie on active paths neither maintain any routing information nor participate in any periodic routing table exchanges. Further, a node does not have to discover and maintain a rout to another node until the two need to communicate, unless the former node is offering services as an intermediate forwarding station to maintain connectivity between two other nodes.

The primary objectives of the algorithm are to broadcast discovery packets only when necessary, to distinguish between local connectivity management and general topology maintenance and to disseminate information about changes in local connectivity to those neighboring mobile nodes that are likely to need the information.

The node which needs the connection broadcasts a route request RREQ to its neighbors who re-route the message and safeguard the node from which they received the message. Along with its own sequence number and the Route Request ID, the source node includes in the RREQ the most recent sequence number it has for the destination. The path cost comparison of packets with the same RREQ allows choosing best path at the moment and discarding anything worse. When a node receives a message and it has an entry corresponding to the destination in its routing table, it returns a RREP through the reverse path to the requesting node. So, the source sends its data through this path to the destination with the minimum number of hops.

It is possible to summarize the rout finding using AOVD in the following steps and example:

1. Ad hoc On-demand Distance Vector Routing
2. On-demand $\rightarrow$ Reactive $\rightarrow$ Construct a route when needed
3. Routing table: Path is not stored. Only next hop.

   - Entry = <destination, next node, "sequence #" (timestamp)>

4. Route Discovery: Flood a route request (RREQ) to all neighbors (Figure 7.24). Neighbors broadcast to their neighbors.
5. Request ID is the RREQ serial number. Used to discard duplicates. Source sequence # is a clock counter incremented when RREQ is sent. Destination sequence # is the most recent sequence from the destination that the source has seen. Zero if unknown.
6. Intermediate nodes can reply to RREQ only if they have a route to destination with higher destination sequence #
7. Route reply (RREP) comes back "unicast" on the reverse path (Figure 7.25).
8. Destination Sequence # is from Destination's counter. Lifetime indicates how long the route is valid.

Src Addr	Req ID	Dest Addr	Src Seq #	Dest Seq #	Hop Count

Figure 7.24   RREQ.

Src Addr	Dest Addr	Dest Seq #	Hop Count	Life Time

Figure 7.25   RREP.

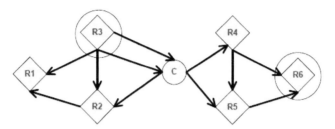

Figure 7.26   Sample route request (broadcast) transmission where R3 is trying to discover a route to R6.

9. Intermediate nodes record node from both RREP and RREQ if it has a lower cost path the reverse path.
10. Backward route to Destination is recorded if sequence number is higher or if sequence number is same and hops are lower.
11. Old entries are timed out.

AODV supports only symmetric links.

Example 7.1: AODV: Rout Discovery

Consider the network of Figure 7.26. In this network, R3 is trying to discover a rout to R6.

As mentioned before, when a source node must discover a route to a destination node, it sends a broadcast route request command. The route request command contains the source network address, the destination network address and a path cost field (a metric for measuring route quality). As the route request command is propagated through the network (refer to the Broadcast Transmission), each node that re-broadcasts the message updates the path cost field and creates a temporary entry in its route discovery table.

Node	Destination Address	Next Hop Address
R3	Router 6	Coordinator
C	Router 6	Router 5
R5	Router 6	Router 6

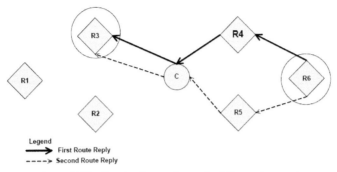

Legend
→ First Route Reply
----→ Second Route Reply

Figure 7.27   Destination node received the rout request.

When the destination node receives a route request, it compares the 'path cost' field against previously received route request commands. If the path cost stored in the route request is better than any previously received, the destination node will transmit a route reply packet to the node that originated the route request. Intermediate nodes receive and forward the route reply packet to the source node (the node that originated route request).

Retries and Acknowledgments: Here R6 can send multiple replies if it identifies a better route. ZigBee includes acknowledgment packets at both the Mac and Application Support (APS) layers. When data are transmitted to remote device, it may traverse multiple hops to reach the destination. As data are transmitted from one node to its neighbor, an acknowledgment packet (Ack) is transmitted in the opposite direction to indicate that the transmission was successfully received. If the Ack is not received, the transmitting device will retransmit the data, up to 4 times. This Ack is called the Mac layer acknowledgment.

In addition, the device that originated the transmission expects to receive an acknowledgment packet (Ack) from the destination device. This Ack will traverse the same path that the data traversed, but in the opposite direction. If the originator fails to receive this Ack, it will retransmit the data, up to 2 times until an Ack is received. This Ack is called the ZigBee APS layer acknowledgment.

Example 7.2: AOVD
Consider the network of Figure 7.28 in which node 1 is requesting rout to node 10.

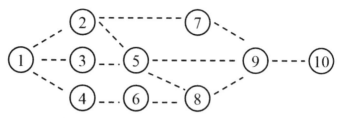

Figure 7.28   Node 1 requesting rout to node 10.

The following steps will be followed.

1. Node 1 broadcasts RREQ to nodes 2, 3, and 4. The request message looks as follows:

   "Any one has a rout to 10 fresher than 1. This is my broadcast #1."

2. Node 2 broadcasts RREQ to nodes 1, 5, and 7.
3. Node 3 broadcasts RREQ to nodes 1 and 5.
4. Node 4 broadcast RREQ to nodes 1 and 6.
5. Node 7 broadcast RREQ to nodes 2 and 9.
6. Node 5 broadcast RREQ to nodes 3, 2, 9, and 8.
7. Node 6 broadcast RREQ to nodes 6 and 8.
8. Node 9 broadcast RREQ to nodes 8, 5, 7, and 10.
9. Node 8 broadcast RREQ to nodes 6, 5, and 9.

Rout Replay

10. Node 10 broadcast PREP to node 9.
11. Node 9 broadcast PREP to node 7.
12. Node 7 broadcast PREP to node 2.
13. Node 2 broadcast PREP to node 1.

Now node 1 has the fresh rout to node 10.

Route Maintenance in AODV

The following sequence are used for route maintenance in AODV:

1. Each node keeps a list of active neighbors (replied to a hello within a timeout).
2. If a link in a routing table breaks, all active neighbors are informed by "Route Error (RERR)" messages.
3. RERR is also sent if a packet transmission fails.
4. RERR contains the destination sequence # that failed.

5. When a source receives an RERR, it starts route discovery with that sequence number.

Disadvantage of this technique: Intermediate nodes may send more up to-date but still stale routes.

7.2.2.2 AODVjr

AODVjr removes form the AODV specification the sequence numbers, gratuitous RREP, hop count, Hello message, RREP, precursor lists. In AODVjr, if communications are unidirectional, the destination sends Connect messages to the source. If data traffic is bidirectional, no additional messages are used. In any case, a source detects a link break in a route when it receives no messages from the destination. An AODVjr routing protocol with multiple feedback policy by means of processing main message during route discovery. The source node starts the route discovery, the destination node processes every Route Request (RREQ) and sends every Route Reply (RREP) as feedback information to the source node, and the source node processes multiple RREP messages for routing decision. The improved AODVjr changes the routing decision commander from destination to source and make a proactive routing decision on the basis of multiple feedback information.

7.2.2.3 Multicast route
- Similar to unicast route discovery
- If a node receives an RREQ but is not a member of the group or does not have the route to any member of the group, it creates a reverse-route entry and broadcasts the request to other neighbors.
- If the node is a member of the group, it sends a RREP message to the source and forwards to other neighbors. Intermediate nodes make a note of this and set up a forward path.

Multicast discovery examples are as follows:

- D and E are members
- A concludes that the paths are ABD and ACE

7.2.2.4 Dynamic source routing (DSR)

Dynamic Source Routing (**DSR**) is a routing protocol for wireless mesh networks. It is similar to AODV in that it forms a route on-demand when a transmitting node requests one. However, it uses source routing instead of relying on the routing table at each intermediate device.

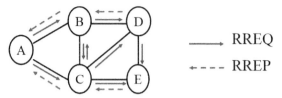

Figure 7.29   Multicast discovery.

To avoid using source routing, DSR optionally defines a flow id option that allows packets to be forwarded on a hop-by-hop basis.

DSR is based on source routing whereby all the routing information is maintained (continually updated) at mobile nodes. It has only two major phases, which are Route Discovery and Route Maintenance. Route Reply would only be generated if the message has reached the intended destination node (route record which is initially contained in Route Request would be inserted into the Route Reply).

To return the Route Reply, the destination node must have a route to the source node. If the route is in the Destination Node's route cache, the route would be used. Otherwise, the node will reverse the route based on the route record in the Route Request message header (this requires that all links are symmetric). In the event of fatal transmission, the Route Maintenance Phase is initiated whereby the Route Error packets are generated at a node. The erroneous hop will be removed from the node's route cache; all routes containing the hop are truncated at that point. Again, the Route Discovery Phase is initiated to determine the most viable route.

DSR is designed to restrict the bandwidth consumed by control packets in ad hoc wireless networks by eliminating the periodic table-update messages required in the table-driven approach. The major difference between this and the other on-demand routing protocols is that it is beacon-less and hence does not require periodic hello packet (beacon) transmissions, which are used by a node to inform its neighbors of its presence. The basic approach of this protocol (and all other on-demand routing protocols) during the route construction phase is to establish a route by flooding *RouteRequest* packets in the network. The destination node, on receiving a *RouteRequest* packet, responds by sending a *RouteReply* packet back to the source, which carries the route traversed by the *RouteRequest* packet received.

The main characteristics of DSR are:

- On-demand (reactive) routing using "Source Route."
- Source Route = List of routers along the path in the packet.

- Routing database: Complete route to recent destinations.
- Each entry has an expiration period and is timed out.
- If a route is not available, send "route request" to all neighbors.

Src Addr	Broadcast 255...255	RREQ	Req ID	Dest Addr	Route Record

- Each neighbor adds itself to the route in the request and forward to all its neighbors (only first receipt). Does not change source address.
- If a node knows the route it appends the rest of the route and returns the "route reply (RREP)"
- RREP goes back along the recorded path.
- All nodes record paths in RREP and RREQ. Multiple routes cached.

Advantages and Disadvantages of DSR: This protocol uses a reactive approach which eliminates the need to periodically flood the network with table update messages which are required in a table-driven approach. In a reactive (on-demand) approach such as this, a route is established only when it is required and hence the need to find routes to all other nodes in the network as required by the table-driven approach is eliminated. The intermediate nodes also utilize the route cache information efficiently to reduce the control overhead. The disadvantage of this protocol is that the route maintenance mechanism does not locally repair a broken link. Stale route cache information could also result in inconsistencies during the route reconstruction phase. The connection setup delay is higher than in table-driven protocols. Even though the protocol performs well in static and low-mobility environments, the performance degrades rapidly with increasing mobility. Also, considerable routing overhead is involved due to the source-routing mechanism employed in DSR. This routing overhead is directly proportional to the path length.

Example 7.3: DSR

Consider the same network of the AODV example (given again in Figure 7.30). In this case, the route discovery will be as follows:

1. Node 1 sends RREQ to 2, 3, 4: "Any one has a route to 10."
2. Nodes 2 send RREQ to 5, 7. Note: RREQ not sent to 1.
3. Node 3 sends RREQ to 5.
4. Node 4 sends RREQ to 6.

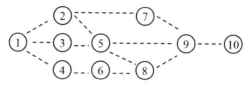

Figure 7.30   Node 1 requesting route to node 10.

The steps will continue as given in Table 7.5.

Route Maintenance in DSR
- If a transmission fails, route error (RERR) is sent to the source. It contains hosts at both ends of the link.
- Intermediate nodes remove or truncate all routes with that link.
- Source may re-initiate the route discovery.
- Caching multiple routes results in a faster recovery but the routes may be stale resulting in cache poisoning at other nodes.
- Not suitable for high-mobility environments.
- Source-route overhead in each packet.

AODV versus DSR
- In DSR a single RREQ can result in routes to several destination.
- In DSR RERR messages are sent to the source not broadcast Many nodes are unaware of failure.
- In DSR, route discovery is delayed until all cached entries have been tried Not good for high mobility.

Feature	DSR	AOVD
Routing table	Rout	Next hop
Packet	Rout	No rout
Replies	Multiple	First only
Rout deletion	Fast	Slow
	Local	Global

7.2.2.5 Tree hierarchical routing protocol (HERA) or ZBR (ZigBee Routing)

In Section 6.3.4 "ZigBee Address Assignment" we mentioned that two schemes are in use for allocating the addresses of the nodes: Distributed

Table 7.5 Route table when using DSR

Pkt # In	Pkt # Out	From Node	To Node	Message Type	Req ID	Hops	Action at Recipient	Route Record in Packet
	1	1	2	RREQ	1	1	New RREQ. Record and forward.	1-2
	2	1	3	RREQ	1	1	New RREQ. Record and forward.	1-3
	3	1	4	RREQ	1	1	New RREQ. Record and forward.	1-4
1	4	2	5	RREQ	1	2	New RREQ. Record and forward.	1-2-5
1	5	2	7	RREQ	1	2	New RREQ. Record and forward.	1-2-7
2	6	3	5	RREQ	1	2	Duplicate ID. Same hops. Record and forward.	1-3-5
3	7	4	6	RREQ	1	2	New RREQ. Record and forward.	1-4-6
4	8	5	8	RREQ	1	3	New RREQ. Record and forward.	1-2-5-8
4	9	5	9	RREQ	1	3	New RREQ. Record and forward.	1-2-5-9
5	10	7	9	RREQ	1	3	New RREQ. Same hops. Record and forward.	1-2-7-9
6	11	5	8	RREQ	1	3	Duplicate ID. Longer Path. Discard.	1-3-5-8
6	12	5	9	RREQ	1	3	New RREQ. Record and forward.	1-3-5-9
7	13	6	8	RREQ	1	3	New RREQ. Same hops. Record and forward.	1-4-6-8
8	14	8	5	RREQ	1	4	Duplicate ID. Longer Path, Discard.	1-2-5-8-6
8	15	8	9	RREQ	1	4	Duplicate ID. Longer Path. Discard.	1-2-5-8-9
9	16	9	3	RREQ	1	4	Duplicate ID. Longer Path. Discard.	1-2-5-5-9
9	17	9	7	RREQ	1	4	Duplicate ID. Longer Path. Discard.	1-2-5-9-7
9	18	9	10	RREQ	1	4	New RREQ. Respond through route 10-9-5-2-1	1-2-5-9-7
10	19	9	10	RREQ	1	4	New RREQ. Respond through route 10-9-7-2-1	1-2-7-9-10

(Continued)

Table 7.5 Continued

Pkt # In	Pkt # Out	From Node	To Node	Message Type	Req ID	Hops	Action at Receipient	Route Record in Packet
10	20	9	8	RREQ	1	4	Duplicate ID. Longer Path. Discard	1-2-7-9-8
10	21	9	5	RREQ	1	4	Duplicate ID. Longer Path. Discard.	1-2-7-9-5
12	22	9	10	RREQ	1	4	New RREQ. Respond through route 10 9 5 3 1	1-3-5-9-10
12	23	9	8	RREQ	1	4	Duplicate ID. Longer Path. Discard.	1-3-5-9-8
12	24	9	7	RREQ	1	4	Duplicate ID. Longer Path. Discard.	1-3-5-9-7
13	25	8	5	RREQ	1	4	Duplicate ID. Longer Path. Discard	1-4-6-8-5
13	26	8	9	RREQ	1	4	Duplicate ID. Longer Path. Discard.	1-4-6-5-9
18	27	10	9	RREP	1	1	Record and forward along return path	10-9 (1-2-5-9-10)
19	28	10	9	RREP	1	1	Record and forward along return path	10-9 (1-2-7-9-10)
22	29	10	9	RREP	1	1	Record and forward along return path	10-9 (1-3-5-9-10)
27	30	9	5	RREP	1	2	Record and forward along return path	10-9-5 (1-2-5-9-10)
28	31	9	7	RREP	1	2	Record and forward along return path	10-9-7 (1-2-7-9-10)
29	32	9	5	RREP	1	2	Record and forward along return path	10-9-5 (1-3-5-9-10)
30	33	5	2	RREP	1	3	Record and forward along return path	10-9-5-2 (1-2-5-9-10)
31	34	7	2	RREP	1	3	Record and forward along return path	10-9-7-2 (1-2-7-9-10)
32	35	5	3	RREP	1	3	Record and forward along return path	10-9-5-3 (1-3-5-9-10)
33	36	2	1	RREP	1	4	Record and forward along return path	10-9-5-2-1 (1-2-5-9-10)
34	37	2	1	RREP	1	4	Record and forward along return path	10-9-7-2-1 (1-2-7-9-10)
35	38	3	1	RREP	1	4	Record and forward along return path	10-9-5-3-1 (1-3-5-9-10)

scheme and Stochastic scheme. Hierarchical Routing Protocol (HERA) is based on distributed scheme of assigning the nodes addresses. In the distributed scheme, during the establishment of the network, the ZigBee coordinator determines maximum number of children routers a parent may have as children. In addition, each node has a "depth" which is the minimum number of hops to reach the coordinator using only parent-child link. For constructing the path, the source node checks if the recipient is one of its descendants because it knows the network address in the block of its child node. Otherwise, the source sends the data to its parent. The parent node also sends the data to its parent until we get to a parent node of the destination node. The downlink of information is provided through a technique for determining the successor based on the ZigBee router address, depth and address of the node. In the integrated routing, a node falls into one of the following two classes: routing node plus (RN+), which has enough memory to perform AODVjr routing; routing node minus (RN−), which has limited memory and only performs cluster-tree routing. While an RN node always follows the cluster-tree, an RN+ node can either follow the cluster-tree or dynamically discover an AODV route, depending on various factors such as session duration and tolerable route discovery delay.

Example 7.4: Hierarchical Routing

Consider the tree topology shown in Figure 7.31 what is the sequence to get the rout between A12 to A30.

In this routing protocol:

- All leaf nodes send the request packet to their parent
- Each parent checks the address to see if it is in its subrange:

    o If yes, it sends to the appropriate child.
    o If not, it sends to its parent.

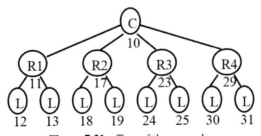

Figure 7.31 Tree of the example.

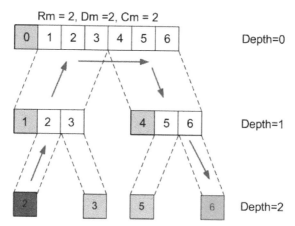

Figure 7.32 Example 2 on hierarchy routing.

Accordingly, the rout between A12 to A30 will be:
Rout from A12 to A30:

1. A12 → R1
2. R1 → Coordinator
3. Coordinator → R4
4. R4 → R30

Example 7.5: Hierarchical Routing

This example is for a frame originating from device 2 with a destination address of 6. Each of the arrows are based on comparing the destination address with the device's address space (except the horizontal one that used to show the flow). If the destination address lies outside the device's address space, the frame gets routed up. Otherwise it gets routed down.

The Problems with Tree Routing: Tree routing is an effective and simple way of routing frames without a complex mechanism like AODV or some of the other routing algorithms mentioned in this section. The big benefit is that it doesn't require tables so it is possible to use tree routing in networks where the devices are very contained on resources, that is, memory. However, there is a fatal flaw in tree routing that renders it almost completely useless in a real world network.

The big problem is that this routing mechanism is fundamentally unreliable because the addresses are static. The position in the tree hierarchy of a device is set once that device joins the network and receives its address and afterwards, it is fairly inflexible to any changes to the network structure. So if a device joins the network and fails or moves out of range, then it will also take out all the devices underneath it in the hierarchy. It is possible to recover from this situation if all of the devices are able to rejoin the network with another parent successfully; however, this can be a painful and error-prone process and potentially involve many devices.

Another fatal flaw is that there is no way to recover if the coordinator fails. It is the single point of failure for this routing method. In mesh routing, if the coordinator fails, things are still okay because from a mesh point of view, the coordinator is just another router in the routing tables.

However in tree routing, the coordinator is the vertex of the tree, where all frames must pass if the destination device is not located on the same branch as the source device. If the coordinator fails, then all the main branches become isolated and your network is pretty much useless. Currently, there's no place in the ZigBee spec (that I'm aware of) that describes how to recover from this failure.

From a reliability standpoint, tree routing is fairly dangerous. In fact, it was probably dangerous enough that the ZigBee Alliance decided to remove tree routing and addressing in the ZigBee Pro spec.

7.2.2.6 ZigBee cluster label (ZiCL)

In ZiCL (ZigBee Cluster Label), the ZigBee network is divided into one or more logical clusters and then a unique Cluster Label is assigned to each cluster where the Cluster Label represents addresses of all nodes in a logical cluster. Within each cluster, nodes can communicate with each other in at most four hops. Every node is associated with a Cluster Label and cluster heads are connected with each other via gateway nodes, which are used to communicate with an adjacent cluster. According to one hop neighbor information, nodes which are connected with other clusters can identify own role as a gateway. In the shortcut tree routing, remaining hops from an arbitrary source to the destination are calculated using the hierarchical addressing scheme in ZigBee, and each source or intermediate node forwards a packet to the neighbor node with the smallest remaining hops in its neighbor table.

7.2.2.7 ZigBee multipath hierarchical tree routing (Z-MHTR)

ZigBee Multipath Hierarchical Tree Routing (Z-MHTR) is a multipath extension of the ZBR and the built paths are node disjoint, used simultaneously to route data. Data packets are routed to the sink (tree root) on up to three disjoint paths. These routing decisions are performed on-the-fly thanks to the

ZigBee tree topology properties without requiring the traditional discovery phase using RREQs.

7.2.2.8 Neighbor table-based routing techniques

A shortcut tree routing (SCT) algorithm is which reduces the routing cost of ZigBee tree routing with the help of a neighbor table, originally defined in ZigBee standards. While following the ZigBee tree routing algorithm, the algorithm suggests, if one can decrease the expense of the routing to particular destination, the most preferred method is forwarding the packet usually to the very near and closest node. However, the links in their method were still invariable and may lead to rapid decline of energy for some nodes. An enhanced routing protocol for ZigBee/IEEE 802.15.4 wireless networks is proposed. For making the ZigBee Tree-based routing algorithm much more efficient, neighbor nodes are considered from where a local shortest path to the destination are calculated and the specific node having the shortest value is selected as next hop node. So, such model is taken on the basis of Greedy algorithm, which is never become mandatory for us to make that we will receive altogether a very shortest path finally. The problem associated with this method is that the lifetime of nodes is not considered and load balancing over nodes is not evaluated. A novel modified tree routing mechanism distinguished with an introduction of neighbor table. A successive improvement on the routing path is done by estimating the cost via each neighbor. It shows a comparable good performance system and commendably very low per packet transfer consumption and an excellent durability. But as this is also on the basis of two-hop neighbor scenario, same cannot be guaranteed in ZigBee networks; so this may lead to high memory overhead and energy.

The DFG-TR (Destination Family Group Tree Routing) was introduced to determine the relationship of forwarding and destination nodes that are closely neighbors. To decrease the routing cost to the definite destination, this technique will suggest the neighbor node as the next hop node. Since it can find neighbor node from which the destination family group exists. However, the life time of nodes and load balancing over nodes are still questionable. Another routing system called ESTR (Energy-Efficient Shortcut Tree Routing) which also a tree based new type is suggested to reduce hop counts as well as to energy balancing in a network with the aid from the available neighbor tables. This also gives optimum low delay route on the basis of balancing load with the associated nodes. But this may lead to computational overhead25. The shortcut tree routing (STR) claims to significantly enhance the ZTR path efficiency with addition of a simple one-hop neighbor data. At the same time, ZTR uses tree links connecting the parent and child nodes only, STR exploits the neighbor nodes by allowing them for shortcut the routing pathway of tree model. Putting in other way, the smaller tree hops which are remained in the next hop node to the destination, which ever may be the type, as is a parent one, the children and or neighbor nodes, is always selected by a source or an intermediate node. This path selection in STR is done in sequential method by individual node and completely compatible with the standards for ZigBee. That accepts various routing protocols according to each node current condition. Another highlight being the feature that it does not requires any extra cost or any alteration in ZigBee standards like creation and maintenance mechanism of one-hop neighbor details, etc....

7.2.2.8.1 Neighbor table

The neighbor table contains a list of the devices that are within transmission/reception range and provide a convenient single hop transmission to the destination. It is also used during the discovery or rejoin process to see if the joining device was previously a child of the node (Tables 7.6 and 7.7). According to the specification, the neighbor structure contains both mandatory and optional fields. However in actual usage, the optional fields are required since they will be needed by some of the ZDO functions. Just a little gotcha for those implementing their own stack:

The neighbor table is initially populated during device discovery when a device is searching for a parent to join to get on the network. As mentioned in Section 7.1.5, when a device tries to join a network, it will first perform

Table 7.6   Neighbor table entry–mandatory

Field	Description
Extended Address	64-bit device address
Network Address	16-bit network address
Device Type	Coordinator, router, end device
Rx On When Idle	Flag to mark sleepy end devices
Relationship	Parent, child, sibling, no relationship
Transmit Failure	Transmission failure counter
LQI	Link quality indicator
Outgoing Cost	Cost of outgoing link as measured by neighbor. *Only required if symmetrical links are used*
Age	Time since link status command was received. *Only required if symmetrical links are used*

Table 7.7   Neighbor table entry–optional

Field	Description
Extended PAN ID	64-bit unique PAN ID
Channel	Operating channel
Depth	Tree depth of device
Beacon order	802.15.4 Beacon order. ZigBee uses no beacons so this should be 0x0F
Permit joining	Flag to indicate whether device is accepting join requests
Potential parent	Flag to indicate if neighbor satisfies parent criteria

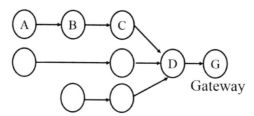

Figure 7.33   Many-to-one routing.

a device discovery where it broadcasts a beacon request. All routers within earshot will respond with a beacon frame containing information about themselves. This information will get stored into the neighbor table. Unfortunately, the spec is a bit light on details about populating the neighbor table after the initial join procedure. In order to keep the table up-to-date, any beacons that are seen should be compared to the neighbor table and added to it if an entry doesn't exist.

7.2.2.9 Many-to-one routing

- Used for sensor data collection. All data goes to a concentrator or a gateway.
- Gateway has a large memory and can hold complete routes to all nodes.
- But each node only remembers the next hop towards gateway (Figure 7.33).

7.2.2.10 ZigBee RF4CE

- Radio Frequency for Consumer Electronics (RF4CE) consortium developed a protocol for remote control using wireless (rather than infrared which requires line of sight).
- RF4CE merged with ZigBee and produced ZigBee RF4CE protocol.
- Operates on channels 15, 20, and 25 in 2.4 GHz.

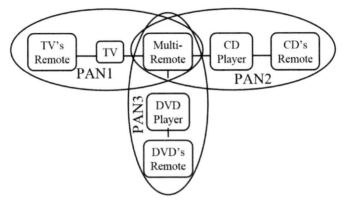

Figure 7.34 RF4CE multi-star topology.

- Maximum PHY payload is 127 bytes.
- Two types of devices: Remotes and Targets (TVs, DVD Player,. . .).
- Status Display: Remote can show the status of the target.
- Paging: Can locate remote control using a paging button on the target.
- Pairing: A remote control works only with certain devices.

7.2.2.11 RF4CE multi-star topology
- Each target device (TV, CD, . . .) forms a PAN with its remote.
- Example: 3 PANs. Multi-function remote can control all 3 devices and is a member of all 3 PANs.

7.2.2.12 ZigBee RF4CE pairing process
- Allows a remote to be associated with the target
- Ensures specific remote will work with only specific TV
- Each remote and the target has a device ID, profiles that it supports, vendor ID
- When target receives a pairing command, it checks if the remote is listed in its pairing table. If yes, it accepts the pairing request.

7.2.2.13 ZigBee smart energy V2
- Monitor, control, automate the delivery and use of energy and water
- Adds plug-in vehicle charging, configuration, and firmware download
- Developed in collaboration with other smart grid communication technologies: HomePlug, Wi-Fi, . . .
- IP based Incompatible with previous ZigBee.

8

ZigBee PRO and ZigBee SECURITY

This chapter considers in detail two topics: ZigBee Pro Feature sets and Zig-Bee security. Besides the two main topics, the chapter discusses: Applications of ZigBee technology, Advantages of ZigBee Technology and Future Scope of ZigBee.

8.1 ZigBee and ZigBee PRO Feature Sets

(Reference: Microchip Note AN1255)
On October 2007, the ZigBee Alliance introduced an important update to the ZigBee specification leveraging the experience gained from the millions of ZigBee nodes deployed by multiple vendors since the specification was first published in 2004.

The specification now includes two distinct Feature Sets– "ZigBee" and "ZigBee PRO"–that define key aspects of how a mesh network operates. The new ZigBee PRO Feature Set offers significant improvements in network scalability, resiliency, security, and ease-of-use, especially for larger, more complex networks.

The ZigBee Feature Set is essentially the same as the 2006 version ZigBee specification, with some new optional features adopted from the ZigBee PRO capabilities. Networks are generally built using one Feature Set or the other, though end devices implementing either Feature Set may operate in any ZigBee network.

ZigBee PRO offers several significant improvements such as network scalability, resiliency, security and the capability to self-form and self-heal the network. It is targeted for use in building automation and environmental and industrial applications that contain more than 30 nodes. ZigBee PRO is based on a mesh topology and is a beaconless network. Networks are generally built using one Feature Set or the other, though end devices implementing either Feature Set may operate in any ZigBee network.

Table 8.1   Comparison of ZigBee and ZigBee PRO feature sets

Feature	ZigBee Feature Set	ZigBee PRO Feature Set
Addressing	Tree-based	Stochastic
Routing	Tree and Mesh	Mesh
Rout Aggregation (i.e., may-to-one and Source Routing)	NO	YES
Asymmetric Link Handling	NO	YES
Frequency Agility	Optional	YES
PAN ID Conflict Resolution	NO	YES
Basic Security included:	Residential	Standard
APS-level encryption	Optional	Optional
"High Security" mode	NO	Optional
Fragmentation	Optional	Optional
Commissioning Cluster	YES	YES
Secure Commissioning	NO	YES

Table 8.1 compares between ZigBee and ZigBee PRO feature sets. This is followed by detailed discussion.

The following are the details of ZigBee PRO feature set characteristics.

8.1.1 Management and Addressing

8.1.1.1 Link management

In a mesh topology, each node can communicate with its neighbor. The node has the ability to evaluate the quality of its neighbor's links and to select the best one for transmission of the packet.

8.1.1.2 Group addressing

ZigBee PRO provides group addressing; a single packet can reach a group of devices.

8.1.1.3 Compatibility

The ZigBee PRO stack identifier is two and is advertised in its beacon frame. Any end device can join ZigBee PRO if it uses the standard security mode.

8.1.1.4 Power management

In ZigBee PRO, only the end devices are powered by batteries; the routers and coordinator use main power. ZigBee PRO allows end devices to go into sleep

Table 8.2 ZigBee versions' features

Feature	ZigBee 2006	ZigBee Feature Set	ZigBee PRO
Coordinator can change channel during operation	No	Yes	Yes
Distributed Address Assignment	Yes	Yes	No
Stochastic Address Assignment	No	No	Yes
Group Addressing	Yes	Yes	Yes
Many-to-one routing	No	No	Yes
AES-128	Yes	Yes	Yes
Trust Center	Coordinator	Coordinator	Any device
Network Scale limited by address assignment scheme	Yes	Yes	No
Fragmentation and Reassembly	No	Yes	Yes
Commissioning Tool	Yes	Yes	Yes
Keep Neighbor Link Quality	No	No	Yes
High Security Mode	No	No	Yes
Topologies	Tree & Mesh	Tree & Mesh	Mesh

Ref. A Elahi and A. Gschwender, "ZigBee Wireless Sensor and Control Network," Prentice Hall, 2009, ISBN: 0137134851, Safari Book.

mode so that they consume less power. While the end device is in sleep mode, the node will miss any network key updates from the trust center. When the device wakes up, it uses its link key to send a message to the trust center to obtain an updated network key.

8.1.2 Improved Security

Security: Since the beginning, ZigBee has included integrated security functions leveraging AES128 encryption for protecting network communications. ZigBee PRO now provides significant enhancements to the security choices available for secure wireless networks. ZigBee PRO offers two different security modes: "Standard" security and "High" security, as follows:

Standard mode: In standard mode, the device is permitted to use the network key and link key.

Network Key: The "network key" is a single network-wide key and all devices share it. It is a standard 128-bit key used to encrypt all network communications using AES128 algorithm. Devices in the network may have the network key pre-configured, individually commissioned, or distributed by a centralized trust center.

Link keys: The ZigBee PRO Standard Security mode further provides for the additional encryption of *application-level* communications between node pairs through the use of individual "link keys". While this is also optional in the ZigBee Feature Set as well the earlier 2006 ZigBee specification, it is expected to be more widely deployed going forward.

This allows specific node-to-node application level data to be protected from other nodes in the network. This is important for the collection of potentially sensitive private data or the validation of potentially disruptive commands.

Link keys and sleeping nodes: The security enhancements in ZigBee PRO are also important to providing secure basic network operation. For example, certain types of ZigBee nodes can "sleep" (power down for a period of time to conserve power use), and hence miss a "change network key" message from the security Trust Center.

When a sleeping device wakes up, it will try to communicate with the network using its old network key, and will be unable to participate in the network. ZigBee PRO provides the ability for the newly awoken node to use a specific link key to encrypt a message that validates itself with the Trust Center and gets the new network key in a similarly encrypted message.

Otherwise, the validation message and network key would need to be sent "in the clear", potentially compromising the security of the overall network. These same mechanisms can be used to wirelessly commission (i.e., configure for use) individual devices securely, never requiring important data to be transmitted without some level of encrypted protection.

In the standard security mode, devices do not require authentication to join a network. The trust center, master key, and SKKE are optional. The trust center for ZigBee PRO standard security mode is used for transporting the network key.

High-security mode: This mode adds some further protections for device authentication and key management and distribution including the use of SKKE. In this mode, three keys are permitted for use: the network, link, and master keys. The trust center and SKKE are mandatory. The trust center uses the transport key command to transport the link and network keys to the devices in the network. The device is required to perform authentication with its parent, and it is required to perform authentication between neighbors.

This mode also provides some additional protections against security replay attacks.

Note: It is important to say here that the additional security capabilities inherent in ZigBee PRO are critical as ZigBee is used in increasingly important applications. The control of critical systems infrastructure, whether in a commercial building, utility grid, industrial plant, or a home security system, must not be compromised.

8.1.2.1 Trust center
ZigBee PRO should have a trust center; the trust center can be a router, coordinator, concentrator, or specific device.

8.1.3 Scalability

One of ZigBee's greatest merits is scalability, allowing networks of thousands of nodes. While this capability is obviously important for large building automation and neighborhood utility metering networks, it has also proven important in home-oriented networks that may grow to 100 nodes or more.

ZigBee has emerged as the choice for advanced Home Automation networks in large part because proprietary wireless technologies aimed at the home have failed to deliver such scalability. The ZigBee PRO Feature Set significantly improves ZigBee's ability to scale through a number of new capabilities, including:

- *Stochastic Addressing and*
- *Route Aggregation.*

8.1.3.1 Stochastic addressing
Stochastic Addressing is a new method for assigning the addresses used by individual nodes for routing in the network. The original ZigBee scheme used a "cluster-tree" routing algorithm where a single ZigBee Coordinator node acts as the root of a network and address tree, and each node's address is assigned based on its position within the tree.

While having the node's location implicit in the address can allow relatively simple routing algorithms, this also limits the practical address space, leading to the possibility of address exhaustion down long branches of the tree structure. Also, changes to the tree topology may cause potentially disruptive re-addressing of significant portions of the network.

By contrast, ZigBee PRO uses a mesh topology. Any device joining the network requires a network address. ZigBee PRO uses a stochastic addressing method, which means a ZigBee device randomly picks up an address when joining the network. Also, the device announces its address on the network by using the device-annc command. The ZigBee network layer provides address conflict resolution based on the MAC address if two nodes have the same network addresses.

Thus the entire 16-bit address space is available to all nodes, anywhere in the network, and the assignment is persistent even if RF conditions change (such as placement of new obstacles in a building or if a node moves).

Stochastic addressing eliminates the need for the parent to maintain an address table for assignment to children.

8.1.3.2 Route aggregation

Another new ZigBee PRO scalability feature is the ability to aggregate routes in the network through the use of new two types of routing: many-to-one routing and multicast (source) routing techniques.

In most wireless sensor networks, many or most nodes in the network need to communicate with one or more centralized nodes, such as a centralized home automation controller, a sensor data collection gateway, or a security trust center.

For illustration purposes, assume many devices are communicating to a centralized gateway. Each device issues a route discovery broadcast to find the gateway, and all the intervening nodes form a routing table entry for each device (see Figure 8.1).

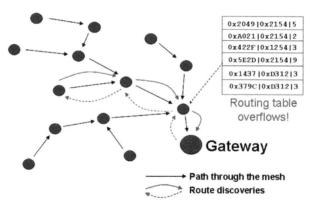

Figure 8.1   In larger networks, many routes to a centralized mode may cause routing table overflows near the central node.

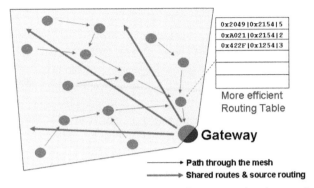

0x2049	0x2154	5
0xA021	0x2154	2
0x422F	0x1254	3

More efficient
Routing Table

Gateway

→ **Path through the mesh**
→ **Shared routes & source routing**

Figure 8.2   Through shared many-to-one routes and source routing, less routing table space is used.

The nodes near the gateway will see many, many such requests, likely overflowing the relatively constrained routing tables available within the small RAMs typical of low-cost ZigBee silicon. Such overflows cause the devices to issue new route request broadcasts, causing continued routing table churn and additional network broadcast traffic.

ZigBee PRO mitigates this behavior through a "many-to-one" route discovery mechanism that allows each device seeking a route to the gateway (continuing the example above) to share the same route and associated routing table entry.

For traffic returning from the gateway to the devices, a source routing technique is used, where the gateway remembers the path used from the device to the gateway, and embeds that path in the returning packets enabling the intervening nodes to forward the packets without requiring a routing table entry (see Figure 8.2).

These mechanisms dramatically improve the efficiency and stability of the routing tables and lower the amount of broadcast traffic in the network.

8.1.4 Improved Ease-of-Use and Operation

The ZigBee PRO Feature Set also provides some additional features that simplify the use and deployment of ZigBee applications. Two features are sharing in improving ZigBee operation:

- Fragmentation
- Commissioning

8.1.4.1 Fragmentation

The IEEE 802.15.4 defines a maximum packet size of 128 bytes, which is optimum for short, efficient control and sensor data messaging. However, there may be times where an application desires to send a larger message.

ZigBee PRO and also ZigBee feature set now provides a means, fragmentation, for the stack to automatically fragment a larger message into smaller ZigBee packets and reassemble the message at the receiving node, relieving the application of this task.

8.1.4.2 Commissioning

It is a tool that is used by the ZigBee device installer to install ZigBee devices:

Commissioning a ZigBee network (like any control network), refers to the task of configuring the individual control or sensor devices with the information they need to perform the desired function. This information may include: what network to join, what ZigBee Feature Set is in use, provision of security keys, and establishing command/control binding relationships (i.e., what lights a given switch turns on or off).

Commissioning typically takes place using a dedicated tool (laptop, PDA, etc.) or through a physical interface on the device (perhaps a simple button).

The updated ZigBee specification provides a new "Commissioning Cluster" that standardizes command sets for usual configuration parameters. ZigBee PRO, with its additional security capabilities, allows this commissioning to be performed in a secure manner.

8.1.5 Improved Resiliency

ZigBee-based networks are inherently self-forming and self-healing in their operation. ZigBee PRO provides a number of new features that enhance network resiliency in certain situations. The new features are:

- Asymmetric link handling
- frequency agility
- PAN ID conflict resolution

8.1.5.1 Asymmetric link

The link between two nodes is usually asymmetric, meaning the quality of the link in both directions is not the same. The quality of the link is represented by the link cost. Figure 8.3 shows a ZigBee network. The link cost from node A to B is different from B to A (asymmetric link) because A and B do not have the same transmission powers and receiver sensitivities. When the nodes

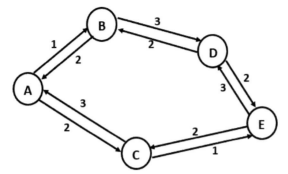

Figure 8.3 Asymmetric link.

A and E use the same route B–D to exchange information, it is considered symmetric. In asymmetric routing, the destination node uses a different path to transmit to the source. In Figure 8.3, the source A uses the B–D path to send a packet to E, and E use the path C to send the packet to A.

8.1.5.2 Frequency agility
One of the advantages of 2.4 GHz-based ZigBee networks is the choice of 16 different operating channels, compared with a single channel for some competing lower-frequency proprietary systems.

ZigBee PRO's Frequency Agility function makes utilization of these channels easier. It selects the best available channel during startup of a network but, during network operation, if any node detects interference due to frequency conflict or noise, it will report the occurrence to the channel manager. (A channel manager can be a dedicated device or trust center.) When the channel manger receives the reports from several nodes on the network, it selects another channel for network operation and informs the nodes of the switch.

Frequency Agility to Combat Interference
Frequency agility can be used to combat interference and accordingly helps to stop loosing of service in the radio channel of operation or across the whole frequency band. To achieve that, it is worthwhile implementing a network manager application on the network coordinator, where this application is responsible for periodically assessing the operating channel and traffic by means of various ZigBee messages, such as Management Network Update notifications. If interference is present in the operating channel, the network

manager can initiate a channel scan to find a safe channel and move the network to this channel (frequency agility). If no suitable channel is found, the network manager node may declare a loss of service.

8.1.5.3 PAN ID conflict resolution

When a coordinator attempts to establish a ZigBee network, an energy scan is initiated to find the best RF channel for its new network. When a channel has been chosen, the coordinator assigns a PAN-ID which will be applied to all the devices that join the network.

PAN-ID is a 16-bit number that is used as a network identifier. A node is allowed to communicate on a network only when it undergoes the association process. The association function is used to join a node to a parent.

PAN-ID Conflict Resolution: ZigBee networks use the IEEE 802.15.4 PAN ID as their basic identifier, allowing different logical ZigBee networks even on the same channel. However, if two networks overlap with the same PAN ID (such as devices using the same defaults), confusion may result. ZigBee PRO provides automatic detection and resolution of conflicting PAN IDs, including application notification.

Table 8.3 shows a comparison of ZigBee 2006, ZigBee, and ZigBee PRO feature sets.

8.2 ZigBee Security

ZigBee protocol found its way in many applications and is growing more and more over the years. In parallel, security attacks on ZigBee-enabled devices have grown incrementally as well. The adoption of ZigBee in personal home and hospital care, commercial building automations, and many others, let the need for assessing and strengthen the security of ZigBee protocol to be crucial. Such high need for assessing the security of ZigBee is clear if we remember that a hack on a critical device such as the pacemaker can be life threatening. The previous sections of the chapter provided an overview of the ZigBee protocol and its network components, a knowledge that serves as a pre-requisite to understanding ZigBee security which is the subject of this section.

ZigBee security architecture is built on the security services provided by IEEE 802.15.4 standard. The IEEE 802.15.4 MAC layer implements several features which are used by the ZigBee protocol in the network and application layers. One of these features is the security services. **IEEE 802.15.4** sets the encryption algorithm to use when cyphering the data to transmit; however,

Table 8.3   Comparison of 2006, ZigBee, and ZigBee PRO feature sets

ZigBee Features	2006 Specification	ZigBee Feature Set Specification	ZigBee PRO Feature Set Specification
Network coordinator selects best channel at startup.	Yes	Yes	Yes
During operation can detect interference and change channel operation.	No	Yes	Yes
Distributed address assignment.	Yes	Yes	No
Stochastic address assignment.	No	No	Yes
Supports group addressing.	Yes	Yes	Yes
Many-to-one routing.	No	No	Yes
128-bit Advanced Encryption Standard (AES) with message integrity code (MIC).	Yes	Yes	Yes
Trust center can be any device or coordinator in the network.	Coordinator	Coordinator	Any device
Network scale up limited to address assignment scheme.	Yes	Yes	No
Message fragmentation is permitted.	No	Yes	Yes
Supports buffering for message fragmentation.	No	Yes	Yes
Supports commissioning tool.	Yes	Yes	Yes
Device keeps information about its neighbor devices.	No	No	Yes
Offers high-security mode.	No	No	Yes
Network topologies.	Tree and mesh	Tree and mesh	Mesh

the standard does not specify **how** the keys have to be managed or what kind of **authentication policies** have to be applied. These issues are treated in the upper layers of the **ZigBee stack.**

Security services provided by ZigBee range from secure key establishment, secure key transportation, frame protection via symmetric cryptography, to secure device management.

This section starts by discussing the security services offered by 802.15.4 followed by the security services offered by ZigBee.

8.2.1 802.15.4 Security: Overview

IEEE 802.15.4 uses **AES** (**Advanced Encryption Standard**) with a **128-bit** key length (16 Bytes) as encryption algorithm. The algorithm is used to achieve Data Security (Data Confidentiality), Data Integrity and authenticity.

Data Security (Data Confidentiality): This term covers two related concepts:

- **Data confidentiality:** Assuring that private or confidential information is not made available or disclosed to unauthorized individual/node.
- **Privacy:** Assures that individuals control or influence what information related to them may be collected and stored and by whom and to whom that information may be disclosed.

The importance of confidentiality is crucial in some fields as health care where the records of the patient must be kept secret and private.

Data confidentiality is performed by encrypting the data payload field with the 128-bit Key.

Data Integrity and Authenticity: This is used to validate the data sent, mainly the integrity of the MAC header and payload data attached. Data integrity covers:

- **Data integrity:** Assures that information is changed only in a specified and authorized manner.
- **System integrity:** Assures that a system (the network) performs its intended function in an unimpaired manner, free from deliberate or inadvertent unauthorized manipulation of the system.

Concerning Authenticity, it means:

- **Authenticity:** The property of being genuine and being able to be verified and trusted: confidence in the validity of a transmission, a message, or message originator.

Data Integrity is achieved using a **Message Integrity Code** (**MIC**) and data authenticity is achieved by **Message Authentication Code (MAC). The MAC** is appended to the message. This code ensures integrity of the MAC header and payload data attached. The system uses the network key to generate the MAC. Upon receiving message, the receiving side generates MAC for the received message using the same key and compares the calculated

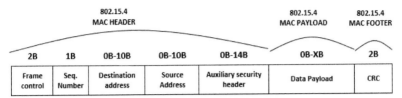

Figure 8.4 IEEE MAC frame.

MAC with that attached with received message. If the two values match, the receiver accepts the message, if the two values does not match the receiver discard the message.

The MAC can have different sizes: 32, 64, and 128 bits; however, it is always created using the **128-bit AES** algorithm. Its size is just the bits length which is attached to each frame. Larger size means more secure. However, increasing the MAC size will be reflected on the payload: larger MAC size means less payload the message can take (Figure 8.4).

MAC Frame

Figure 8.4 shows the MAC frame. There are three fields in the IEEE 802.15.4 MAC frame which are related to security issues:

- **Frame Control** (located in the MAC Header)
- **Auxiliary Security Control** (in the MAC Header)
- **Data Payload** (in the MAC Payload field)

a. Auxiliary Security Frame

The **Auxiliary Security Frame** is only enabled if the **Security Enabled** subfield of the **Frame Control Frame** is turned on. This special header has 3 fields (Figure 8.5):

- **Security Control** field. It is 1B (1 Byte) length. This field specifies which kind of protection is used (see Table 8.3).
- **Frame Counter field. It is** 4B (4 Bytes = 32 bits) length. This field is a counter given by the source of the current frame in order to protect the message from replaying protection. For this reason, each message has a unique sequence ID represented by this field. This means that this field is used to protect from the replay attack.
- **Key Identifier: The length of this field varies** 0B and 9B (0 bit to 72 bits). The field specifies the information needed to know what key we are using with the node we are communicating with.

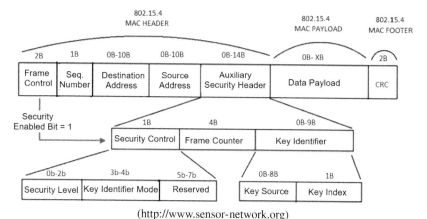

(http://www.sensor-network.org)

Figure 8.5   MAC frame: Auxiliary security frame.

i. **Security Control Subfield: Security Control** specifies the type of protection provided by the network. It is the place where global Security Policy is set. The choice of security level determines the length of the key and what is to be encrypted, that is, each security level provides a certain degree of frame encryption and integrity checks. ZigBee defines 8 different security levels available to the NWK and APS Layer as summarized in Table 8.4.

The security level identifiers in the table mean:

o The **0x00** value sets no encryption. This means that (1) data will not be encrypted: no data confidentiality, (2) the data authenticity will not be validated.

o From the **0x01 to 0x03,** the data are authenticated using the encrypted Message Authentication Code (**MAC**).

o The value **0x04** encrypts the payload ensuring **Data Confidentiality**.

o The **0x05 to 0x07** range ensures both data confidentiality and authenticity.

ii. **Key Identifier Subfield:** The **Key Identifier Mode** subfield sets the kind of key (**implicit or explicit**) which should be used by the sender and the receiver. Possible values are:

• **0**–the key id is known implicitly by the sender and the receiver (It is not specified in the message)

• **1**–the key id is determined explicitly by the 1Byte **Key Index** from the **Key Identifier** Field and the *macDefaultKeySource*.

Table 8.4 ZigBee security levels

Security Identifier	Security Attributes	Data Encryption	Frame Integrity (length of MIC)
0x00	None	OFF: Data are not encrypted. Data authenticity is not validated.	NO (M = 0)
0x01	MIC-32	OFF: Data are not encrypted. Data authenticity is validated.	YES (M = 4)
0x02	MIC-64	OFF: Data are not encrypted. Data authenticity is validated.	YES (M = 8)
0x03	MIC-128	OFF: Data are not encrypted. Data authenticity is validated.	YES (M = 16)
0x04	ENC	ON: Data are encrypted. Data authenticity is not validated.	NO (M = 0)
0x05	ENC-MIC-32	ON: Data are encrypted. Data authenticity is validated.	YES (M = 4)
0x06	ENC-MIC-64	ON: Data are encrypted. Data authenticity is validated.	YES (M = 8)
0x07	ENC-MIC-128	ON: Data are encrypted. Data authenticity is validated.	YES (M = 16)

- **2**–the key id is determined explicitly by the 1Byte **Key Index** and the 4Byte **Key Source** both subfields from the **Key Identifier** Field.
- **3**–the key id is determined explicitly by the 1Byte **Key Index** and the 8Byte **Key Source** both subfields from the Key Identifier Field.

As pointed before, the **Key Identifier** field is set when the **Key Identifier Mode** subfield is not zero. The **Key Source** subfield specifies the group Key originator. The **Key Index** subfield helps to identify different Keys from an specific **Key Source**.

b. Data Payload Field
The **Data Payload** field can have three different configurations depending on the previously defined security fields:

- **AES-CTR:** All the data are encrypted using the defined 128b key and the AES algorithm. The Frame Counter sets the unique **message ID**, and the **Key Counter** (Key Control subfield) is used by the application layer if the **Frame Counter** max value is reached.
- **AES-CBC-MAC:** The Message Authenticity Code (**MAC**) is attached to the end of the data payload. Its length depends on the level of security

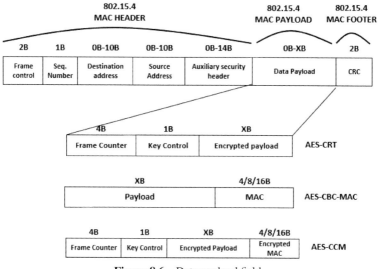

Figure 8.6 Data payload field.

specified in the Security Policy field. The MAC is created encrypting information from the 802.15.4 MAC header and the data payload.

- **AES-CCM:** It is the mixture of the previously defined methods. The subfields correspond with the AES-CTR mode plus the extra AEX-CBC-MAC subfield encrypted.

iii. **Frame Counter:** Frame counter is a counter given by the source of the current frame in order to protect the message from replaying protection.

8.2.1.1 802.15.4 Access control list

Each **802.15.4 transceiver** has to manage a **list** to control its *"trusted brothers"* along with the security policy. For this reason, each node has to control its own **Access Control List (ACL)** which stores the following fields:

- **Address:** The address of the node we want to communicate with.
- **Security Suite:** The security police which is being used (AEC-CTR, AES-CCM-64, AES-CCM-128,...).
- **Key:** The 128b key used in the AES algorithm.
- **Last Initial Vector (IV) and Replay Counter:** Both are the same field. The Last IV is used by the source and the Replay Counter by the destination as a message ID in order to avoid reply attacks.

When a node wants to send a message to a specific node or receives a packet, it looks at the **ACL** to see if it is a **trusted brother** or not. In the case it is, the node uses de data inside the specific row apply the security measures. In the case the node is not in the list or its message is rejected or an authentication process starts.

The ACL is used to prevent unauthorized devices from participating in the network. The ACL is stored in MAC PAN Information Base (PIB) and is accessed and modified similar to other MAC attributes.

8.2.2 ZigBee Security

The security measures provided by IEEE 802.15.4 and the use of ACL have a main shortage: they do not specify how the keys have to be managed or the type of authentication policies to be applied. These issues are managed by ZigBee. ZigBee handled all these issues by the **two extra security layers** on top of the 802.15.4 one: The **Network** and **Application** security layers. The ZigBee security provided includes methods for key establishments, key transport, frame protection, and device management. These services form the building blocks for implementing security policies within a ZigBee device.

All the security policies rely on the AES 128-bit encryption algorithm so the hardware architecture previously deployed for the link level (MAC layer) is still valid.

8.2.2.1 ZigBee security services

As one of its defining features, ZigBee provides facilities for carrying out secure communications, protecting establishment and transport of cryptographic keys, cyphering frames (frame protection), and controlling and managing devices. It builds on the basic security framework defined in IEEE 802.15.4. This part of the architecture relies on the correct management of symmetric keys and the correct implementation of methods and security policies.

The ZigBee standard supports many optional security services as (will be discussed in detail latter):

- Encryption/decryption
- Replay Protection
- Device Authentication
- Secure over-the-air (OTA) firmware upgrades
- Logical link-based encryption
- Runtime key updates
- Network interference protection

The security features of ZigBee are based on certain inherent assumptions:

- ZigBee assumes an "open trust" model; that is, the protocol stack layers trust each other and that the layer that originates a frame is responsible for initially securing it.
- The security services cryptographically protect the interfaces between different devices only.
- Interfaces between different stack layers on the same devices are arranged non-cryptographically.
- The secret keys are not inadvertently revealed during key-transport. (An exception to this is during pre-configuration of a new device, in which a single key may be sent unprotected.)
- Availability of almost perfect random number generators.
- Availability of tamper-resistant hardware.

Based on the above assumptions, in the following we are going to explore the security models provided by the ZigBee standard, the various keys used for secure communication, key management methodologies suggested by ZigBee, and other inherent security mechanisms such as authentication, replay protection, etcetera.

8.2.3 ZigBee Security Models

The basic mechanism to ensure confidentiality is the adequate protection of all keying material. Trust must be assumed in the initial installation of the keys, as well as in the processing of security information. For an implementation to globally work, its general conformance to specified behaviors is assumed.

Keys are the cornerstone of the security architecture; as such their protection is of paramount importance, and keys are never supposed to be transported through an insecure channel. A momentary exception to this rule occurs during the initial phase of the addition to the network of a previously unconfigured device. The ZigBee network model must take particular care of security considerations, as ad hoc network may be physically accessible to external devices. Also, the state of the working environment cannot be predicted.

Within the protocol stack, different network layers are not cryptographically separated, so access policies are needed, and conventional design assumed. The open trust model within a device allows for key sharing, which notably decreases potential cost. Nevertheless, the layer which creates a frame

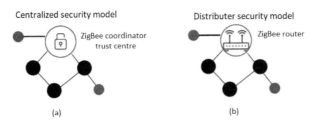

Figure 8.7 The two ZigBee security models: Centralized and distributed.

is responsible for its security. If malicious devices may exist, every network layer payload must be ciphered, so unauthorized traffic can be immediately cut off. The exception, again, is the transmission of the network key, which confers a unified security layer to the grid, to a new connecting device.

In general, ZigBee standard supports two types of security models as shown in Figure 8.7:

- Centralized security model
- Distributed security model.

The two models mainly differ in how they admit new devices into the network and how they protect the messages on the network.

Centralized Security model: This model is complex but, it is the most secure model and involves a third logical device, the Trust center (network coordinator). The Trust center, as mentioned in Section 8.1.2.1 could be the coordinator device, or a device designated as the Trust center by the coordinators of the ZigBee Network. The role of the trust center is given before while discussing the nodes type (Section 6.4.2). The Trust center, as discussed before, is responsible for:

- Configuring and authenticating routers and end devices that join the network,
- Generating network key to be used for encrypted communication across the network,
- Periodically or as required switching to a new network key. Thus, if an attacker acquires a network key, it will have a limited lifetime before expiring,
- Establishing a unique Trust Center link key for each device as they join the network to securely communicate with the Trust Center, and
- Maintaining the overall security of the network.

Distributed security model: This model is simple, but less secure. This model supports only routers and end devices. Routers form the distributed network and are responsible for enrolling other routers and end devices. Routers issue network keys (used to encrypt messages) to newly joined routers and end-devices. All the nodes in the network use the same network key for encrypting messages. Also, all nodes are pre-configured with a link key (used to encrypt the network key) prior to being enrolled in the network.

Security Assumptions

Aside from the open trust model between layers, the security of ZigBee ultimately depends on the following assumptions:

1. The safekeeping of symmetric keys. ZigBee assumes that secret keys are not available outside of the device in an unsecured way, meaning that all transmission of keys must be encrypted. An exception to this is during pre-configuration of a new device, in which a single key might be sent unprotected, creating a brief vulnerability. Here, if the keys are stolen because the adversary has physical access to the devices, many information then become available. ZigBee's security policy does not protect against attack to hardware due to its low-cost nature.

2. The protection of mechanism employed. All Router and End Device nodes should support both centralized security and distributed security by adapting to the security scheme employed by the network that they join.

3. The proper implementation of cryptographic mechanism and associated security policies involved. Here, ZigBee developers are assumed to follow the complete protocol in practice. ZigBee also assumes the availability of almost perfect random number generators.

8.2.3.1 ZigBee security architecture

The security architecture includes security mechanisms at two layers of the stack: the network and the APS layers. The two layers are responsible for secure transport of their frames. APS, also, provides services for the establishment and maintenance of security relationships. The ZDO manages the security policies and the security configuration of a device.

ZigBee uses 128-bit keys to implement its security mechanisms. A key can be associated either to a network, being usable by both ZigBee layers and the MAC sublayer, or to a link, acquired through pre-installation, agreement or transport. Establishment of link keys is based on a master key which controls link key correspondence. Ultimately, at least, the initial master key must be obtained through a secure medium (transport or pre-installation), as the security of the whole network depends on it. Link and master keys are only visible to the application layer. Different services use different one-way variations of the link key to avoid leaks and security risks.

Key distribution is one of the most important security functions of the network. A secure network will designate one special device which other devices trust for the distribution of security keys: the trust center. Ideally, devices will have the center trust address and initial master key preloaded; if a momentary vulnerability is allowed, it will be sent as described above. Typical applications without special security needs will use a network key provided by the trust center (through the initially insecure channel) to communicate.

Thus, the **trust center** maintains both the network key and provides point-to-point security. Devices will only accept communications originating from a key supplied by the trust center, except for the initial master key. The security architecture is distributed among the network layers as follows:

- The MAC sublayer is capable of single-hop reliable communications. As a rule, the security level it is to use is specified by the upper layers.
- The network layer manages routing, processing received messages and being capable of broadcasting requests. Outgoing frames will use the adequate link key according to the routing if it is available; otherwise, the network key will be used to protect the payload from external devices.
- The application layer offers key establishment and transport services to both ZDO and applications.

The security levels infrastructure is based on CCM*, which adds encryption- and integrity-only features to CCM.

8.2.3.2 Network layer security

When a frame originating at the network layer needs to be secured, or when a frame originates at a higher layer and the NWK secure all frame attribute in the NIB is TRUE, ZigBee shall use the frame-protection mechanism, unless the security enable parameter of the NLDE-DATA. Request primitive is FALS, explicitly prohibiting security.

The NWK layer is responsible for the processing steps needed to transmit outgoing frames and securely receive incoming frames securely. Similar to the MAC layer, upper layers set up the appropriate keys and frame counter and establish which security level to use. The NWK layer sometimes broadcast route request messages and process received route reply messages. In

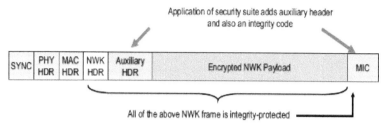

Figure 8.8   ZigBee frame with security at the NWK layer.

doing so, the NWK layer uses link keys if available; otherwise, it uses its active network key. The Network layer's frame-protection mechanism shall make use of the AES (Advanced Encryption Standard) and use CCM AS Specified. The security level applied to a network frame shall be determined by the NWK security level attributed in NIB. Upper layers manage network layer security by setting up active and alternate network keys and by determining which security level to use.

The frame format explicitly indicates the key used to protect the frame. Figure 8.8 shows an example of an encrypted network layer.

8.2.3.3 APL layer security
All the security related with the APL layers is handled by the APS (application support) sublayer. The APS layer is responsible for the processing steps needed to securely transmit outgoing frames, securely receive incoming frames, and securely establish and manage cryptographic keys. Upper layers control the security level or the management of cryptographic keys by issuing primitives to the APS layer. Figure 8.9 shows an example of encrypted APS layer.

The APS layer is also responsible for providing applications and the ZDO within key establishment, key transport, and device management services.

	PHY	MAC	NWK	APS	Auxiliary	Encrypted APS	
SYNC	HRD	HRD	HRD	HRD	HRD	Payload	MIC

In ZigBee 3.0, ZigBee protocol can also create an application-level secure link between a pair of devices in the network by establishing a unique set of AES-128 encryption keys between a pair of devices. This supports the virtual private links between a pair of devices which needs higher security. An example is in a functional network of home area network that connects

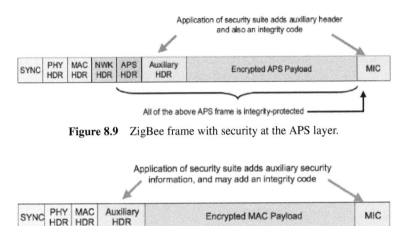

Figure 8.9 ZigBee frame with security at the APS layer.

Figure 8.10 ZigBee frame with security at the MAC layer.

many devices (lights, thermostats, occupancy sensors, door locks, window sensors, and garage door openers), an extra layer of security qualification is established between door locks and garage door openers to limit the ability of an attacker acquiring the network key to inject messages that would open the door lock; in this case, the attacker would also need the link key between door locks and garage door openers.

[Ref. ZigBee 3.0 Task Force, ZigBee: Securing the Wireless IoT (2017)].

8.2.3.4 MAC Layer Security

The MAC layer security is based on the security of IEEE 802.15.4 (based on its specification) augmented with CCM∗. CCM is an enhanced counter with CBC-MAC mode operation encryption scheme, while CCM∗ is CCM with encryption-only and integrity-only capabilities. The MAC layer uses a single key for all CCM∗ security levels (CCM∗ throughout the MAC, NWK, and APS layers) [5]. As part of the open trust model, the MAC layer is responsible for its own security processing, but the upper layers determine which keys or security levels to use. The upper layer sets the MAC layer default key to coincide with the active network key and the MAC layer link keys to coincide with any link keys from the upper layer [5]. MAC layer link keys (which are set by the upper layer are preferred. Figure 8.10 shows an outgoing MAC frame in ZigBee protocol with its security processing.

8.2.3.5 Trust center role

As discussed before while discussing the types of nodes, for security purposes, ZigBee defines the role of Trust Center. The trust center is the device trusted by devices within a network to distribute keys for the purposes of network and end-to-end application configuration management. All members of the network shall recognize exactly one Trust center, and there shall be exactly one trust center in each secure network. In high security, commercial applications a device can be pre-loaded with the Trust center address and initial master key. Alternatively, if the application can tolerate a moment of vulnerability, the master key can be sent via an in-band unsecured key transport. If not pre-loaded, a device's Trust center defaults to the ZigBee coordinator or a device is designated by the ZigBee Coordinator. In low security, residential applications a device securely communicates with its Trust Center using the current networks key, which can be preconfigured or sent via an in-band unsecured key transport. For purposes of trust management, a device accepts an initial master or active network key originating from its trust center via unsecured key transport. For purposes of network management, a device accepts initial active network key and updated network keys only from its Trust Center. For purposes of configuration, a device accepts master keys or link keys intended for establishing end-to-end security between two devices only form its Trust Center. Aside from the initial master key or network key, additional link, master, and network keys are generally only accepted if the originate form a device's Trust Center via secured key transport.

8.2.4 ZigBee Security Keys

There are three types of symmetric keys (each of length 128-bit) used in the ZigBee standard:

- Network key
- Link key
- Master key

Network Key

Network key: It is a unique 128-bit key shared among all the devices in the network. is used in broadcast communication and applied by NWK and APL layers of ZigBee. Each node requires the network key in order to communicate securely with other devices on the network. The trust center generates the network key and distributes it to all the devices on the network. The trust center regenerates this key at different intervals. A device on the network acquires a network key **via key-transport** (used to protect transported network keys) or **pre-installation**. There are two different types of network keys: **standard** (sending network key in the open)**, and high-security** (network key is encrypted). The type of network key controls how a network key is distributed and may control how network frame counters are initialized. However, the type does not affect how messages are secured.

The Trust Center regenerates the network key at different intervals. Each node has to get the Network Key in order to join the network. Once the trust center decides to change the Network Key, the new one is spread through the network using the **old Network Key** (see Figure 8.12 about *"ZigBee Residential Mode"*). Once this new key is updated in a device, its **Frame Counter** (see in the previous sections) is initialized to zero.

Link Key

Link key: is used in unicast communication and applied by the APS of the ZigBee stack. A device acquires link keys either **via key-transport** (key-load key is used to protect transported link keys), **key-establishment** (based on the "master" key and other network parameters)**, or pre-installation** (e.g., during factory installation). Usually, link keys related to the Trust Center are pre-configured using an out-of-band method, for instance, QR code in the packaging, whereas the link keys between nodes are generated by the Trust Center and encrypted with the network key before sending it to the node.

ZigBee defines two types of link keys: global and unique (can in turn be of two types–the first is the trust center link key, where the key is established between the trust center and the device; and the second is the application link key that is established between two devices in the network other than the trust center). The type of link key determines how the device handles various trust center messages (APS commands), including whether to apply APS encryption or not.

A global link key has the advantage that the memory required by the Trust Center does not grow with the number of devices in the network. On the other hand, a unique link key has the advantage of being unique for each device on the network and application communications can be secured from other devices on the network, but more memory resources are needed in each device.

Both types of keys may be used on the network, but a device shall only have one type in use per device-key pair. Furthermore, each node may also have the following pre-configured link keys which would be used to derive a

Trust Center link key (derivation is done by means of Certificate-Based Key Establishment protocol (if SE security is enabled), APS request key method, Touchlink commissioning, or by using Matyas-Meyer-Oseas hash function):

- A **default global trust center link key** defined by the ZigBee Alliance. It has a default value of 5A 69 67 42 65 65 41 6C 6C 69 61 6E 63 65 30 39 (ZigBeeAlliance09) and is used or supported by the device if no other link key is specified by the application at the time of joining.
- A **distributed security global link key**, a manufacturer specific key used for interaction between devices from the same manufacturer.
- **Install code** is a preconfigured link key. All ZigBee devices can contain a unique install code, a random 128-bit number protected by a 16-bit cyclic redundancy check (CRC). The Trust Center may require that each new device use a unique install code to join a centralized security network and the install code must match a code previously entered into the Trust Center out-of-band (i.e., QR code). Once the install code is verified, the joining device and the Trust Center derive a unique 128-bit Trust Center Link Key from the install code using the Matyas-Meyer-Oseas (MMO) hash function
- Touchlink preconfigured link key.

Master key

Master key: forms the basis for long-term security between two devices and is used only by the APS. Its function is to keep the link key exchange between two nodes in the Symmetric-Key Key Establishment protocol (SKKE) confidential. A device acquires a master key via **key-transport** (key-load key is used to protect transported master keys), **pre-installation** or **user-entered data** such as PIN or password. Besides the above three symmetric keys, ZigBee uses another two keys:

Key-transport key: The *key-transport key* is a key used to secure the transmission of any key, other than the master key, from the trust center to the requesting device.

Key-load key: The *key-load key* is a key used to secure the transmission of the master keys.

Commercial Mode and Residential Mode

Each pair of devices can have set **both** Network and Link Keys. In this case, the Link key is always used (more security although more **memory**

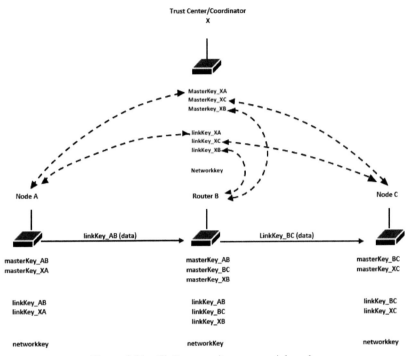

Figure 8.11 ZigBee security commercial mode.

is needed). There are two kinds of security policies which the Trust Center can follow:

- **Commercial mode, also high security mode**, is designed for high security commercial applications. In this mode, the Trust Center maintains a list of devices, master keys, link keys and network keys that it needs to control and enforce the policies of network key updates and network admittance. It also mandates the implementation of key establishment using SKKE and entity authentication (see Figure 8.11). This mode requires high memory resources. This mode offers a complete centralized model for the Key Security control.

- **Residential or Standard mode:** designed for residential applications. In this mode, Trust Center shares just the **Network Key** (it is the ideal mode when embedded devices have to cope with this task due to the low resources they have). It maintains a standard network key and controls policies of network admittance. In this mode, each device that joins the network securely shall either have a global link key or a unique link key

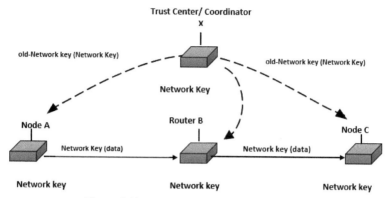

Figure 8.12 ZigBee security residential mode.

depending upon the application in use. It is required that the trust center have prior knowledge of the value of the link key and the type (global or unique) in order to securely join the device to the network.

  o A global link key has the advantage that the memory required by the Trust Center does not grow with the number of devices in the network.
  o A unique link key has the advantage of being unique for each device on the network and application communications can be secured from other devices on the network.

Both types of keys may be used on the network, but a device shall only have one type in use per device-key pair.

This is the mode normally chosen for the **Wireless Sensor Network** model (see Figure 8.12 about the *"ZigBee Residential mode"*).

The keys used by a centralized security model of ZigBee protocol are summarized in Figure 8.13.

8.2.5 Key Management

One characteristic of ZigBee security model is the fact that it has a variety of key management mechanisms:

Pre-installation: The manufacturer installs the key into the device itself. In the case when the device has more than one preinstalled key, the user can select one of the installed keys by using a series of jumpers in the device.

Key establishment: This is a local method of generating link keys based on the master key. Different security services of the ZigBee Network use a

Security Key	Description	
Network-level Security		
Network Key	o Essential key used to encrypt communications between all nodes of the network o Randomly generated by Trust Centre o Distributed to joining nodes, encrypted with a pre-configured link key	
Application –level Security		
Global link key (pre-configured)	o Used between the Trust Centre and all other nodes o Pre-configured in all nodes (unless a unique link key is pre-configured) o Also used in joining to encrypt network key transported from Trust Centre to joining node	
	o If ZigBee –defined, allows nodes from all manufacturers to join the network o If manufacturer-defined, allows only nodes from one manufacturer to join the network o Touchlink Pre-configured Link Key is a key of this type o Distributed Security Global Link Key is a key of this type	
Unique link key	Optional key used to encrypt communications between a pair of nodes- may be one of:	
	Pre-configured unique link key	o Used between the Trust Centre and one other node o Pre-configured in Trust Centre and relevant node o Also used in joining to encrypt network key transported from Trust Centre to joining node o Install Code-derived Pre-configured Link Key is a key of this type.
	Trust Centre Link Key (TCLK)	o Used between the Trust Centre and one other node o Randomly generated by the Trust Centre o Distributed to node encrypted with network key and pre-configured link key o Replaces pre-configured link key (if any) but application must retain the pre-configured key in case it needs to be reinstated
	Application link key	o Used between a pair of nodes, not including the Trust Centre o Randomly generated by the Trust Centre o Distributed to each node encrypted with network key and pre-configured link key (if any)

Figure 8.13 ZigBee security key summary for centralized model.

key derived from a one-way function (with link key as the input) to avoid security leaks due to unwanted interactions between the services. The use of uncorrelated keys ensures logical separation of the execution of different security protocols. This key-establishment is based on the SKKE (Symmetric Key Establishment) protocol. The devices involved in communication must be in possession of the master key, which may have been obtained through pre-installation or key transport or user-input.

Key transport: The network device makes a request to the Trust Center for a key to be sent to it. This method is valid for requesting any of the three types of key in commercial mode (High security mode), whereas, in residential mode (Standard mode), the Trust Center holds only the network key. The key-load key is used by the Trust Center to protect the transport of the master key.

Certificate-Based Key Establishment: In addition to the above, in the centralized model, keys can be distributed using Certificate-Based Key Establishment protocol (CBKE). CBKE provides a mechanism to negotiate symmetric keys with the Trust Center based on a certificate stored in both devices at manufacturing time and signed by a Certificate Authority (CA).

From the certificate, it is possible to generate a public key and other security elements. The CBKE method provides a mechanism to safely identify a device and to allow it to start communicating. A key establishment procedure involves the following four steps:

1. Exchange static data (certificate validation)and ephemeral data
2. Generate the key
3. Derive a Message Authentication Code(MAC) key and key data
4. Confirm the key using the MAC

For the second and third steps, the key establishment procedure refers to the Elliptic Curve Menezes-Qu-Vanstone (ECMQV) key agreement scheme and a key derivation function respectively. At the end of this process, the Trust Center and the authenticating device share a new link key that will be used to protect data communications between them.

8.2.5.1 Security key modification

In a centralized security model, the TC periodically creates, distributes, and switches the network key to limit the time that an attacker acquires a network key. The new network key is encrypted with the TC-generated Trust Center Link Key. When the new key first reaches the nodes, the transported key is automatically saved but not activated. A node can store more than one network key while identifying the current one with a unique 'key sequence number' assigned by the TC. Similarly, application link key can also be replaced with the new link key generated by the TC. There is also over-the-air (OTA) updates that allow a manufacturer to add new features, fix defects in the product, and apply security patches as new threats are identified. OTA updates create potential security vulnerability if the protocol does not provide enough protection or the device manufacturer does not use all the available protection. ZigBee provides multi-layered security to update devices and assure that updated code images are not malleable. It encrypts all image transfers OTA with a unique key, signs the OTA image with another unique key, then encrypts the image during manufacturing so that only the end product can decrypt it. The image might be stored in on-chip memory that is configured with the debug read-back feature disabled preventing reverse engineering with standard debugging tools, which is a common vulnerability of other solutions. Once the encrypted image is received, its secure bootloader decrypts the image, validates the signature, and updates the device. The bootloader also checks the validity of each image each time the device boots to prevent it from updating and return to using the previous known good image if the image is invalid (detecting image corruption quickly).

8.2.6 ZigBee Stack Security Measures

The ZigBee standard supports the following optional security services:

Encryption/decryption: ZigBee frames can be optionally protected with the security suite AES-CCM* to provide data confidentiality, data authentication and data integrity. AES-CCM* is a minor variation of AES (Advanced Encryption Standard) with a modified CCM mode (Counter with CBC-MAC).

Figure 8.14 below shows the role of AES-CCM* in **data authentication and confidentiality**. On the transmitter side, the plaintext in the form of 128-bit blocks of data enters the AES-CCM*. The responsibility of the AES-CCM* is to encrypt the data and generate an associated MIC, which is sent to the receiver along with the frame. The receiver uses the AES-CCM* to decrypt the data and generate its own MIC from the received frame to be compared with the received MIC (**data integrity**). A MIC provides stronger assurance of authenticity compared to the CRC. The MIC generated by the CCM* detects intentional and unauthorized modifications of the data as well as accidental errors.

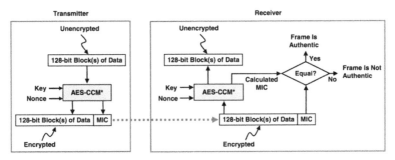

Figure 8.14   Role of AES-CCM* in data authentication and confidentiality.

The CCM* is referred to as a generic mode of operation that combines the data encryption, data authentication and, data integrity. The CCM* offers encryption-only and integrity-only capabilities as shown in Figure 2 above. The nonce used in the process is a 13-octet string constructed using the security control, the frame counter, and the source address fields of auxiliary header. The size of the MIC can be 32 bits, 64 bits, or 128 bits.

Replay Protection: Each node in the ZigBee network contains a 32-bit frame counter that is incremented at every packet transmission. Each node also tracks the previous 32-bit frame counter of each of device (node) that it is connected to. If a node receives a packet from a neighboring node with the same or lesser frame counter value than it had previously received, the packet is dropped. This mechanism enables replay protection by tracking packets and dropping them if they were already received by the node. The maximum value that a frame counter can be 0XFFFFFFFF, but if the maximum value is reached, no transmission can be made. The only time the frame counter is reset to 0 is when the network key is updated.

Device Authentication: The ZigBee standard supports both device authentication and data authentication. Device authentication is the act of confirming a new device that joins the network as authentic. The new device must be able to receive a network key and set proper attributes within a given time to be considered authenticated. Device authentication is performed by the trust center. The authentication procedure is different in residential and commercial modes.

In residential mode, if the new device that joins the network does not have a network key, the trust center sends the network key over an unprotected link, which causes a moment of vulnerability. If the new device already has the network key, it must wait to receive a dummy (all-zero) network key from the trust center as part of authentication procedure. The new device does not know the address of the trust center and uses the source address of this received message to set the trust center address. The joining device is then considered authenticated for residential mode.

In commercial mode, in contrast, the trust center never sends the network key to the new device over an unprotected link. But the master key may be sent unsecured?in commercial mode if the new device does not have a shared master key with the?trust center. After the new device receives the master key, the trust center and the new device start the key establishment protocol (SKKE). The new device has a limited time to establish a link key with the trust center. If the new device cannot complete the key establishment before the end of the timeout period, the new device must leave the network and retry the association and authentication procedure again. When the new link key is confirmed, the trust center will send the network key to the new device over a secured connection. The joining device is now considered authenticated for commercial mode.

Furthermore, ZigBee also supports device-unique authentication at joining such as Touchlink commissioning-which is an easy-to-use proximity mechanism for commissioning a device into a network. This method works by the Touchlink 'initiator' determining the proximity of the target device (to be commissioned) and negotiating/transferring network parameters.

Secure over-the-air (OTA) firmware upgrades: OTA updates allow a manufacturer to add new features, fix defects in its product, and apply security patches as new threats are identified. However, OTA updates also represent a potential security vulnerability if the protocol does not provide ample protections, or the device manufacturer does not use all available safeguards. ZigBee devices and associated silicon platforms provide multi-layered security to update devices in the field and assure that updated code images have not been modified maliciously:

- First, the ZigBee standard provides a method to encrypt all image transfers over the air with a unique key.
- Second, the standard provides a method to sign the OTA image with another unique key.
- Third, the image may be encrypted during manufacturing so that only the end product contains the key to decrypt it.
- Finally, the image may be stored in on-chip memory that is configured with the debug read-back feature disabled, preventing reverse engineering with standard debug tools, which is a common vulnerability of other solutions.

During an OTA upgrade, once a device receives an encrypted image, its secure bootloader decrypts the image, validates the signature, and then updates the device. Furthermore, the bootloader checks the validity of the active image each time the device boots. If the image is invalid, the bootloader prevents it from updating and returns to using the previous known good image. Thus, image corruption will be quickly detected and the system operator can take action.

Logical link-based encryption: Another key security tool is the ability to create an application-level secured link between a pair of devices in the network. This is managed by establishing a unique set of AES-128 encryption keys between a pair of devices. This allows logical, secured links between any two devices in the network, thus supporting "virtual private links" between a pair of devices in a network with many others. This measure limits the ability of an attacker that acquires the network key from intercepting or injecting messages that other devices would act upon.

Runtime key updates: periodically or as when required, the trust center takes the initiative to change the network key. The Trust center generates a new network key and distributes it throughout the network by encrypting it with the old network key. All devices continue to retain the old network key for a short period of time after the update until every device on the network has switched to the new network key. Also, the devices, on receiving the new network key initialize their frame counter to zero.

Network interference protection: in low-cost ZigBee nodes, using a band-select filter might not be even an option due to cost or node size limitations to protect the network from interference. However, basic properties of IEEE 802.15.4 and ZigBee network such as low RF transmission power, low duty cycle, and the CSMA/CA channel access mechanism help reduce the effect of the presence of a ZigBee wireless network on other nearby systems and vice-versa. There are two approaches to improving the coexistence performance of ZigBee networks: collaborative and non-collaborative.

In collaborative methods, certain operations of the ZigBee network and the other network (e.g., an IEEE 802.11b/g network) are managed together. Every time one network is active, the other network stays inactive to avoid packet collisions. In this method, there must be a communication link between the ZigBee network and the other network to implement and manage the collaboration.

The non-collaborative methods are the procedures any ZigBee network can follow to improve its coexistence performance without any knowledge regarding the operating mechanism of the nearby interfering wireless devices. This method is based on detecting and estimating interferences and avoiding them whenever possible. Some of the non-collaborative methods that can be used in ZigBee wireless networking include:

- Carrier Sense Multiple Access with Collision Avoidance (CSMA/CA).
- Signal spreading-spreading method such as DSSS allows the desired signal to have the advantage of processing gain over any interferer that resides in the same frequency band. Therefore, the signal spreading generally improves the robustness of a network against interferers.

- Dynamic RF output power control-adjust the RF output power of the transmitter based on the channel condition and the distance between the nodes. Reducing the transmitter output power decreases the interference with other nearby wireless devices, but the recipient of the signal becomes more susceptible to interference.
- Mesh networking and location-aware routing-If a certain router node in the network is constantly in the presence of strong interferers that cause frequent failure in packet delivery to the next hop, a mesh network may have the option of selecting an alternative path to carry the message to the final destination and avoid the router located close to a major source of interference. This is sometimes referred to as path diversity. In location-aware routing, the information regarding the areas of possible high interference, if known, can be taken into account in calculating link-cost functions. In this way, the packet traffic flow is directed away from the high-interference areas whenever possible. But the interferers still affect the transmissions initiated by, or intended for, the nodes within the high-interference areas.
- Frequency Channel Selection-Changing the frequency channel when the energy of the interferer signal in the desired channel is unacceptable can be a simple way of addressing the interference problem. ZigBee provides frequency agility capability that allows the entire network to change channels in the face of interference. If the frequencies of operations and bandwidths of the interfering signals in the nearby networks are known, the frequency channel of the ZigBee network can be selected accordingly to minimize the effect of interfering signals. This is referred to as channel alignment.
- Adaptive Packet Length Selection-is based on channel condition. Reducing the size of the packet is normally considered a way of improving the PER in presence of interferers. Generally speaking, a smaller packet has a better chance of receiving the destination before an interferer appears in the same frequency channel. However, some experiments have shown that reducing the packet length does not always result in better PER performance.

8.2.7 ZigBee Vulnerabilities

The security issues mentioned above, can achieved the confidentiality, integrity and, authentication but they cannot be considered as a comprehensive security suite that can stop different attacks and vulnerabilities. This section explores ZigBee vulnerabilities and some practical attacks that could be performed on a ZigBee network.

Vulnerabilities in a ZigBee network can be attributed to protocol issues (certain combination of protocol feature choices renders the network vulnerable) or poor implementation of the protocol by the developers. Some vulnerabilities include the following.

8.2.7.1 Implementation vulnerabilities

Insecure key storage: ZigBee protocol's security is based on the assumption that its keys are stored securely. Usually, the coordinator is preconfigured with the network key and, other devices such as routers and end devices are pre-configured with the link key. If the keys (network, link, or master) are insecurely stored; extraction of the keys from the device's firmware becomes easy. Hence, when a node is compromised this way, a hacker can extract the network key and compromise the whole network or extract the link key and obtain all the unicast communication of the that device as well.

Insecure key transportation: In standard security mode, any node that joins a ZigBee network obtains its network key over-the-air, sometimes in plaintext. As a result, the key can be obtained by merely eavesdropping on the network. As the network key is shared among all devices, compromise of the network key compromises the whole network.

Reusing Initialization Vector (IV) value with the same key is a security vulnerability inherited by ZigBee from 802.15.4. Reusing nonce values (Initialization Vector (IV)) with the same key introduces a vulnerability that enables an attacker to recover two plaintexts using their cipher-texts in the AES-CTR mode (recovering the two plaintexts is a simple operation; XOR operation on the two cipher-texts has been encrypted with the same keys and nonce values). This attack is known as the same-nonce attack. There are 2 occasions that allow the reuse of nonce with the same key:

- *Multiple Independent Access Control Entries*: 802.15.4 radio devices have an Access Control List (ACL) for configuring the security suite that is to be used. Since each chip has multiple independent ACL, the same IV or nonce could be reused with the same key.
- *Repopulating Access Control Table*: this happens when a ZigBee device is inadvertently powered off, resulting in the loss of ACL entries and the need for the entries to be repopulated. If the last nonce states are unknown after the power failure, the system might reset the nonce states

to a default value. This reset action increases?the chance of reusing the same nonce with a key that has been used before the power failure and could lead to the same-nonce attack.

Sending security headers in clear text: Auxiliary frames provide semantic security and replay protection. When an adversary crafts an invalid security header without knowing the key generating the MIC, although the integrity attack fails, the recipient device in fact expends certain amount of energy receiving and processing those bogus messages. If an adversary sends scads of such crafted messages to the victim device, a significant amount of energy will be dispensed by the device leading to its battery depletion. This attack is capable of reducing the battery life of a device from years to days [ghost attack or ghost-in-wireless].

Predictable sensor polling rates: ZigBee end devices remain in sleep mode for certain periods of time to conserve energy. They wake up at regular intervals to poll the coordinator for data inputs or wake up when the coordinator beacon so. However, knowing the polling rate will help an attacker to relay crafted messages to the end device at regular intervals even if it wasn't required, thereby forcing the end device to dispense excess energy for polling (energy is dispensed when the device navigates from sleep mode to active mode) and processing the received messages. (Vidgren et al. demonstrates how it is possible to discharge the batteries of a sensor if the attacker knows the adopted sensor polling rate.)

8.2.7.2 Protocol Vulnerabilities
Vulnerability has many reasons. It starts by the protocol itself followed by, as in Section 8.2.7.1, vulnerability in implementation and others. In this section, ZigBee protocol vulnerability is discussed.

Default link key values are used by manufacturers to provide interoperability for all ZigBee devices. For example, a case where a user has a new ZigBee device to be added to his/her network, but the new device has no specific authorization associated to it, the ZigBee network will allow a default link key to fall back to at the startup time. Therefore, an attacker can join the network by using an unknown type device and collect the data needed to temper the network's functionality. SEC Consult's research has also revealed that millions of IoT devices are directly accessible via the Internet, and attribute it to insecure default configuration.

Unauthenticated acknowledgment packets (ACK): The 802.15.4/ZigBee specification does not provide integrity and confidentiality protection for acknowledgment packets. Hence, an unauthenticated remote attacker can spoof acknowledgment packets to cause a remote node to believe an acknowledgment was received by the intended node. This attack (association flooding) is a direct consequence of man-in-the middle attack.

Man-in-the-middle attack instance: A sender device sends a packet to the receiver and waits for an ACK confirming that the receiver has received the packet in order to continue its operation. In case an ACK was not received, the sender will send the packet again. However, in case the network is compromised and the ACK packet is not authenticated, an attacker can intercept the packet and send fake ACK to the sender, thereby forcing the sender to send all subsequent packets to the attacker.

CSMA/CA trade-off: At the MAC layer, an attacker can flood a channel with frames, thereby forcing the network to deny any communication among devices. This is made possible because ZigBee uses CSMA/CA (if it is running in non-beacon mode) and devices that use CSMA/CA communication always back off if a channel is busy, thereby resulting in an inadvertent DDoS. Exposed and hidden node problems in a network can also be considered shortfalls of CSMA/CA mechanism.

Unencrypted keys: When a non-preconfigured device joins a network, a single network key (default link key) is sent unencrypted by the Trust Center (at least in residential mode). This one-time transmission of the unprotected key results in a short timeframe of exploitability. If the key could be sniffed by an attacker, the active network key stands compromised and consequently the whole ZigBee network communication is compromised.

Vidgren, et al. demonstrated a sniffing attack on a ZigBee network that sends the network key unencrypted by having a node (Atmel RZRAVENUSB) interact with a ZigBee device and collecting their wireless transmissions/interactions. The captured packets were later analyzed using open source tools (KillerBee) to obtain the network key.

Predictable PAN IDs and limited channels: In ZigBee protocol, a rejoining device joins a network when its saved parameters match the PAN ID and the channel that the network is operating on. The PAN ID is a 64-bit value, and can be brute-forced (It will take some processing time to find the right PAN ID, but eventually an attacker can find its value). Also, the channel numbers that ZigBee operates on is limited to 16 which makes it easy to find the correct network channel as well.

Insufficient replay protections: You might argue that ZigBee has replay attack protection mechanisms as part of its specification; however, 802.15.4 specification has limited replay protection mechanisms that even an encrypted message cannot complement. Hence, a crafty attacker can replay any previously observed traffic (ZigBee has the advantage of preventing the attacker from modifying the packets) until key rotation (provided 802.15.4 has no authentication mechanisms enabled). There are several other attacks that accompany replay attack:

- DDoS: A malicious user can receive packets at one point in the network and then replays these packets in other areas to interfere with the overall network functionality (wormhole attack).
- Temporal Key Integrity Protocol (TKIP): MIC attack in IEEE 802.11 networks is the one in which an attacker decodes the payload one byte at a time by using multiple replays and observing the response over the air on MIC failures.

Over the years, researchers have also succeeded in implementing the replay attack using the "Killer Bee" tool made by Joshua Wright and have clearly illustrated the severe problems it can cause. For example, medical devices such as Fitbit, pacemakers and insulin pumps which replay old messages can lead to erroneous outputs and dangerous events (loss of life).

Signal interference: Although ZigBee has certain network interference protection techniques established, they are however not comprehensive and slow to execute, allowing an attacker to take control of the frequency channel the network is operating in. This phenomenon is prominent in large networks, where transmissions are much more frequent and the signals can be detected quite efficiently.

Unauthorized network commissioning: The ZigBee device connects to the first network that is made available to it without any further interaction from the ZigBee device's user. Hence, it is possible to force a ZigBee device to join a fake network (even without the knowledge of the active secret keys expected by the ZigBee device if it were rejoining a network). An attacker could do this by sending a "reset to factory default" command to the device and wait on the device to look for a ZigBee network to connect to.

Lack of DDoS Protection Mechanisms: A denial-of-service (DoS) attack causes a node to reject all received messages. In a ZigBee network, DoS can be achieved by:

- Maximizing the frame counter: The attacker composes a message that includes random content (legit is fine too) as the encrypted payload (without knowing the security key), sets the frame counter to the maximum, and sends it to the node. By setting the frame counter to the maximum, any legitimate frame that arrives at the node after the attack will be automatically rejected by the recipient device because the frame counter of the received message will be less than the one recorded by the node at the time of the attack. This attack is made possible in situations where MIC of the packet is not verified.
- Flooding the network with messages (legit or spoofed packets) or frames. Krivtsova et al. proposed the broadcast storm attack that involved clogging the network by sending numerous broadcast packets.
- Altering routing tables to redirect all the traffic of the network to a bogus device (sinkhole attack) by purposely sending messages to construct artificial routing paths or introduce loops to the routing process of legitimate sensors. As a consequence, transmission of packets among devices is hindered, resulting in a DDoS.
- Using jamming techniques (signal jamming, reflexive jamming, changing the Power Spectral Density, etc.) to trick the user to initiate a factory reset or preventing the devices from communicating. This technique can also force a rejoin and reestablish the attack time-frame to sniff the network key.

Re-using link key: ZigBee allows link keys to be re-used for rejoining the network. Hence, it would make it possible for an attacker to copy a device's addressing credentials and spoof a network layer insecure rejoin using a separate device. This would result in the Trust Center passing the network key encrypted with the previously used link key to the cloned device. As a consequence, an attacker could gain complete access to the network key and hence the entire network.

TouchLink Factory reset: It is legitimate for a Touchlink initiator to send a factory-reset command to a target node already?in the network, allowing the target node to be removed from the current network and moved to another network. The process is aptly known as stealing. However, it provides a window of opportunity for an unauthorized person to move a device into another network of his/her for malicious intent.

Privacy issues: ZigBee protocol/network, similar to other wireless networks is susceptible to statistical attacks. Gleaning network traffic can reveal, with certain accuracy, the nature of the functionality implemented, thereby aiding the attacker to work around it to perform his or her malicious deed. However, a statistical attack on a ZigBee network can compromise ones Privacy as well. For instance, in home automation systems, an attacker can gather information to ascertain if tenants are at home or not. This information can aid an attacker to perform an informed robbery.

8.2.7.3 Security risks
The main security risks for a ZigBee wireless network are:

- Theft of sensitive data from a node–This may be user data that can be used for criminal purposes but is more likely to be network security data, such as encryption keys, that will allow access to the node and network. It is therefore important to carefully select and protect the security keys used by the network.
- Theft of a node–A node may be stolen in the sense that it is removed from the current network and moved to another network where it can be accessed and controlled. This is a particular risk when Touchlink commissioning has been used (described later), but there are ways to prevent it.
- Unauthorized control of a node–This may result from the above thefts or from replay attacks in which genuine over-the-air command frames are captured and re-sent to the node at a later time to achieve subversive control. The frame counters incorporated in ZigBee security help to defeat this kind of breach but other measures, such as regularly changing the network key, can also help.
- Loss of network service–This may result from the jamming of the radio channel or whole radio band by interference. Applications can employ frequency agility to overcome this kind of impediment by moving the network to a quieter channel.

8.2.7.4 Security key weaknesses
The least secure element of a ZigBee network is the network key. It is used by all nodes of the network and although it is securely passed to a joining node encrypted with a pre-configured link key, this link key is likely to be of the 'global' type and susceptible to exposure. Therefore, depending only on the network key leaves a network vulnerable. A ZigBee application profile (such as Home Automation) defines the level of security that is appropriate

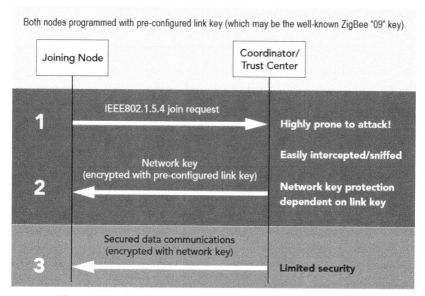

Figure 8.15   Security problems start with pre-configured link key.

for itself. For example, Home Automation uses the well-known ZigBee 09 global pre-configured link key to authenticate devices onto the network and distribute the network key. ZigBee 3.0 has no application profiles but defines a base device behavior (BDB) that adopts and adapts the security functionality of the Home Automation profile. It permits use of the ZigBee 09 key, as this allows backward compatibility with legacy Home Automation devices. However, the use of this global key will create a hole in network security (see Figure 8.15). To achieve a high level of security, ZigBee 3.0 allows the use of install codes (see Figure 8.16).

8.2.8 Measures for Maximizing ZigBee Security

This section introduces some of the measures that can be used to increase the security of the keys .The work done by NXP Laboratories UK is the base of the discussions.

8.2.8.1 Protecting the network key

The basic security provided by the ZigBee network is the encryption of data using the network key. This key is transported to the joining device during the authentication process. The network key is never sent over-the-air unencrypted. It is always encrypted with a pre-configured link key, but having

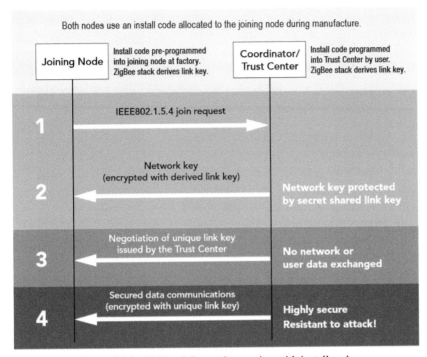

Figure 8.16   ZigBee 3.0 security starting with install code.

knowledge of the link key makes obtaining the network key possible by capturing over-the-air packets using an IEEE802.15.4 packet sniffer application. Exposing the network key could have a very large impact on the security of the network and its vulnerability to security attacks and hacks. Therefore, the choice and distribution of the pre-configured link key are critical. The ZigBee specification does not mandate how to distribute the link keys and it is left to each vendor to use a mechanism for securing the keys that is suitable for their products. To properly protect the network key over-the-air, the user has the choice of using an explicit 16-byte link key which is unique for each device or using install codes, described later. The use of a pre-configured link key to protect the network key is illustrated in Figure 8.15. This scheme is typical in a network based on a legacy ZigBee application profile, such as Home Automation, and may result in compromised security.

Install Codes

In ZigBee 3.0, an install code can be used to create the link key used to authenticate a node into a centralized security network and pass the network key to the node. A random install code is assigned to the node in the factory and programmed into the node. The ZigBee stack within the node derives a link key from the install code using a Matyas-Meyer-Oseas (MMO) hash function. The install code must also be communicated (by unspecified means) for the purpose of commissioning the node into a network. During commissioning, the install code is entered into the Trust Center and the ZigBee stack again derives the same link key from the install code. The Trust Center and node can subsequently use the link key in joining the node to the network. An install code is made up of 6, 8, 12 or 16 bytes with a 2-byte CRC appended to the end. Therefore, from the user's viewpoint, the install code consists of 8, 10, 14 or 18 bytes. It is recommended that install codes are 16bytes long (18bytes including the CRC). They should be randomly generated and not tied to any other credentials of the node, such as its IEEE/MAC address. Protecting the install code is vital and it should never be put on the outside of the product packaging or directly on the product. ZigBee 3.0 also allows a unique link key to be negotiated between the Trust Center and node in order to enhance the security of communications between them. The use of an install code and derived link key to protect the network key is illustrated in Figure 8.16, which also includes the negotiation of a unique link key. This scheme is available for networks that employ the ZigBee 3.0 standard and will result in very high level security.

8.2.8.2 Rolling the network key

Once a device has left a network, it is possible to obtain the network credentials from the device, including the network key and any link keys that it had for peers in the network. To prevent this from compromising the network, it is recommended that when a device leaves the network, the Trust Center rolls the network key to a new key, as follows:

1. The Trust Center broadcasts a new network key (with a new key sequence number) to all the nodes of the network.
2. The Trust Center then broadcasts a Switch Key command (with the new key sequence number) to all the nodes of the network.
3. Each device in the network switches to the new network key. Any devices that do not receive the new network key will try to find the network again and will receive the new network key encrypted with the last negotiated Trust Center link key.

It is possible to prevent replay attacks on devices on the network by simply rolling the network key at regular intervals. However, this would be a huge overhead if done every time a device joins the network and it is more feasible to do it when network commissioning is opened and/or when the network is closed for any further joins. During normal network operation, rolling the key every fourth night would be a reasonable time frame.

8.2.8.3 Preventing spoof leave notifications key

It is possible to repeatedly issue a spoof Leave request from a non-network device for a device in the network. This would result in the Trust Center and/or the parent of the victim device losing all network information for the device, including its keys. This renders the device stranded, outside the network. This situation can be prevented by the Trust Center or parent not clearing network data for the device on receiving a Leave request that supposedly originates from the device, but only when a Leave request has been issued by the parent of the device. If a device does attempt a (genuine) self-leave then it is prudent for the removal of the network credentials to require manual interaction by an authorized person. In order to avoid the unnecessary or accidental removal of a device from the network:

- The management leave server should be disabled, since the Management Leave command is only network key encrypted.
- Leaves should be disabled on routing devices so that a network layer Leave command is ignored, since this command is only network key encrypted.

To cause a device to leave the network, it is safest to allow this action only through Remove commands issued by the Trust Center, as this requires link key encryption.

8.2.8.4 Preventing rejoin security attacks

When a device joins the network, a pre-configured link key is used to join the network and decrypt the network key received from the Trust Center. If the same link key is used for every join attempt, it opens up the system to rejoin security attacks. Therefore, the same link key should not be used for any subsequent rejoins. In more detail, if the same link key is allowed to be used for rejoins, it is possible to copy a device's addressing credentials and spoof a network layer insecure rejoin using a separate device. This would result in the Trust Center passing the network key encrypted with the previously used link key to the cloned device. If the network key finds its way into the wrong hands, it compromises the entire network.

To prevent this type of attack, the ZigBee 3.0 mechanism of Trust Center link key negotiation should be used. In this scheme, once the device has joined the network, it is issued with a new Trust Center link key which is randomly generated by the Trust Center and replaces the original link key. Thus, the new link key (instead of the original link key) is subsequently used for transporting the network key during future rejoins. It is also recommended that rejoins are disabled by setting the Trust Center policy to disallow automatic rejoins with the pre-configured link key. Then if a device needs to rejoin the network, it will perform a fresh join with the pre-configured link key but, in this case, it is advised that the applications implement a mechanism for manual intervention at the Trust Center by the user in order to authorize the join.

8.2.8.5 Persisting essential data when using cloud services

The storage of network parameters in a remote location or in the Cloud provides huge benefits in terms of security, memory management and remote access. The weakness of this approach is that access to the network parameters is lost when the connection to the Cloud or remote location goes down. It is recommended that only non-essential data are stored away from the network nodes. Essential data should be persisted on the nodes themselves, so that they continue to operate through the loss of backhaul links or network issues. Essential data include the following:

• Network key and link key
• PAN and channel information
• Authentication and commissioning data
• Factory-default settings

8.2.8.6 Key establishment cluster and security certificates

The ZigBee Smart Energy profile includes a Key Establishment cluster which provides the functionality to authenticate devices using very strong Elliptic Curve Cryptography (ECC) based security. This security scheme generates and employs unique link keys for communications between pairs of nodes. It does, however, require digital security certificates for the joining nodes and Trust Center, and is reliant on the availability of these certificates. There are a number of certification authorities that can issue these certificates and this brings its own complications. Certificates issued by different authorities are likely to be different and a device may need to hold certificates and root keys from multiple certification authorities. This has huge memory implications for the devices, which have small memory footprints. The only practical ways of managing the use of security certificates on devices with very limited memory are to either use only devices that have certificates issued from the same authority or rely on installing the certificates into the devices using a custom validation and certification method.

8.2.9 ZigBee Security–Conclusion

[NXP Laboratories UK member of ZigBee came in their document number: MAXSECZBNETART REV 0 with the next conclusions]

The ZigBee protocol defines a highly secure operating environment but does not mandate the level of security that a product needs to provide. This is a choice for the product manufacturer, depending on how sensitive the handled data are, as well as how easily and securely the product should interoperate with products from other manufacturers.

- If data security is important then care must be taken to avoid eavesdropping as well as rejoin and replay attacks, and it is absolutely critical to protect the network key that is used to secure all network communications.
- If backwards compatibility and ease-of-use are requirements then there is a possible vulnerability introduced by using well-known keys to securely distribute the network key. The use of install codes and Trust Center link keys is highly advisable to ensure that the network key is secured when passing it to joining nodes.
- If the intention is to produce highly-secured network applications, as a minimum they should be designed to employ link key security and not just network key security. In addition, use of the Key Establishment cluster should be given due consideration, with its support of ECC-based key establishment, digital certificate management and choice of certification authority.
- If Cloud services are used by low-power wireless network products, these devices will also be vulnerable if the connection to the Internet and Cloud is not properly secured. So the end-to-end security of the system should also be considered.
- The ZigBee network should also not be fully reliant on such backhaul connections; otherwise, the loss of a connection could render the ZigBee network unusable. There should be some autonomy for network recovery and maintenance in the event of the loss of backhaul links.
- The wireless medium can be contaminated with interference and it is possible to cause sufficient interference for two devices not to be able to communicate effectively. It may be possible to overcome such interferers by building frequency agility into the network and it is strongly

recommended that this is considered for the final implementation of the system.

• Outside the scope of the ZigBee protocol itself, the encryption of program memories and firmware as well as read/write protection on devices should be considered. Erasing the security credentials of any device that is no longer in use is also advisable.

The level of security that is realistically required depends on the nature of the product but manufacturers should not overlook security for the sake of cost. The security measures implemented in a product should form part of the product specifications and features list in the advertising, marketing, documentation and packaging.

8.3 Applications of ZigBee Technology

The applications of this technology include the following:

1. Collection of Medical Data

This technology is used in home patient monitoring where collection of medical data is crucial. In this system, a patient wears a ZigBee device which collects the information like the pulse rate, temperature of the body, blood pressure etc.

2. Intelligent Smoke Alarms

This technology is used in intelligent smoke alarms. This system consists of a smoke detection module, a wireless communication module, data collection module and an intelligent identification module which differentiates the types of smokes, that is, smoke arising from cooking, from accidental fire, foggy air etc. thus improving the safety of the user.

3. Home Automation System

This technology is used in home automation system leading to create smarter homes that has enhanced the comfort, convenience, security of the users. Smart Air conditioners, smart TV's, smart refrigerator, smart home alarms can be made inter operable using this technology.

4. Wireless Sensor Networks

An intelligent smart wireless sensor network system helps to gather information and to monitor air quality, weather conditions, traffic etc.

Figure 8.17 Applications of ZigBee technology.

8.4 Advantages of ZigBee Technology

8.4.1 ZigBee Technology Advantages

- ZigBee is a popular technology for smart home, because it offers some distinct advantages over other wireless technologies:
- For starters, ZigBee communication is two-way: devices confirm commands given to them. This makes ZigBee devices very reliable. It makes sure devices never miss a command, and allows your sensors to dub as a smart security system.
- ZigBee is also relatively standardized, especially in lighting. Carried by big brands such as Philips Hue and IKEA, ZigBee has a universal basis for lighting control. That is why you can pair your lights from IKEA to a ZigBee hub from a different brand like Hue.
- Another advantage is the fact that ZigBee uses the same wireless band as Wi-Fi and Bluetooth do: 2.4 GHz. This means that devices can be certified relatively simple and can immediately be sold worldwide. Z-Wave devices on the other hand need different frequencies, certifications and requirements for different regions.
- The technology also requires very little power. It allows very small devices like a Xiaomi Door Sensor to run on a single battery for years.

It's also fairly mainstream and affordable–ZigBee devices usually cost less than their counterparts on Z-Wave.

- ZigBee is built as a mesh network. Mesh networks allow all the devices to talk to each other and repeat commands. In practice, it means range and reliability of the whole network *improve* with every ZigBee device you add.
- It is more stable and reliable.
- It is less complex than Bluetooth.
- Installation is easy and consumes less power.
- Its protocol patent is free and hence cost effective.
- Easier to control and monitor home devices with a touch of a button.

8.4.2 Disadvantages

- Every technology has its trade-offs. For ZigBee, this is mainly range. This issue is solved partly because ZigBee uses mesh networking, as described above. When you do not have a mesh, however, the point-to-point range of ZigBee is fairly limited, especially when compared to technologies like Z-Wave or 433 MHz. As ZigBee works on the 2.4 GHz band and is energy-efficient, the direct range is often limited to 15-20 meter in open air.
- Another thing in ZigBee is the quality and compatibility difference between devices. Where ZigBee light bulbs and plugs are pretty evolved, some other products are less mature. An example is the 5 button dimmer from IKEA, which cannot be paired properly to a ZigBee hub like Homey.
- You cannot connect ZigBee devices directly to your phone. Your phone does not feature a ZigBee antenna. So you always need a smart home hub to connect with ZigBee devices. Luckily, Homey includes ZigBee, so if you have Homey you have nothing to worry about.
- It operates on short range, that is, 10–100 m line of sight.
- Low transmission rate.
- It has certain limitations in the area of memory size, processing speed of data.
- Less secure compared to Wi-Fi based security systems.

8.4.3 Conclusion

To sum up, ZigBee is a powerful, energy-efficient and affordable smart home technology. It is widely used for lighting by big brands like IKEA and Philips

Hue. Next to lighting, ZigBee is found in sensors, plugs, roller blinds and many more attractive products from a variety of brands.

ZigBee has less range than Z-Wave, but this is easily offset by building a good ZigBee mesh network in your home which will relay signals. And because ZigBee devices are cheaper, you can automate your home for less. More and more brands and devices are adopting ZigBee, but not everything is on ZigBee yet. That is why we recommend to combine ZigBee with another technology like Z-Wave if you want to fully automate your home.

8.5 Future Scope Of ZigBee

{Review on Selecting Topologies in ZigBee Networks, International Journal of Science, Engineering and Technology Research (IJSETR) Volume 5, Issue 6, June 2016}

1. Various types of fields, that is, national security, defense, monitoring, and control, are provided by devices depending on ZigBee standard. To achieve the highest throughput from a ZigBee network, the nodes position required to be considered particularly. A significant role has been played by the nodes movements in a ZigBee network from the point of view of the throughput and network efficiency.

2. In a ZigBee network, a wireless sensor network is often built by utilizing a tree topology. These wireless sensor networks are utilized for applications that need the data delivery. Utilizing the tree topology for the nodes position in the network and by configuring the nodes movement according to the group mobility model, network provides the highest throughput and possess least data dropped in the network.

3. This is an important obstacle to ZigBee which means that most of the wireless devices should have a level of automated intelligence introduced in them so as to enable easy commissioning and reliable usage. Approaching this challenge will need advanced planning to automate issues faced by the people who is actually responsible for installing the ZigBee application.

4. In the WSNs, the coordinator is responsible for forwarding the signal to all the sensor nodes linked to it. As the distance from the coordinator increases, the signal strength becomes weak causing interaction with destination nodes complex as a result the link breaks with all the broadly distributed wireless sensor nodes. Thus, it is complicated to perform reliable and stable communication with broad range nodes.

5. Chipcon is utilizing ZigBee to create a road map product that decreases the system and chip costs and increases integration level with low power consumption. Sensors are currently being utilized in agricultural and environmental applications, but the main target–home automation. ZigBee technology is also being utilized and examined in applications related to health monitoring.

6. A Wireless sensor network (WSN) contains spatially distributed wireless nodes at the bottom most layer. These sensor nodes are often battery powered devices and consume very low power. Data are forwarded from these devices to the gateway by the support of the subsystem on the top of devices. These sensor nodes can behave both as a full functional device and decreased functional device. But one of the major issue in WSN is the hot spot such as all the FFD's linked to the coordinator are responsible for forwarding the data to the coordinator on the behalf of nodes linked to them. So these nodes use more power which increases the load on them and die out much sooner in comparison of other nodes which breaks the link with the whole network. For overcoming the hot-spot issue, utilizing the mobile coordinator is the best solution.

7. Control overhead is a very significant indicator for evaluating Zig-Bee routing protocol performance. More control overhead will append network energy consumption and decrease network survival time. 8) Energy consumption is also a ZigBee challenge. ZigBee was particularly planned for home applications and have some characteristics, that is, high latency because of the low power consumption, long battery life, low cost etc. The configuration is the important issue and it should be performed very carefully. Routing method selection and type of topology plays a significant role because ZigBee objectives to obtain greater efficiency. A combined routing technique of AODV and Tree Routing is also planned in the ZigBee specification. But there is no mechanism designed to balance these two Routing modes for achieving better performance. So selecting the best mechanism is also a ZigBee challenge.

CONCLUSION ZigBee will play a significant part in the future in the fields, that is, smart lighting, home automation, war fields and smoke and intruder warning traffic management and will build computer and communication technology more accessible and usable to the subscribers. These networks are easy to deploy which is cheaper in comparison of other technologies. It would cost only $5 for a single chip. But the smaller memory size of protocol stack will further lower. ZigBee price to around $2 per chip. ZigBee technology is very helpful from security point of view as the devices manage a list of trusted devices inside the network and frame integrity to secure data from being altered by parties without cryptographic keys. Because of its emergence, researchers are facing several challenges in the deployment and development of the technology but because of the intensive research, all the issues are being sorted out day by day and the technology is becoming less vulnerable to the issues and also becoming more sustainable and reliable. The wireless communication technologies are frequently spreading to several new fields, involving the wireless sensors and the significance of the usage of wireless technologies in building control, data acquisition, monitoring systems and automation of manufacturing procedures will develop in future So, ZigBee has a very promising future in front of it.

References

[1] ZigBee Alliance, ZigBee Specification Document 053474r17, 2008
[2] Daintree Network, "Comparing ZigBee Specification Versions," www. daintree.net/resources/spec-matrix.php
[3] "How Does ZigBee Compare with Other Wireless Standards?" www.st g.com/wireless/ZigBee-comp.html
[4] Craig, William C. " ZigBee: Wireless Control That Simply Works," ZigBee Alliance, 2003
[5] Ember Corporation, EmberZNet Application Developer's Reference Manual, 2008
[6] IEEE Std 802.15.4 2003
[7] IEEE Std 802.15.4 2006
[8] Harsh Dhaka, –ZigBee Technology: Wireless Control that Simply Works?, Communications Design Conference, 2 October 2003
[9] Jorjeta G. Jetcheva, Yih-Chun Hu, Amit Kumar Saha, and David B. Johnson. "Design and Evaluation of a Metropolitan Area Multitier Wireless Ad Hoc Network Architecture in Wireless Sensor Networks." In Proceedings of the Fifth IEEE Workshop on Mobile Computing Systems & Applications, Monterey, CA, pp. 32–37, 2004
[10] ZigBee Alliance. ZigBee Overview. September 2003
[11] Aaron Alva, DMCA security research exemption for consumer devices, October 28, 2016. https://www.ftc.gov/news-events/blogs/techftc/2016 /10/dmca-security-research-exemption-consumer-devices.

[12] Wireless Mesh Networking ZigBee vs. DigiMesh. https://www.digi.com.

[13] About ZigBee Protocol. https://sites.google.com/site/xbeetutorial/xbee-introduction/zigbee.

[14] Inc. ZigBee Alliance, ZigBee Specification053474r20 (September 7, 2012).

[15] ZigBee Specification053474r06 (December 4, 2004).

[16] ZigBee 3.0 Task Force, ZigBee: Securing the Wireless IoT(2017).

[17] Luke Tutty, Guide to ZigBee 3.0, March 21, 2017. https://mmbnetworks.atlassian.net/wiki/display/ITZ/Guide+to+Zigbee+3.0.

[18] Ricky A. Melgares, ZigBee Analysis and Security (May 31, 2011).

[19] Bjorn Stelte and Gabi Dreo Rodosek, Thwarting Attacks on ZigBee–Removal of the KillerBee Stinger.

[20] Jeff John Roberts, Light Bulbs Flash "SOS" in Scary Internet of Things Attack, November 03, 2016. http://fortune.com/2016/11/03/light-bulb-hacking/.

[21] Darren Pauli, IoT worm can hack Philips Hue light bulbs, spread across cities, November 10, 2016. https://www.theregister.co.uk/2016/11/10/iotwormcanhackphilipshuelightbulbsspreadacrosscities/.

[22] Internet of Things Map. https://p16.praetorian.com/iotmap.

[23] Inc. ZigBee Alliance, ZigBee 3.0 Stack User GuideJN-UG-3113 (October 5, 2016).

[24] ZigBee 3.0 Devices User GuideJN-UG-3114 (December 1, 2016).

[25] Tobias Zillner, ZigBee Exploited: the good, the bad, and the ugly (August 16 2015).

[26] Joshua Wright, KillerBee API Documentation, 2009.

[27] Michael Ashbridge, Eddystone-EID, 2016. https://github.com/google/eddystone/blob/master/eddystone-eid/tools/eidtools.py

Section IV

Bluetooth

9

Bluetooth

9.1 Introduction: Overview

Bluetooth is a short range wireless communication technology designed to replace cables connecting portable unit and maintaining high levels of security. Bluetooth products communicate over the radio (Figure 9.1) and therefore no cables or optical connection is needed (which is not the case for IrDA). It started within Ericsson concern as means to replace the cables to peripheral devices for Ericsson's mobile telephones. It has since then grown to a short-range wireless ad hoc networking solution.

The key features of Bluetooth are low cost, robustness and low power consumption. Bluetooth operates in the universally unlicensed Industrial, Scientific and Medical (ISM) band at 2.4 GHz. To avoid possible interference from other devices using the same band, Bluetooth uses a frequency hopping scheme, where devices alternate rapidly among the 79 available frequencies in a pseudo-random fashion, which changes the transmission frequency 1600 times per second in a pseudo random way.

The range of the Bluetooth radio is approximately 10 m with a standard Bluetooth device. The range can be extended to 100 m by increasing the output power. Maximum communication speed is around 700 kbps. Bluetooth technology is based on **Ad hoc technology**. When two or more Bluetooth devices form an ad hoc network it is called a piconet. Several piconets can be linked together and the resulting network is called a scatternet, which is a local area network with a very limited coverage (see Section 9.1.2).

Bluetooth wireless technology is currently one of the most common means of communication for embedded systems. Bluetooth is described under IEEE 802.15.1 standard.

All products using Bluetooth wireless technology must follow the communication protocols documented in the Bluetooth specification. The advantage of this is interoperability; all products from different manufacturers function well together.

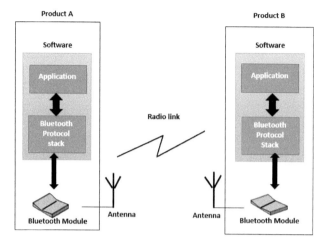

Figure 9.1 Two products communicating over a wireless Bluetooth radio link.

Figure 9.2 Bluetooth cards.

The Bluetooth specification is a highly technical document (over 1500 pages) describing all the functionality that must be implemented in order to build a product that complies with the Bluetooth specification.

In order to communicate with another device using Bluetooth technology, a special Bluetooth card is needed (Figure 9.2). It is important to note that the regular Wi-Fi card that normally used on laptop or smartphone is for 802.11 technology, and it is not compatible with Bluetooth based on the 802.15 standard.

9.1.1 History of Bluetooth

WLAN technology enables device connectivity to infrastructure based services through a wireless carrier provider. The need for personal devices to communicate wirelessly with one another without an established infrastructure has led to the emergence of **Personal Area Networks (PANs)**.

- Ericsson's Bluetooth project in 1994 defines the standard for PANs to enable communication between mobile phones using low power and low cost radio interfaces.
- In May 1988, Companies such as IBM, Intel, Nokia and Toshiba joined Ericsson to form the Bluetooth Special Interest Group (SIG) whose aim was to develop a defacto standard for PANs.
- IEEE has approved a Bluetooth based standard named IEEE 802.15.1 for Wireless Personal Area Networks (WPANs). IEEE standard covers MAC and Physical layer applications.

Bluetooth specification details the entire protocol stack. Bluetooth employs Radio Frequency (RF) for communication. It makes use of **frequency modulation** to generate radio waves in the **ISM** band.

The usage of Bluetooth has widely increased for its special features.

- Bluetooth offers a uniform structure for a wide range of devices to connect and communicate with each other.
- Bluetooth technology has achieved global acceptance such that any Bluetooth enabled device, almost everywhere in the world, can be connected with Bluetooth enabled devices.
- Low power consumption of Bluetooth technology, an offered range of up to 10 m, has paved the way for several usage models.
- Bluetooth offers interactive conference by establishing an ad hoc network of laptops.
- Bluetooth usage model includes cordless computer, intercom, cordless phone and mobile phones.

9.1.2 Piconets and Scatternets

In order to communicate, Bluetooth devices organize themselves into small networks called *piconets*, comprising one *master* and up to 7 *slave* devices, in which the frequency hopping sequences are synchronized and controlled by the master. The device that initializes establishment of the Piconet becomes the **master**. In general the devices forming piconett has the ability to act either as a master or slave. The specification allows a mechanism for **master** and **slave** to switch their roles. Point to point configuration with one master and one slave is the simplest configuration.

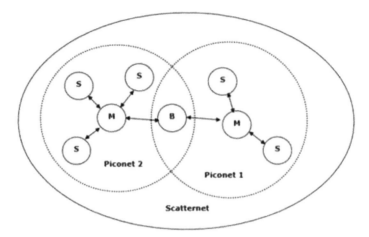

Figure 9.3 Piconets and scatternets.

The master is responsible for transmission control by dividing the network into a series of time slots amongst the network members, as a part of **time division multiplexing** scheme which is shown below.

The features of Piconets are as follows:

- Within a Piconet, the timing of various devices and the frequency hopping sequence of individual devices is determined by the clock and unique **48-bit address** of master.
- Each device can communicate simultaneously with up to seven other devices within a single Piconet.
- Each device can communicate with several piconets simultaneously.
- Piconets are established dynamically and automatically as Bluetooth enabled devices enter and leave piconets (Bluetooth is a mobile ad hoc network).
- There is no direct connection between the slaves and all the connections are essentially master-to-slave or slave-to-master.
- Slaves are allowed to transmit once these have been polled by the master.
- Transmission starts in the slave-to-master time slot immediately following a polling packet from the master.
- A device can be a member of two or more piconets, jumping from one piconet to another by adjusting the transmission regime-timing and frequency hopping sequence dictated by the master device of the second piconet.
- It can be a slave in one piconet and master in another. It however cannot be a master in more than once piconet.

- Devices resident in adjacent piconets provide a bridge to support inner-piconet connections, allowing assemblies of linked piconets to form a physically extensible communication infrastructure known as **Scatternet**.

9.1.3 Bluetooth Spectrum

Bluetooth technology operates in the unlicensed industrial, scientific and medical (ISM) band at 2.4 to 2.485 GHZ, using a spread spectrum hopping, full-duplex signal at a nominal rate of 1600 hops/s. The 2.4 GHZ ISM band is available and unlicensed in most countries.

The frequency range from 2.4 GHz to 2.485 GHz is the same as that of WLAN; therefore, using the Bluetooth communication would interfere with WLAN networks, if both of them are used in the same area.

9.1.4 Bluetooth Frequency and Connectivity Ranges

Bluetooth enables, low power communication between devices that are in close proximity of each other. It operates in the unlicensed Industrial, Scientific, and Medical (ISM) radio frequency (RF) 2.4 GHz spectrum and has a range from between 0.5 m and 1 m to 100 m.

The technology has three classes of devices, which offer three different connectivity ranges. Class 1 devices offer a range of 100 m and transmit at 100 mW. Class 2 devices, the most common devices, have a range of 10 m and transmit at 2.5 mW. Class 3 devices have a range of about 1 m and transmit at 1 mW.

One main advantage of the technology is its ability to transmit both voice and data simultaneously; however, the 2.4 GHz radio frequency spectrum is shared with many consumer appliances (i.e., microwaves, baby monitors, and cordless phones), which could cause interference. According to research, coexistence in wireless technologies resulted in Bluetooth signals being harmed significantly by other wireless technologies. For this reason, Bluetooth technology uses hops in the Bluetooth frequency at 1600 hops per second and a technology called spread spectrum to avoid interference.

9.1.5 Data Rate and Modulation

Bluetooth 1.1 and, the later improved version of Bluetooth, Bluetooth 1.2 are known as Basic Rate (BR). They allow transmission speeds up to 1 Mbps. Bluetooth version 2.0, known as Enhanced Data Rate (EDR), allows transmission speeds up to 3 Mbps. Bluetooth version 3.0, known as High Speed (Hs), provides data rates up to 24 Mbps. Bluetooth version 4.0, known as Low Energy (BLE), is simple and more efficient, offers 1 Mbps, and achieves lower power consumption for use in medical devices. Bluetooth 4.1 and 4.2 are the most updated versions. The technology, developed for coin cell battery-powered devices (e.g., medical devices), provides both Low Energy and BR/EDR/HS. Bluetooth 4.1 offers speeds in range of 1–3 Mbps, while Bluetooth 4.2 offers 1 Mbps using Gaussian frequency shift keying (GFSK). GFSK is a method of modulation used for digital communication and is one method used in Bluetooth technology. It is able to send information by shifting carrier frequencies by Gaussian filtering square-wave signals and by increasing or decreasing the duration of carrier frequencies based on symbols (i.e., decreasing the frequency for 0 and increasing the frequency for 1).

Bluetooth 5 has been announced as the next version of Bluetooth. It will also be known for Low Energy mode.

9.2 Bluetooth Stack Architecture

A Bluetooth protocol stack is a software library that implements the communication protocol as defined in the Bluetooth specification.

Bluetooth stacks can be roughly divided into two distinct categories:

1. *General-purpose* implementations that are written with emphasis on feature-richness and flexibility, usually for desktop computers. Support for additional Bluetooth profiles can typically be added through drivers.
2. *Embedded system* implementations intended for use in devices where resources are limited and demands are lower, such as Bluetooth peripheral devices.

Some of the General-purpose implementations are as follows:

FreeBSD: The FreeBSD Bluetooth stack is implemented using the Netgraph framework. A broad variety of Bluetooth USB dongles are supported by the ng_ubt driver.

NetBSD: NetBSD has its own Bluetooth implementation, committed in 2006, and first released with NetBSD § 4.0.

OpenBSD: OpenBSD has had the implementation from NetBSD for some time, but it was removed in 2014 due lack of maintainership and code rot.

DragonFly BSD: DragonFly BSD has had NetBSD's Bluetooth implementation since 1.11 (2008), first released with DragonFly BSD § 1.12

Linux: The Linux operating system currently has BlueZ included with the official Linux kernel distributions, initially developed by Qualcomm.

BlueZ: BlueZ is a Bluetooth stack for Linux kernel-based family of operating systems. Its goal is to program an implementation of the Bluetooth wireless standards specifications for Linux. As of 2006, the BlueZ stack supports all core Bluetooth protocols and layers. It was initially developed by Qualcomm, and is available for Linux kernel versions 2.4.6 and up. In addition to the basic stack, the bluez-utils and bluez-firmware packages contain low level utilities such as dfutool which can interrogate the Bluetooth adapter chipset to determine whether its firmware can be upgraded.

BlueZ is licensed under the GNU General Public License (GPL), but reported to be on its way toward switching to the GNU Lesser General Public License (LGPL).

Some of Embedded System Implementations are as follows:

Apache Mynewt NimBLE: Apache Mynewt NimBLE is a full-featured, open source Bluetooth Low Energy 4.2 and 5.0 protocol stack written in C for embedded systems. NimBLE is one of the most complete protocol stacks, supporting 5.0 features including high data rate and extended advertising. The implementation supports all layers of the Bluetooth protocol. The first ports for the Controller part are to nRF51 series and nRF52 SoCs from Nordic Semiconductor. NimBLE also supports standard HCI interfaces to work with controllers, including ST, Dialog and Em Micro chipsets. It leverages the open source Apache Mynewt OS which is designed to support multiple microcontroller architectures. NimBLE can also run with FreeRTOS and is portable to other real-time operating systems. The implementation allows for the Mynewt NimBLE Controller part to be used with a non-Mynewt NimBLE Host.

BlueCode+: BlueCode+ is the portable higher layer Bluetooth protocol stack from Stollmann E+V GmbH. BlueCode+ 4.0 is qualified to Bluetooth version 3.0. The protocol stack is chipset and operating system independent and supports any Bluetooth HCI chips available. The APIs offer control of the profiles and stack functions, as well as direct access to lower level functions. BlueCode+ 4.0 supports the protocols L2CAP, eL2CAP, RFCOMM, SDP Server and Client, MCAP, HCI-Host Side and AVDTP. Supported profiles are Generic Access (GAP), Service Discovery Application (SDAP), Serial Port Profile (SPP), Health Device Profile (HDP), Device Identification Profile (DID), Dial-up Networking (DUN), Fax, Headset (HSP), Handsfree (HFP),

SIM Access (SAP), Phone Book Access (PBAP), Advanced Audio Distribution Profile (A2DP), Audio/Video Remote Control (AVRCP) and OBEX. The stack has been ported to a wide range of different microcontrollers and operating systems.

BlueCore Host Software (BCHS) a.k.a. CSR Synergy: CSR's BCHS or BlueCore Host Software (now called CSR Synergy) provides the upper layers of the Bluetooth protocol stack (above HCI, or optionally RFCOMM)–plus a large library of Profiles–providing a complete system software solution for embedded BlueCore applications. Current qualified Profiles available with BCHS: A2DP, AVRCP, PBAP, BIP, BPP, CTP, DUN, FAX, FM API, FTP GAP, GAVDP, GOEP, HCRP, Headset, HF1.5, HID, ICP, JSR82, LAP Message Access Profile, OPP, PAN, SAP, SDAP, SPP, SYNC, SYNC ML.

BlueLet: It is also a product of IVT Corporation. DUN, FAX, HFP, HSP, LAP, OBEX, OPP, PAN and SPP are currently supported.

BlueMagic: BlueMagic 3.0 is Qualcomm's (formerly Open Interface North America's) highly portable embedded Bluetooth protocol stack which powers Apple's iPhone and Qualcomm-powered devices such as the Motorola RAZR. BlueMagic also ships in products by Logitech, Samsung, LG, Sharp, Sagem, and more. BlueMagic 3.0 was the first fully certified (all protocols and profiles) Bluetooth protocol stack at the 1.1 level.

Zephyr Project: The Zephyr Project RTOS includes a complete, open source Bluetooth Low Energy 5.0 protocol stack written in C for embedded systems. It contains both a BLE Controller and a BLE and BR/EDR capable Host running on nRF51 Series and nRF52 SoCs from Nordic Semiconductor.

9.2.1 Relation between Classic Bluetooth, OSI model, and 802.11 Wi-Fi

Figure 9.4 demonstrates the relationship between the Open Systems Interconnection (OSI) model, selected 802.11 (Wi-Fi) layers and the components of the classic Bluetooth v1.1 protocol stack. The diagram shows that parts of the Bluetooth stack can be mapped to the Wi-Fi stack, primarily in the areas encompassing the Data Link and Physical OSI layers. Although the diagram does not include the Bluetooth LE stack for the purposes of simplicity, the same mapping principles would still apply. The depicted relationship is significant because it implies that techniques that have been evaluated as effective in the context of other protocols such as Wi-Fi may also be applicable to Bluetooth.

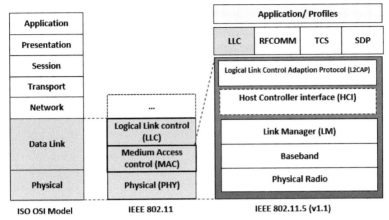

Figure 9.4 Mapping classic Bluetooth (IEEE 802.11.5 v1.1) and IEEE 802.11 (Wi-Fi) to the data link and physical layers of the OSI model. *(The figure is adapted from Marks, Gifford, and O'Hara (2001, p. 52); Patil (2003).)*

Figure 9.4 shows that when using Wi-Fi communication (that is based on an 802.11 Protocol), all the layers of its OSI model are involved in the communication. This means that Wi–Fi is using layer 1 which is now a wireless physical layer (modulation and coding). Layer 2 is needed to have 802.11 header and then on layer 3, all the IP information, and so on.

With the Bluetooth protocol stack it is different as devices do not have to use all the protocols in the stack (all the layers of the communication model). It is because, the Bluetooth was developed to be used by a variety of communication applications, and it is the application, that designates which part of the Bluetooth stack is used by the communication.

Bluetooth versus Wi-Fi (IEEE 802.11)

Bluetooth and Wi-Fi (Wi-Fi is the brand name for products using IEEE 802.11 standards) have some similar applications: setting up networks, printing, or transferring files. Wi-Fi is intended as a replacement for high-speed cabling for general local area network access in work areas or home. This category of applications is sometimes called wireless local area networks (WLAN). Bluetooth was intended for portable equipment and its applications. The category of applications is outlined as the wireless personal area network (WPAN). Bluetooth is a replacement for cabling in a variety of personally carried applications in any setting, and also works for fixed location applications such as smart energy functionality in the home (thermostats, etc.).

Wi-Fi and Bluetooth are to some extent complementary in their applications and usage. Wi-Fi is usually access point-centered, with an asymmetrical client-server connection with all traffic routed through the access point, while Bluetooth is usually symmetrical, between two Bluetooth devices. Bluetooth serves well in simple applications where two devices need to connect with a minimal configuration like a button press, as in headsets and remote controls, while Wi-Fi suits better in applications where some degree of client configuration is possible and high speeds are required, especially for network access through an access node. However, Bluetooth access points do exist, and ad hoc connections are possible with Wi-Fi though not as simply as with Bluetooth. Wi-Fi Direct was recently developed to add a more Bluetooth-like ad hoc functionality to Wi-Fi.

9.2.2 Classical Bluetooth Stack

Figure 9.5 shows the Bluetooth Stack, and Figure 9.6 shows Bluetooth 4 stack.

The Bluetooth protocol layers, together with their associated protocols are as follows:

- **Bluetooth Core Protocol Baseband** – LMP, L2CAP, SDP
- **Cable Replacement Protocol** – RFCOMM
- **Telephony Control Protocol** – TCS Binary, AT-Commands
- **Adopted Protocols** – PPP, UDP/TCP/IP, WAP.

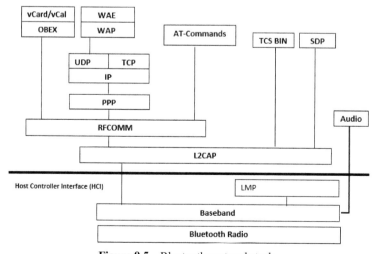

Figure 9.5 Bluetooth protocol stack.

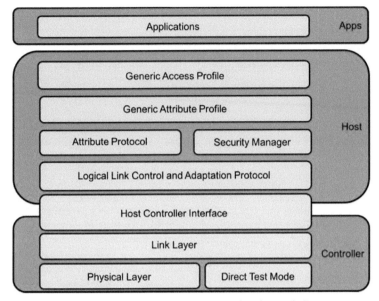

Figure 9.6 Bluetooth protocol stack (Bluetooth 4).

One additional element that can be seen on the stack is the **Host Controller Interface** (HCI). This HCI provides a command interface to the baseband controller, link manager, hardware status, registers. Due to this fact, all the names of the Linux tools that are used for Bluetooth communication are starting from "hci"; for example–"hciconfig," "hcidump", "hcitool."

9.2.2.1 Bluetooth stack as core and profile

A Bluetooth protocol stack can also be divided into two parts: the core and the profiles.

i. **Bluetooth stack core consists of:**
- Host Controller Interface (HCI) which provides a standardized communications protocol between the stack and the Bluetooth module. HCI communication packets can be transmitted over UART, RS232, USB, or PC Card interface.
- Logical Link Control and Adaptation Protocol (L2CAP) allows multiple communication channels to share a single Bluetooth link (multiplexing). L2CAP also handles segmentation and assembly of long messages.
- Radio Frequency Communication Protocol (RFCOMM) implements the functionality needed for a virtual RS232 link, including modem control

signals. Most of the application profiles use RFCOMM to send and receive data.

- Service Discovery Protocol (SDP) provides functionality to publish supported Bluetooth functionality (SDP server), as well as to query other Bluetooth products for Bluetooth functionality (SDP client).

ii. **Bluetooth profile**

To use Bluetooth wireless technology, a device must be able to interpret certain Bluetooth profiles, which are definitions of possible applications and specify general behaviors that Bluetooth-enabled devices use to communicate with other Bluetooth devices. These profiles include settings to parameterize and to control the communication from the start. Adherence to profiles saves the time for transmitting the parameters anew before the bi-directional link becomes effective. There are a wide range of Bluetooth profiles that describe many different types of applications or use cases for devices.

The Bluetooth specification defines how Bluetooth should be used within specific application fields (each application usage is defined in its own profile). A number of profiles have already been defined, and more are expected to be developed in the future. Some of the most important profiles are:

- Generic Access Profile (GAP) is mandatory and is used to manage connections between different Bluetooth products. GAP exposes a high-level API that can be used by the application program to configure the stack and manage connections to different Bluetooth products.
- Service Discovery Application Profile (SDAP) is mandatory and is used to query other Bluetooth products for supported services and functionality. SDAP exposes a high-level API that can be used by the application program to query other Bluetooth products for supported functionality.
- Serial Port Profile (SPP) provides emulated virtual serial ports over a Bluetooth radio link. SPP exposes a high-level API that can be used by the application program to transmit and receive data in a way that is very similar to a standard serial port.
- Headset Profile (HSP) provides functionality for linking sound between an audio gateway (usually a GSM phone) and a headset. HSP exposes a high-level API that can be used by the application program to establish and manage the sound link.

There are also many other profiles for different purposes, for example, file transfer, network access, and dial-up network connections. New profiles are being specified for, for example, cars, audio/video, digital cameras and

Table 9.1 Examples of bluetooth profiles

Profile	API Function	Description
Serial Port Profile	SPP_AddPortRq()	Creates a virtual serial port
	SPP_RemovePortRq()	Deletes a virtual serial port
	SPP_OpenRq()	Opens a virtual serial port
	SPP_CloseRq()	Closes a virtual serial port
	SPP_WriteRQ()	Prints data to a virtual serial port
	. . .	. . .
Headset Profile	HSP_OpenVoiceRq()	Opens a sound channel
	HSP_CloseVoiceRq()	Closes a sound channel
	HSP_VolumeRq()	Adjusts the volume
	. . .	. . .

printer management. This means that a Bluetooth device must not support the entire specification but rather a segment of the protocol stack.

When a Bluetooth protocol stack has been purchased an application program must be developed to use the stack functionality by calling the different API functions. The stack takes care of all the complex communication rules that are defined in the specification and manages the communication with the Bluetooth module.

Table 9.1 contains a few examples of API functionality provided by a stack (in this case the IAR Systems stack) to an application program for a few different profiles.

9.3 Bluetooth Stack Layers

Bluetooth is defined as a layer protocol architecture consisting of core protocols, cable replacement protocols, telephony control protocols, and adopted protocols. Mandatory protocols for all Bluetooth stacks are LMP, L2CAP and SDP. In addition, devices that communicate with Bluetooth almost universally can use these protocols: HCI and RFCOMM.

In the following, we present the different layers.

9.3.1 Bluetooth Radio

The most common Bluetooth radio has an output power of 0 dBm. These devices are called class 3 devices. There are two more classes: class 2 which has a maximum output power of 4 dBm and class 1 which has a maximum output power of 20 dBm. Channel spacing is 1 MHz and a guard range is used in the top and bottom of the frequency band which makes for 79 channels. There is a 23-channel radio defined for countries with special radio frequency regulations. The modulation technique used is Gaussian Frequency Shift Keying (GFSK).

9.3.2 Baseband

The Baseband (BB) sets up connections, defines links and packet types and provides error correction. It sets up the hopping sequence of the piconet based on the master's unique device address (BD ADDR). Two different connections are supported: Asynchronous Connection Less (ACL) and Synchronous Connection Oriented (SCO). Typically, the ACL link transmits data and the SCO link is used only for voice.

The physical channel is divided into time slots of 625 µs in length, each time slot corresponds to a RF frequency which is determined by the masters BD ADDR. A time division duplex TDD scheme is used where the master and slave transmits alternatively. Normally a BB packet covers one time slot but there are packets covering 3 and 5 time slots as well (Figure 9.7).

Each active member in a piconet is given a three bit address called the Active Member Address (AM ADDR) which acts as a MAC address. There can only be seven active slaves participating in a piconet at the same time, AM ADDR 000 is reserved for broadcast. There can, however, be a number of slaves in so-called parked mode, these slaves are not participating in the piconet traffic but remain synchronized with it.

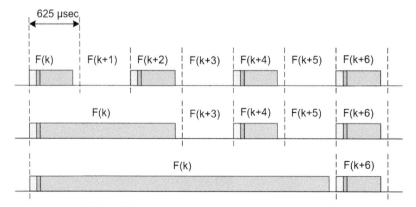

Figure 9.7   Packet lengths supported by baseband.

The master starts its transmission in even-numbered time slots only, the slave starts its transmission in odd-numbered time slots only. The time slots are numbered according to the Bluetooth clock (CLKN) of the piconet master. In an ACL connection a slave can only transmit a packet if it has received a packet containing its AM ADDR in the previous time slot. However, a slave with an SCO connection may transmit even though it has not received its AM ADDR in the previous time slot.

Each packet carries an access code which is the first 72 or 68 bits, the latter if no packet header exists. The access code is used for synchronization, DC offset compensation and identification. There are three access codes defined within Bluetooth, these are as follows.

- Channel Access Code: identifies which piconet the packet belongs to. It is included in every packet transmitted on a piconet.
- Device Access Code: used during the device discovery process to page a device. It is derived from the BD ADDR of the device.
- Inquiry Access Code (IAC): used in the device discovery process when a device inquires about which devices are in range. There are two kinds of Inquiry Access Codes, *General* IAC (GIAC) and *Dedicated* IAC (DIAC). GIAC is used in a general case when a device wishes to know all devices in range. DIAC is used when a device only wishes to know about devices with special characteristics.

The BB determines the hopping sequence of the Bluetooth device. There are five hopping sequences for a standard 79 channel device. These are described below:

- Channel Hopping Sequence: distributes the hopping frequencies pseudo randomly but equally over the 79 channels. It has a very long period so as not to repeat itself in the short term. It is calculated from the BD ADDR of the master of the piconet.
- Inquiry Hopping Sequence: a sequence of 32 wakeup frequencies. It is calculated from the GIAC or the DIAC. The inquiry hopping sequence switches between 2 inquiry trains of 16 frequencies each. The phase is determined from the native clock (CLKN) of the unit.
- Inquiry Response Sequence: also uses two trains of 16 frequencies each, which are exactly the same as the 2*16 frequencies of the Inquiry Hopping Sequence.
- Page Hopping Sequence: a sequence of 32 frequencies calculated from the device being paged's BD ADDR. The phase is determined from an estimate of the paged device's clock. The paging device will calculate

the frequency at which the paged device will be listening on and start transmitting on that frequency to speed up the process. Two trains of 16 frequencies each are used.

- Page Response Sequence: uses two trains of 16 frequencies each, which are exactly the same as the 2*16 frequencies of the Page Hopping Sequence.

9.3.3 Link Manager Protocol (LMP)

The Link Manager Protocol (LMP) provides means for setting up secure, power efficient links for both voice and data. It has the ability to, whenever necessary, update the link properties to ensure optimum performance. The Link Manager also terminates connections, either on higher layers request or because of various failures. The Link Manager Protocol basically consists of a number of PDUs (Protocol Data Units) that are sent from one device to another. The following is a list of supported services:

- Transmission and reception of data.
- Name request
- Request of the link addresses.
- Establishment of the connection.
- Authentication.
- Negotiation of link mode and connection establishment.

The LMP also handles security (authentication is one of security measures) and the different low-power modes *sniff*, *hold* and *park*:

- Active: a device participates in the piconet by listening (in the master-to-slave time slots) for packets containing its own AM ADDR.
- Sniff: in sniff mode, a device acts similar to an active device. Upon entering sniff mode, the master and slave decide a sniff interval, the time between two time slots where the slave will listen for packets.
- Hold: a device that enters hold mode, does so for a specified time. During this time ACL packets are not supported but SCO packets can still be transmitted.
- Park: when a slave enters park mode, it gives up its AM ADDR but remains synchronized to the piconet. It is given a Parked Member Address (PM ADDR) that the master uses for un-parking the slave and an Access Request Address (AR ADDR) that the slave can use to request to be un-parked.

9.3.4 Host Controller Interface (HCI)

The Host Controller Interface provides a command interface for the controller and for the link manager, which allows access to the hardware status and control registers. This interface provides an access layer for all Bluetooth devices. The HCI layer of the machine exchanges commands and data with the HCI firmware present in the Bluetooth device. One of the most important HCI tasks that must be performed is the automatic discovery of other Bluetooth devices that are within the coverage radius.

9.3.5 Logical Link Control and Adaptation Protocol (L2CAP)

The *Logical Link Control and Adaptation Protocol* (L2CAP) supports protocol multiplexing, packet segmentation and reassembly, the maintenance of Quality of Service and group management. L2CAP is placed above the Baseband Protocol and interfaces with higher protocols.

L2CAP has a number of modes.

Basic mode: In *Basic* mode, L2CAP provides packets with a payload configurable up to 64 kB, with 672 bytes as the default MTU, and 48 bytes as the minimum mandatory supported MTU.

Retransmission and Flow Control Mode: In *Retransmission and Flow Control* modes, L2CAP can be configured either for isochronous data or reliable data per channel by performing retransmissions and CRC checks.

Bluetooth Core Specification Addendum 1 adds two additional L2CAP modes to the core specification. These modes effectively deprecate original Retransmission and Flow Control modes:

Enhanced Retransmission Mode (ERTM): This mode is an improved version of the original retransmission mode. This mode provides a reliable L2CAP channel.

Streaming Mode (SM): This is a very simple mode, with no retransmission or flow control. This mode provides an unreliable L2CAP channel.

Reliability in any of these modes is optionally and/or additionally guaranteed by the lower layer Bluetooth BDR/EDR air interface by configuring the number of retransmissions and flush timeout (time after which the radio flushes packets). In-order sequencing is guaranteed by the lower layer.

Only L2CAP channels configured in ERTM or SM may be operated over AMP logical links.

9.3.6 Radio Frequency Communications (RFCOMM)

RFCOMM is a simple transport protocol intended as a cable replacement protocol. The RFCOMM provides transparent data stream and control channels over the L2CAP channels.

RFCOMM provides for binary data transport and emulates EIA-232 (formerly RS-232) control signals over the Bluetooth baseband layer, that is, it is a serial port emulation.

RFCOMM provides a simple, reliable, data stream to the user, similar to TCP. It is used directly by many telephony related profiles as a carrier for AT commands, as well as being a transport layer for OBEX over Bluetooth.

Many Bluetooth applications use RFCOMM because of its widespread support and publicly available API on most operating systems. Additionally, applications that used a serial port to communicate can be quickly ported to use RFCOMM.

9.3.7 Adapted Protocols

There are a number of protocols that have been adapted to fit into the Bluetooth protocol stack. Adopted protocols are defined by other standards-making organizations and incorporated into Bluetooth's protocol stack, allowing Bluetooth to code protocols only when necessary. The adopted protocols include:

Point-to-Point Protocol (PPP): Internet standard protocol for transporting IP datagrams over a point-to-point link.

TCP/IP/UDP: Foundation Protocols for TCP/IP protocol suite.

Object Exchange Protocol (OBEX): Session-layer protocol for the exchange of objects, providing a model for object and operation representation.

Wireless Application Environment/Wireless Application Protocol (WAE/ WAP): WAE specifies an application framework for wireless devices and WAP is an open standard to provide mobile users access to telephony and information services.

9.3.8 Service Discovery Protocol

The Service Discovery Protocol (SDP) enables a Bluetooth device to inquire what services are available in a piconet and how to access them. For example, when you use a mobile phone with a Bluetooth headset, the phone uses SDP to determine which Bluetooth profiles the headset can use (Headset Profile, Hands Free Profile, Advanced Audio Distribution Profile (A2DP) etc.) and the protocol multiplexer settings needed for the phone to connect to the headset using each of them. The SDP should be able to determine the properties of any service, future or present, of arbitrary complexity in any operating environment. However, it will not provide any access to the services themselves or negotiate service parameters.

SDP uses a server-client model where the server has a list of elements called Service Records. Each Service Record describes the characteristics of one service. There can only be one SDP server in a Bluetooth device, if a device has several services the one SDP server can act on behalf of all of them. Similarly multiple applications may use a single SDP client to query servers of Service Records. If a device only acts as an SDP client it does not need to have an SDP server. A device may operate as both a server and a client at the same time.

The server-client model used by SDP states that each transaction consists of one request protocol data unit (PDU) and one response PDU. To provide some flow control, a client must receive a response to each request before issuing another request on the same L2CAP connection. The SDP database consists of a set of records describing the services of the device. A service is described using Service Attributes. A Service Attribute consists of a 16 bit Attribute ID and a variable length Attribute Value. The Attribute Value PDU has a header describing their type and length to the receiver.

9.3.9 Server-Client Model

An overview of the Bluetooth client and server model is given in this section. A Bluetooth service is an application that acts as a server and provides assistance to client devices via Bluetooth communication. This assistance typically takes the form of a capability or a function unavailable locally on the client device. Examples of some of a Bluetooth server applications are LAN access services, file transfer services, business-card exchange services, and synchronization services. Additional profiles describing standardized services were developed after the Bluetooth profiles specification v1.1. For example, three printing profiles were defined that offer various forms of printing services. We note here that if the service is defined by a Bluetooth profile, then the profile specification describes the requirements for the service record, device security, device discoverable modes, and so on.

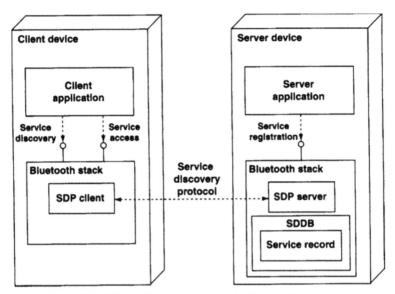

Figure 9.8   Bluetooth components involved in service discovery.

Developers can create Bluetooth server applications to implement one of the Bluetooth profiles or to implement their own custom service. These services are made available to remote clients by the definition of a service record that describes the service and the addition of that service record to the service discovery database (SDDB) of the local device.

Figure 9.8 represents the Client-Server model and illustrates the Bluetooth components involved in service registration and service discovery. The SDP is a Bluetooth protocol for discovering the services provided by a Bluetooth device. A server application adds a service record to the SDDB. The Bluetooth stack provides an SDP server, which maintains this database of service records. Service discovery clients use SDP to query the SDP server for any service records of interest. A service record provides sufficient information to allow an SDP client to connect to the Bluetooth service on the server device.

After registering a service record in the SDDB, the server application waits for a client application to initiate contact with the server to access the service. The client application and the server application then establish a Bluetooth connection to conduct their business.

Service Discovery protocol will be discussed again in more details in Section 9.6.

9.4 Device Discovery

Bluetooth, as mentioned before, is a mobile ad hoc network. In mobile ad hoc environments, devices initially have no information about their surrounding environment or the devices that operate within their range. There is no centralized instance to query about the environment. Therefore, a protocol must exist that provides means for detecting devices and enables devices to set up a connection, Bluetooth uses the Baseband protocol for this task. Two procedures are used in the device discovery procedure: *inquiry* and *page*: firstly, the *inquiry* process, where a master device discovers neighboring slave devices; and secondly, the *page* process, where connections between them are established. During the first step, information about slave clock times is exchanged for the purposes of synchronization. This can be used during the second step, which is hence much faster. We therefore concentrate on the inquiry process. We now describe in more detail the procedure executed by an *inquiring device* (a master trying to discover slaves) and a *scanning device* (a potential slave device who wants to be discovered). Next the two steps are considered in details.

9.4.1 Inquiry

In order to set up a connection, a device must detect what other devices are in range. This is the goal of the inquiry procedure (Figure 9.9). The process is initiated by the unit that wishes to collect device information or create a connection. To conserve power and coexist with other link activity, inquiry is always initiated by higher level control protocols. The inquiry procedure must overcome the initial frequency discrepancy between devices. Therefore, inquiry only uses 32 of the 79 hop frequencies. Typically a device enters inquiry mode periodically. Similarly, a device that wishes to be visible to inquiring units enters inquiry scan in certain intervals. In order to find each other, one device must be in Inquiry state and one (or more) device must be in Inquiry Scan substate simultaneously (Figure 9.10).

Like all Bluetooth devices, the inquiring device has a 28 bit free-running clock, which ticks every $312.5\mu s$. On two consecutive $312.5\mu s$ time slots, it sends on two sequential frequencies, that is, it sends two ID packets every $625\mu s$. During the next two time slots, the device scans for a reply on these same two frequencies, that is, each scan occurs $625\mu s$ after the corresponding send (in fact, a $10\mu s$ margin is added to the start and end of the scan in case replying devices are not completely synchronized). The device now proceeds

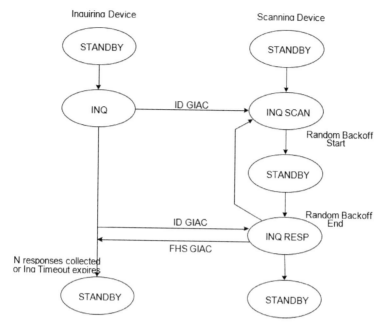

Figure 9.9   State transitions during the inquiry process.

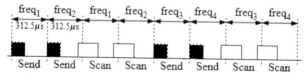

Figure 9.10   Timing of the inquiring device's behavior.

to send and scan on the next pair of frequencies in the same fashion. This procedure is illustrated in Figure 9.10.

The inquiring device does so for the duration of the inquiry window denoted $T_{w\ inquiry}$ (Figure 9.11). The device can also be set to exit Inquiry state after a number, denoted $N_{inq\ responses}$, of devices has been found. In periodic inquiry mode, the time between two consecutive inquiries is determined by the inquiry interval, $T_{inquiry}$, (Figure 9.11).

An ID packet contains the Inquiry Access Code. As mentioned earlier in Section 9.3.2, a device can inquire about all units in range using GIAC or a specific type of device using DIAC. The Inquiry Hopping Sequence, however, is always derived from the Lower Address Part (LAP) of the GIAC which is common for all devices and therefore it is the same for all devices. The 32

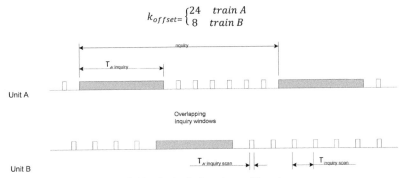

Figure 9.11 Periodic inquiry and inquiry scan.

frequencies are divided into two trains, A and B, each train is 10 ms in length and according to the specification a train must be repeated at least 256 times before a new train is used ($N_{\text{inquiry}} = 256$). The specification also states that in an error free environment at least three train switches must have taken place in order to ensure that all devices in range respond to the inquiry, resulting in $T_{\text{w inquiry}} = 10.24$ s. In an error prone environment more train switches may be necessary to ensure responses from all devices in range.

The phase within the trains is derived from the *CLKN* of the inquiring device and therefore unique to every device. The phase X_i is calculated as follows:

$$X_i = [CLKN_{16-12} + k_{offset} + (CLKN_{4-2,0} - CLKN_{16-12}) \bmod 16]$$
$$\bmod 32$$

where $CLKN_{x-y,z}$ denotes bits x to y and bit z of the native clock. K_{offset} selects the train used:

$$k_{offset} = \begin{cases} 24 & train\ A \\ 8 & train\ B \end{cases}$$

A device wishing to be found by inquiring units periodically enters Inquiry Scan substate. The interval between two consecutive inquiry scans is determined by the inquiry scan interval, $T_{\text{inquiry scan}}$. The time a device stays in Inquiry Scan substate is determined by the inquiry scan window, $T_{\text{w inquiry scan}}$. During this time the device listens to a single frequency of the Inquiry Hopping Sequence. The phase is determined by its native clock:

$$X_{ir} = [CLKN_{16-12} + N] \bmod 32$$

The value of N is increased each time $CLKN_1$ is set to zero, which corresponds to the start of a master transmission slot.

According to the specification, $T_{\text{w inquiry scan}}$ must be greater than 10 ms in order to ensure that a frequency synchronization takes place, assuming that the listening frequency is in the train transmitted. It also states that $T_{\text{inquiry scan}} = 2.56$ s.

Upon reception of an ID packet a device in Inquiry Scan substate will leave Inquiry Scan substate for a random backoff time, uniformly distributed between 0 and 1023 time slots which corresponds to $[0,\ldots,639.375\ ms]$. This is done to reduce the probability that devices will respond to the same ID packet simultaneously, thus colliding. After the random backoff the device will enter Inquiry Response state, in which it listens for a second ID packet. After receiving this, the device responds with a Frequency Hopping Selection (FHS) packet containing its device information, that is, its BD ADDR and its current clock (CLKN).

If the inquiring device receives a response FHS packet in the first half of the receive slot, it cannot receive a response from another device in the second half because the radio channel synthesizer does not permit that. Similarly if it receives a response in the second half of the receive slot, it cannot send the first ID packet in the following transmit slot.

In Inquiry state, a device cannot exchange ACL data with another device. However, SCO links are still supported in the Inquiry and Inquiry Scan substates. If there are SCO links active, the inquiring device must increase N_{inquiry} and $T_{\text{w inquiry}}$ to make up for the time slots lost to the SCO links.

9.4.2 Page

In Bluetooth the connection establishment is handled by the page process (Figure 9.12). The page process requires knowledge of the BD ADDR of the device with which the connection is to be established. Furthermore the device being paged must be in Page Scan substate, that is, listening for page messages. At the end of the page process a connection has been set up, the paging device becomes the master and the paged device becomes the slave. As with inquiry a device typically enters Page state periodically and a device that wishes to be able to connect to paging units enters page scan in certain intervals.

In Page state a device transmits ID packets at twice the normal frequency hopping rate using the 32 frequencies of the Page Hopping Sequence. The Page Hopping Sequence is derived from the paged device's BD ADDR and

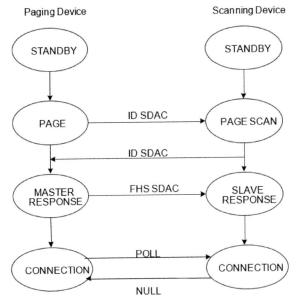

Figure 9.12 State transitions during the page process.

the phase is calculated as follows:

$$X_p = [CLKE_{16-12} + k_{\text{offset}} + (CLKE_{4-2,0} - CLKE_{16-12}) \bmod 16]$$
$$\bmod 32$$

where $CLKE_{x-y,z}$ denotes bits x to y and bit z of the clock estimate of the paged device. The parameter k_{offset} selects the train used. Unlike inquiry the page process always start on train A. This is because the train A starts with the most probable listening frequency of the paged device, based on the *CLKE* received from, for example, an inquiry. It is of course possible to perform a page without the *CLKE* but it will consume more time.

Similar to the inquiry process, the time a device stays in Page state is determined by the page window, $T_{\text{w page}}$. In periodic page mode, the time between two consecutive pages is determined by the page interval, T_{page}.

In periodic page scan mode, the interval between two consecutive page scans is set by the page scan interval, $T_{\text{page scan}}$. The length of the scan is determined by the page scan window, $T_{\text{w page scan}}$. The scanning device listens to frequencies in the Page Hopping Sequence. The phase is determined by it is native clock:

$$X_p = CLKN_{16-12}$$

Table 9.2   Relationship between scan interval, train repetition and paging modes

Mode	T Page Scan	N_{page}
R0	Continous	> 1
R1	< 1.28 sec	> 128
R2	< 2.56 sec	> 256
Reserved	–	–

There are three modes concerning the Page Scan substate, *R0, R1* and *R2*. These modes set the behavior of the scanning device (Table 9.2). Depending of the scanning device's scan mode the paging device must repeat the page train a certain amount N_{page} times, called the repetition number, or until a response is obtained. If the scanning device's scan interval corresponds to R1, then $N_{page} \geq 128$; if the slave scan interval corresponds to R2, then $N_{page} \geq 256$.

When a unit in Page Scan substate receives an ID packet containing its own Access Code it acknowledges the page message with an ID packet and enters Slave Response state. In this state, it freezes the $CLKN_{16-12}$ bits at their current values, then using a value one larger than the preceding slot for determining the phase:

$$X_{prs} = [CLKN^*_{6-12} + N] \, mod \, 32$$

where $CLKN^*_{16-12}$ are the frozen clock bits and N is a counter set to zero in the slot the paged device acknowledges the page and incremented each master TX slot.

After receiving the ACK from the paged device, the paging device enters the Master Response substate. In this mode, the phase is determined as follows:

$$X_{prm} = [CLKE^*_{16-12} + k^*_{offset} + (CLKE^*_{4-2,0} - CLKE^*_{16-12}) mod \, 16] \\ mod \, 32$$

In Master Response substate the device sends a FHS packet containing its CLKN, which will be the piconet clock CLK, and the AM ADDR the paged device shall use. After acknowledging the FHS packet the paged device, now called the slave, switches to the hopping sequence of the pager, now called the master. Now the two units are connected and they both switch to Connection state. To make sure the slave has changed to the master's hop sequence, it sends a Poll packet to the slave which in turn answers with an arbitrary packet. The entire page process is shown in Figure 9.13. The Paging

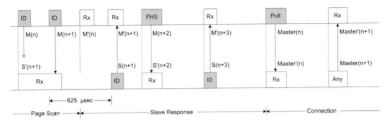

Figure 9.13 Page procedure.

scheme described above is the normal one, all units support it. In addition to supporting this mandatory paging scheme, a unit may support one or more optional paging schemes.

9.4.3 Optional Paging Schemes

To this date only one optional paging scheme has been defined, called optional scheme. It uses the same 32 page frequencies as the mandatory paging scheme and they are split into two trains, A and B. In contrast to the mandatory scheme, the same 32 frequencies that are used for transmitting are also used for reception of page responses, called reception trials. The page train construction differs from the mandatory paging scheme in two ways; the page train consists of 10 slots, or 6.25 ms. The first 8 slots of the train are used to transmit the ID packets, the 9th slot is used to send a marker packet, and the 10th slot is used for the return of a slave response. The marker packet indicates that a response may be sent and it also contains information about what frequency to be used for the response.

9.5 Service Discovery and Access

After the devices in an area are discovered, the next step before connecting to a device is finding the services a device has running on it. Unlike device discovery, the service discovery process involves only a single pair of devices (Figure 9.14). The service discovery process requires the device searching for services to ask a device with services whether it has a service defined by a service record that has a specific set of attributes. If a remote device has a service with the attributes specified, the remote device returns the service record describing the service. The service record has multiple attributes. These attributes provide additional information on a specific service. These attributes may contain anything, including information on how to connect to the service.

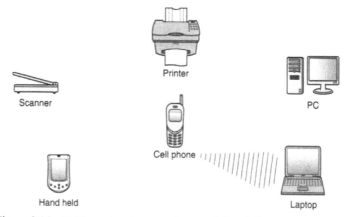

Figure 9.14   Unlike an inquiry, a service search is only between two devices.

Bluetooth Service Discovery Protocol (SDP) provides means for discovering services within remote devices but it does not provide access to them, hence there is the need for other service discovery protocols that addresses this issue such as Jini, Salutation, Universal Plug and Play (UPnP), etc. Universal Plug and Play (UPnP) that developed by MicrosoftOne will be given as example of this techniques

9.5.1 Bluetooth Service Discovery Protocol

The Service Discovery Protocol allows Bluetooth devices to discover what services other devices may offer. It permits service browsing and searching for specific services. But SDP does not provide access to the services themselves.

SDP relies on L2CAP links between the SDP server and client. Following the setup of a L2CAP link, a client may query the SDP server about services available (Figure 9.16). SDP uses a dedicated L2CAP channel known beforehand to all devices. This channel uses a Protocol Service Multiplexer reserved for SDP. After the SDP data has been received, a client must set up a new connection to access the service; the SDP channel cannot be used. Devices supporting the SDP application profile must support pairing and authentication.

9.5.1.1 Format

A SDP PDU is a packet of variable length depending on the parameters of the service record. Every SDP PDU consists of a header followed by

PDU-specific parameters; the header consists of three fields:

- PDU ID (1 byte) identifies the type of PDU, that is, its meaning and the specific parameters.
- Transaction ID (2 bytes) identifies request PDUs and is used to match responses to requests.
- Parameter Length (2 bytes) specifies the length of the parameters carried in the PDU.

As some requests may require responses that are longer than the PDU length, a PDU may contain a Continuation State parameter to indicate that not all parameters were sent in the PDU. Then the client repeats the request with the Continuation State copied from the response. The server then responds with the rest of the parameters.

Every request PDU has a corresponding response PDU. Should there be an error in the request or the server cannot respond to the request, the server will respond with an error PDU (*SDP ErrorResponse*).

9.5.1.2 Service record

All information about a service is stored in a single Service Record within the SDP server. A Service Record consists of a list of Service Attributes. There are 16 Service Attributes defined in the specification v 1.1. A Service Record is identified by a Service Attribute called *ServiceRecordHandle*. Some Service Attributes are common to all service records, but there are reserved fields where vendors can define their own Service Attributes. A Service Attribute consists of two components:

- Attribute ID (16 bits), distinguishes each Service Attribute within a Service Record. It also identifies the semantics of the Attribute Value.
- Attribute Value (variable length), a data element consisting of two fields, a header and data. The meaning of the Attribute Value is defined by the corresponding Attribute ID.

A typical Service Record is made up as follows: The first part is the ServiceRecordHandle followed by the ServiceClassIDList which lists the services offered by the device.

The ProtocolDescriptorList lists the protocols needed to use the service. There may also be protocol specific parameters that are needed to use the protocols listed, such as channel numbers, etc.

The BluetoothProfileDescriptorList lists the profiles which the service supports described by their UUID and a version number. If a profile is

Figure 9.15 Example of a data element containing a 16-bit unsigned int.

dependent of another profile, only the top level profile is listed. GAP is never listed since it is mandatory for all services.

Finally there may be some extra attributes needed to use the service, for example, ServiceName.

9.5.1.3 Representation of data

All data in the Attribute Values is represented as a data element. A data element is a typed data representation of variable length. It consists of a header and data field. The header contains a 5 bit Type Descriptor and a 3 bit Size Descriptor. There are nine types of data defined:

- 0, Null.
- 1, Unsigned int.
- 2, Signed two complement int.
- 3, Universally Unique Identifier (UUID).
- 4, Text string.
- 5, Boolean.
- 6, A sequence of data elements, all of which make up the data.
- 7, Data sequence alternative. A sequence of data elements, of which one must be chosen.
- 8, Uniform Resource Locator (URL).

The Size Descriptor is an index of the size using the three bits left in the byte from the Type Descriptor. Sometimes, however, Service Attributes can be too long to be described with the index, such as text strings, hence the length must be described in another way. Therefore, the Size Descriptor can also indicate where the actual data size can be read. The Size Descriptors and what they indicate are listed below:

- 0, 1 byte (or 0 if Null type).
- 1, 2 bytes.
- 2, 4 bytes.
- 3, 8 bytes.
- 4, 16 bytes.
- 5, Data size in next 1 byte.

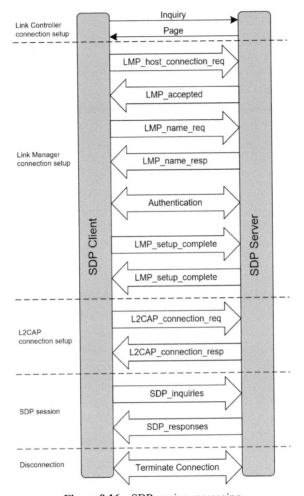

Figure 9.16 SDP session messaging.

- 6, Data size in next 2 bytes.
- 7, Data size in next 4 bytes.

This system saves bandwidth since most attributes fit in the 1–16 bytes of the Size Descriptor index part.

9.5.1.4 Searching and browsing services

To make services easily searched and browsed, they are arranged in a tree structure. The hierarchy of the tree and which services will be browsable is up to the service providers to decide. The top of the tree is called

PublicBrowseRoot. Service Classes are used to identify services, and all services are members of a Service Class. A Service Class record defines all attributes contained in a Service Record of the fore-mentioned class. It always includes a BrowseGroupList attribute, which is a list of UUIDs of all browse groups associated with the service.

A client can search for services with specific attributes contained in their Service Records provided that the attributes are of UUID type. Search for arbitrary values is not supported. In all searches SDP ServiceSearchRequest is used, it includes a ServiceSearchPattern which is a list of UUIDs that are to be searched for by the server in its database. The server responds with a SDP ServiceSearchResponse that includes the ServiceRecordHandle of the matching Service Records. The ServiceRecordHandles are then used to retrieve the attributes of the desired services, done with SDP ServiceAttributeRequest.

To speed up the process of retrieving Service Attributes a client can combine a service search and an attribute request using SDP ServiceSearchAttributeRequest. The server then responds with the ServiceRecordHandles and attributes of the services matching the ServiceSearchPattern.

A client that wishes to browse through the services of a server creates a ServiceSearchPattern containing the UUID of the PublicBrowseRoot. All services that may be browsed from the top level are made members of the root group by including the PublicBrowseRoot UUID in the BrowseGroupList attribute.

9.5.2 Examples of SDP: Universal Plug and Play

Universal Plug and Play (UPnP), developed by Microsoft, is designed to enable simple ad hoc communication between distributed devices and services. UPnP uses the TCP/IP protocol stack which is virtually ubiquitous.

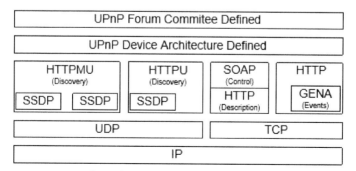

Figure 9.17   The UPnP protocol stack.

It enables a device to dynamically join a network, obtain an IP address, convey its capabilities, and learn about the presence and capabilities of other devices.

9.5.2.1 Architecture overview

A UPnP device, as mentioned earlier, uses the TCP/IP protocol suite as transport. It uses Simple Service Discovery Protocol (SSDP) for announcing its presence to other devices as well as discovering services. SSDP can function with or without a directory service. It uses HTTP over unicast and multicast UDP called HTTPU (unicast UDP) and HTTPMU (multicast UDP). The multicast is sent on a reserved channel to which all devices must listen. XML is used to describe devices and services. Control messages are also expressed in XML using Simple Object Access Protocol (SOAP). Event messages also use XML, the messages are formatted using General Event Notification (GENA). UPnP introduces the term control points as potential clients of services.

9.5.2.2 Operations overview

A device entering the system can use AutoIP to obtain an IP address without explicit administration. It then multicasts a number of discovery messages (ssdp:alive), each corresponding to one of its embedded devices and services to advertise its services to the control points in the network. The messages should include an expiry time for the service or device, if the service or device is still available at the expiry time the advertisement should be resent.

A control point entering the network multicasts a discovery message searching for interesting devices, services, or both. All devices listen to the standard multicast address for these messages and must respond if any of their embedded devices or services match the search criteria in the discovery message. The discovery message may also be directed to a directory service if present. The fundamental exchange in both cases is a discovery message containing a few essential specifics about the device or one of its services, for example, its type, identifier and a pointer (URL) to more detailed information.

After a control point has discovered a device, it will use the URL to retrieve more information about the device and/or how to interact with it. The description for a service is expressed in XML and includes a list of any embedded devices or services, as well as a URL for control, and presentation.

A control point then can send actions to a device's service through control messages. The message is sent to the control URL of the device and the device in turn responds with any action-specific values.

A UPnP service description includes a list of actions the service responds to and a list of variables that model the state of the service at runtime. A service publishes updates when these variables change. A control point may subscribe to receive these updates through event messages. Event messages contain the name of one or more state variables and their values.

If a device or service has a URL for presentation; a control point may display this to the user and, depending on the capabilities of the page, allow the user control and or view of the device status. When a device and its services are to be removed from the network, the device multicasts a message (ssdp:byebye) corresponding to each of the discovery (ssdp:alive) messages it multicasted that have not already expired.

9.6 Bluetooth Security

In any wireless networking setup, security is a concern. Devices can easily grab radio waves out of the air, so people who send sensitive information over a wireless connection need to take precautions to make sure those signals aren't intercepted.

Bluetooth security like that for any other wireless system is of paramount importance as devices are susceptible to a variety of wireless and networking attacking including denial of service attacks, eavesdropping, man-in-the-middle attacks, message modification, and resource misappropriation.

Bluetooth security must also address more specific Bluetooth related attacks that target known vulnerabilities in Bluetooth implementations and specifications. These may include attacks against improperly secured Bluetooth implementations which can provide attackers with unauthorized access.

The latest releases of Bluetooth have increased the levels of security to combat the threat of hackers–any wireless link provides an opportunity for entry into a network.

Not only is the level of security increasing on the hardware elements that often have inbuilt security, but also the level of security built into Bluetooth itself is improving.

In this section, we discuss Bluetooth security. There are two guides and standards for Bluetooth protocols and security, NIST 800-121-R1 and IEEE 802.15.1. NIST 800-121-R1 details the recommended Bluetooth security processes. These recommendations include the authentication and verification of the sender, confidentiality regarding information, and authorization in regard to who has control over access to the information. IEEE 802.15.1 is the standard for Bluetooth Wireless Technology.

9.6.1 Basic Means of Providing Bluetooth Security

There are three basic means of providing Bluetooth security:

- *Authentication:* In this process, the identity of the communicating devices is verified. User authentication is not part of the main Bluetooth security elements of the specification.
- *Confidentiality:* This process prevents information being eavesdropped by ensuring that only authorized devices can access and view the data.
- *Authorization:* This process prevents access by ensuring that a device is authorized to use a service before enabling it to do so.

Besides the three basic means there is a fourth type of Bluetooth security: *Pairing*.

The four security means work together in the following sequence:
Pairing stage: Pairing is the first step in the process of Bluetooth security. When two devices come into contact with one another for the first time and want to use security, the devices must establish a shared secret used for authentication and encryption. Pairing requires the user of each device to input a common code or PIN into each device.

The PIN is then used to do an initial authentication of both devices. After the initial pairing, a shared secret is established and is stored within the Bluetooth device to allow authentication of both devices in the future without the need for the pairing process. Figure 9.18 shows how two devices can retrieve the PIN to complete the pairing process. The pairing process is transparent to the application. It is the responsibility of the BCC to retrieve the PIN from the user or determine what the PIN should be.

Authentication phase: After pairing comes authentication. Bluetooth authentication verifies the identity of one device to another device using a challenge and response scheme. Bluetooth authentication, as mentioned above, does not authenticate users but authenticates devices. When device A wants to authenticate device B, device A sends a challenge to device B. When it receives this challenge, device B applies the shared secret to the challenge and sends the result to device A. Device A then combines the challenge that was sent with its shared secret and compares the result with the result sent from device B. Although it authenticates device B to device A, this process does not authenticate device A to device B. The same process must be used to authenticate device A to device B. To perform authentication, device A and device B must complete the pairing process so that the shared secret can be established.

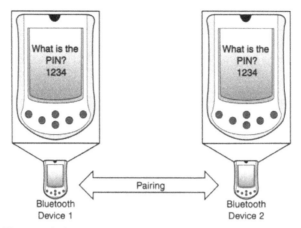

Figure 9.18   For two devices to complete the pairing process, a common PIN must be entered.

Encryption phase: Once the authentication process has been completed, encryption can be turned on. Encryption is used to prevent an eavesdropper, Eve, from intercepting communication between two entities, Alice and Bob. When one device wants to turn on encryption, it must ask the other Bluetooth device to do so also. If the other device accepts the request, all packets between the devices are encrypted. If the other device rejects the request, the connection is closed. Unlike the mechanism of authentication, it is not possible for communications sent from device A to device B to be encrypted while communications sent from device B to device A are unencrypted.

Authorization phase: Another option within Bluetooth security is authorization. Authorization is the process of determining whether a connection request from a specific Bluetooth device should be granted. Authorization is completed on a connection-by-connection basis. The Bluetooth specification has also defined the concept of a trusted device, which is a device that is automatically granted authorization when authorization is requested. In other words, a trusted device is authorized to connect to any service on the local device. When a trusted device connects to a service that requires authorization, the request is automatically accepted without the BCC asking the user if the device is authorized to use the service. The BCC is in charge of maintaining the list of trusted devices. When an authorization request is received by the BCC for a nontrusted device, the BCC requests the user to grant or deny the connection.

The above discussion means that each level of security builds on the previous level. Authentication requires pairing. Encryption and authorization require authentication. JABWT enforces these requirements. If encryption is requested on a link and the link has not been authenticated, the JABWT implementation authenticates the remote device before encrypting the link.

9.6.2 Bluetooth Security and Trust Modes

9.6.2.1 Bluetooth security modes

The various versions of the specifications detail four Bluetooth security modes. Each Bluetooth device must operate in one of four modes:

- *Bluetooth Security Mode 1:* This mode is **non-secure**. The authentication and encryption functionality is bypassed and the device is susceptible to hacking. Bluetooth devices operation in Bluetooth Security Mode 1. Devices operating like this do not employ any mechanisms to prevent other Bluetooth-enabled devices from establishing connections. While it is easy to make connections, security is an issue. It may be applicable to short range devices operating in an area where other devices may not be present. Security Mode 1 is only supported up to Bluetooth 2.0 + EDR and not beyond.

- *Bluetooth Security Mode 2:* **Application/Service Based**–In this mode, once a connection is established, a Security Manager performs authentication, thereby restricting access to the device. For this Bluetooth security mode, a centralized security manager controls access to specific services and devices. The Bluetooth security manager maintains policies for access control and interfaces with other protocols and device users.

  It is possible to apply varying trust levels and policies to restrict access for applications with different security requirements, even when they operate in parallel. It is possible to grant access to some services without providing access to other services. The concept of authorization is introduced in Bluetooth security mode 2. Using this it is possible to determine if a specific device is allowed to have access to a specific service.

  Although authentication and encryption mechanisms are applicable to Bluetooth Security Mode 2, they are implemented at the LMP layer (below L2CAP).

  All Bluetooth devices can support Bluetooth Security Mode 2; however, v2.1 + EDR devices can only support it for backward compatibility for earlier devices. After introducing security mode 3 and mode 4, the role of mode 2 will be discussed in some detail in Section 9.6.2.1.

- *Bluetooth Security Mode 3:* **Link level enforced security**–In Bluetooth Security Mode 3, the Bluetooth device initiates security procedures before any physical link is established. In this mode, authentication and encryption are used for all connections to and from the device.

The authentication and encryption processes use a separate secret link key that is shared by paired devices, once the pairing has been established.

Bluetooth Security Mode 3 is only supported in devices that conform to Bluetooth 2.0 + EDR or earlier.

• ***Bluetooth Security Mode 4:*** This is service level enforced security with encrypted key exchange .Bluetooth Security Mode 4 was introduced at Bluetooth v2.1 + EDR.

In Bluetooth Security Mode 4 the security procedures are initiated after link setup. Secure Simple Pairing uses what are termed Elliptic Curve Diffie-Hellman (ECDH) techniques for key exchange and link key generation.

The algorithms for device authentication and encryption algorithms are the same as those defined in Bluetooth v2.0 + EDR.

The security requirements for services protected by Security Mode 4 are as follows:

  o Authenticated link key required
  o Unauthenticated link key required
  o No security required

Whether or not a link key is authenticated depends on the Secure Simple Pairing association model used. Bluetooth Security Mode 4 is mandatory for communication between v2.1 + EDR devices.

9.6.2.1.1 Security mode 2 role

We refer in to Figure 9.18 during this discussion. Figure 9.19 shows that the Security Manager is cognizant of the goings on in both the L2CAP and RFCOMM layers. When an attempt is made to establish a peer-to-peer connection at either of these layers, the Security Manager is made aware of this and acts as an arbiter. It does not matter if the connection is being initiated by the client's application, or requested by a remote device, the Security Manager has intimate knowledge of what is happening and responds appropriately. It can decide on the course of action, basing its decision on configuration data placed in the service database. The options available to the Security Manager are as follows:

• Do nothing and allow the peer-to-peer connection to establish itself.
• Initiate authentication procedures

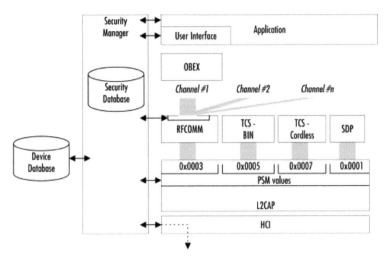

Figure 9.19 Trigger points are located within RFCOMM and L2CAP to invoke mode 2 security.

Table 9.3 Associated protocol service multiplexor specifying the service it represents

Service Module	Protocol Service Multiplexor
SDP	0x0001
TCS bin	0x0005
TCS cordless	0x0007
RFCOMM	0x0003

- Initiate authorization procedures.
- Start encryption once a communications link is established.

With security being triggered at the L2CAP layer, there is the potential for blocking access to services above this layer. Service Discovery Protocol (SDP), Telephony Control Specification (TCS), RFCOMM, and OBEX functions (and all application profiles relying on these underlying building blocks) can be selectively protected. When an L2CAP connection is established, a value called a protocol service multiplexor (PSM) must be specified, identifying which of the modules above this layer is to be accessed. Table 9.3 lists the PSM values along with their corresponding upper layer connection module to give you a view of services that can be protected if security is linked to L2CAP.

Usually, when using the L2CAP layer as the security trigger, the intention of the user will go to protect either the cordless telephony/intercom profile

(TCS) or SDP. Protecting SDP may not be in best interest of the user as this implies that the user is not inclined to provide services to other devices that do not know what the user is doing. It is important to note here that once a remote device passes authentication, and if the link key is stored (bonding completes), authentication will successfully pass in future sessions without user intervention. Perhaps a different strategy would suffice in protecting your device from others that do not know what you do–like configuring your device to be non-discoverable.

In a manner similar to L2CAP, the Security Manager has access to the internal workings of RFCOMM as well and can trigger security based upon connection requests being made at this level. Associating security with the RFCOMM protocol layer protects applications requiring the serial port profile and profiles built upon this foundation such as fax, modem, LAN access, and OBEX.

As was the case with L2CAP, the Security Manager can be selective in determining which applications to protect as well. Peer-to-peer connection establishment at the RFCOMM layer requires that a specific channel (out of a possible 60 channel values) be specified for the connection to complete successfully. This channel number is always associated with a particular service or profile being offered by the Bluetooth server unit. This channel number is made available to client devices through SDP. Therefore, to protect a specific service relying upon serial profile support, you would set up the Security Manager to trigger when a connection attempt is made using RFCOMM and a service-specific channel ID.

There are a few interesting things the reader should be made aware of. First, server applications (such as a LAN Access Point) relying on RFCOMM must register their use of the RFCOMM interface by entering information into the SDP service database; specifically, this equates to a channel number associated with the RFCOMM module along with the service supported, such as LAN access. Devices interested in using this service must query the service database using the SDP facility, extract this information, then make a request to connect to the specified RFCOMM channel number. The Security Manager detecting this request will make a determination if security is required based upon configuration information contained within its own internal service database. It will then take action and invoke security measures as required.

The Security Manager, in accordance with the Bluetooth specification, can also initiate security measures if a particular type of connection (RFCOMM or L2CAP) is initiated by your own application. For instance, assuming for a moment that as a client application, I want to establish a connection to a server offering "FAX" capability (RFCOMM channel #7 as revealed by an earlier SDP session). After establishing a radio connection at the Link Manager level, a connection request would be made to the server unit at the L2CAP layer. Next, before attempting to connect at the RFCOMM layer, authentication would be invoked by my side. My device would be the Verifier. If successful, a connect request to RFCOMM would then proceed. Note that authentication is supported on outgoing (as well as incoming) connection requests. Authorization and encryption are only triggered on incoming connection requests.

9.6.2.2 Bluetooth trust modes

In addition to the security modes discussed above, there are two levels of trust for Bluetooth devices, trusted and untrusted. They are described as follows:

1. **Trusted**–A trusted device has established a fixed relationship with another device and has unrestricted access to all services.
2. **Untrusted**–An untrusted device only has access to a restricted set of services. Although the device has passed authentication successfully, it does not have a fixed relationship with another device.

9.6.3 Discoverability in Devices

Discoverability modes of Bluetooth devices also affect the device's security. Devices in discoverable mode are more vulnerable, as they can be recognized. The device name, class, list of services, and technical information are all exchanged in discoverable Bluetooth devices that are in range (approximately 10 m). In addition, every Bluetooth device has a unique 48-bit address used for identification, known as the BD_ADDR. This address is similar to a MAC address, which is a manufacturer assigned address for hardware that serves as a unique identification number. The BD_ADDR, like a MAC address, is assigned by the manufacturer.

9.6.4 Bluetooth Security Services

The first time that two devices attempt a connection, a trusted relationship needs to be established through authentication. Authentication is performed by using challenge-response, based on BD_ADDR and a link key. The link keys, once established, are kept by both devices to be used for future pairing. In older versions of Bluetooth (v2.0 and earlier), common secret PIN codes, which are passkeys required for first time Bluetooth connections, are used.

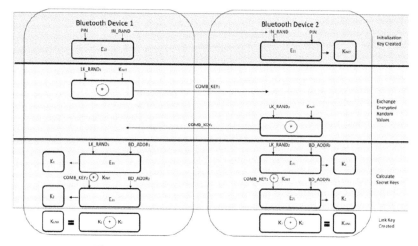

Figure 9.20 Link-key generation with PIN [19].

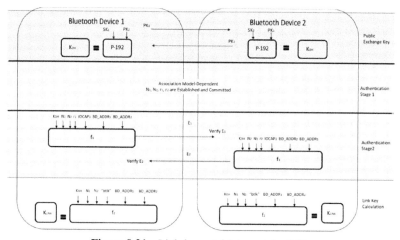

Figure 9.21 Link-key establishment for SSP.

The PINs are used by both devices and consist of between 4 and 16 characters. These codes are specifically used for link-key generation. This is illustrated in Figure 9.20. In some cases, once the PIN is set, it cannot be changed. It is also important to note that two devices cannot communicate or be paired if the devices have fixed PINs. Newer versions of Bluetooth (v2.1 and later) use SSP for the pairing process, which utilizes public key cryptography instead of a PIN. This protocol is illustrated in Figure 9.21.

Authorization begins by first determining whether the device had previously been authorized as a trusted device. If the device database lists it as a trusted device, then access to services is granted. If the device is not listed as a trusted device, trust must first be established before it can be authorized.

Confidentiality is achieved through encryption, more specifically the use of the E0 stream cypher. A link key and the BD_ADDR of the device are used to develop a keystream that when combined with plaintext achieves a cyphered text. Attacks and cryptanalysis attempts on E0 have proven that the stream cipher is vulnerable to attacks.

9.6.5 Built-in Security Features

Bluetooth technology has certain built-in features that help secure the technology. They include:

(1) Adaptive Frequency Hopping: Frequency hopping in Bluetooth uses a 2.4 GHz ISM band with 79 channels to enable hops at 1600 hops per second. During the hopping, existing frequencies are excluded. The ability to frequency hop reduces both jamming and interference.
(2) E0 Cipher Suite: The cipher generally has a key length of 128 bits and uses stream ciphering.
(3) Undiscoverability: This prevents devices from responding to scanning attempts. A device's 48-bit BD_ADDR address is also concealed.
(4) Pairing: Pairing enables devices to communicate. A device's BD_ADDR must be known for a pairing request to be made. The BD_ADDR, which is discussed in the previous two sections, is identified from knowledge of previous pairing or by scanning.

9.7 Bluetooth Vulnerabilities and Threats

In this section, we discuss the vulnerabilities in different versions of Bluetooth. We then discuss Bluetooth attacks, which are a result of vulnerabilities in the technology. Finally, we introduce the Bluetooth threat taxonomy and discuss common Bluetooth attacks.

9.7.1 Vulnerabilities in Bluetooth Versions

The version of Bluetooth that is being used, and the security of communications between devices, which is only as strong as the weakest link (i.e., the device with the oldest (weakest) version) is important when discussing

Bluetooth vulnerabilities. Since many older devices are still being used today, the vulnerabilities in the older versions of Bluetooth continue to be present.

(1) Versions before Bluetooth 1.2: Link keys, which are based on static unit keys, are used for pairing and can be reused. If the key is retrieved, malicious devices can eavesdrop on the original devices, as well as spoof the original device and/or connected devices.

(2) Versions before Bluetooth 2.1 + EDR: Codes that consist of short PINs are permitted. These PINs are easy for attackers to guess due to their short length. These versions are lacking in PIN management, which is a desirable security capability at an enterprise level. In addition, the keystreams in these early versions become vulnerable after being connected for 23.3 h. This is the time period at which the keystream repeats. This increases an adversary's ability to decrypt messages.

(3) Versions 2.1 and 3.0: If Security Mode 4 devices are connecting to devices that do not support Security Mode 4, earlier security modes are used in the connection. For example, it is possible that Security Mode 1, which offers no security, will be used. This rollback in security modes makes versions 2.1 and 3.0 more vulnerable to attacks. In addition, SSP static keys are used in versions 2.1 and 3.0, which increases the device's vulnerability to Man-in-the-Middle attacks.

(4) Versions before Bluetooth 4.0: There is an unlimited number of authentication challenge requests, which enables adversaries to obtain information on many challenge responses. This allows them to gain insight on secret link keys. In addition, the stream cipher E0 function, which is used in early versions, is considered weak.

(5) All versions of Bluetooth: Adversaries can view and potentially modify link keys if they are stored improperly. In addition, encryption key lengths may be small, which can make them vulnerable to attackers. It is possible that encryption keys can be as small as 1 byte. Regarding authentication, there is no user authentication. The Bluetooth standard only includes device authentication. It is important to note that a device can remain in discoverable/connectable mode for an indefinite period of time.

9.8 Reasons for Bluetooth Vulnerabilities

Security in Bluetooth networks depends on the security of the Bluetooth medium, protocols and parameters. Weaknesses in each of these aspects result in vulnerabilities in a Bluetooth network. Bluetooth makes several assumptions about security. Firstly, it assumes that once a connection is established between two devices, it will remain permanently secure with the keys it has stored. Secondly, there is an assumption that short range provides some level of security since adversaries need to be in close proximity (typically 5–30 m). Moreover Bluetooth authenticates devices, not users, so there is an assumption that all users on a particular device should follow the same security protocol. In this section, we look at Bluetooth's vulnerability to eavesdropping and weaknesses in the mechanisms of Bluetooth's security model.

9.8.1 Vulnerability to Eavesdropping

Due to the nature of wireless RF communication, eavesdroppers are often not detected in a Bluetooth network. Unencrypted transmissions obviously make it easy for an eavesdropper to see the contents of any packets. However, the eavesdropper has to be in range of the Bluetooth network. Bluetooth packets consist of an access code, packet header and a payload. Since the access code and packet header are always sent unencrypted, even when encryption is used on the data and payload, an eavesdropper can always see general piconet information pertaining to the devices. Using this information, the eavesdropper could figure out the authorization levels of the legitimate piconet devices. Furthermore, as explained in our attack below 6, low energy legacy pairing provides no passive eavesdropping protection. If successful, eavesdroppers can capture secret keys distributed during low energy pairing.

9.8.2 Potential Weaknesses

9.8.2.1 Encryption mechanisms

The most significant weakness in the Bluetooth encryption mechanism is when 128-bit encryption cannot be used. When two devices communicate the parameters for encryption, the length of the encryption key is restricted by the Bluetooth device that has the shorter maximum encryption key length. Another weakness occurs when the PIN inputted in the Passkey Entry or Just Works association model is used to generate the link keys. This PIN can be brute-forced and used for replication of the link keys by a passive eavesdropping attacker.

9.8.2.2 Association models of SSP

One particular weakness of the association models of SSP is the Just Works model. This pairing method provides no MITM protection between trusted devices.

9.8.2.3 Device configuration

The default settings of Bluetooth provide little security, as the device is set as discoverable and non-secure, meaning that an attacker can discover the BD ADDR of it and perform various attacks.

9.8.2.4 Bad coding during development of RFCOMM stack implementation

- Failures in secure stack implementation may lead to buffer overflows.
- Some manufacturers may not be releasing any patches for the original version of their codes they deploy on end-devices.

9.8.2.5 Re-use of older services for different protocols

- Some highly privileged services are left open.

9.8.2.6 IrMC permissions

- IrMC defines a set of access permissions for common Bluetooth objects.
- Permissions are sometimes not followed or just open, leading to exploitation of open IrMC services.

All the mentioned vulnerabilities do not directly say–what are really the threats of using Bluetooth devices (knowing that those vulnerabilities exist). To name just a few, the attacker might be able to:

- Steal the information.
- Perform a DoS attack on the end-device using Bluetooth.
- Remotely execute code.
- Inject viruses or worms.
- Inject crafted connections to go via a Bluetooth device (working as proxy).

The Bluetooth threat Taxonomy illustrated below in Table 9.4, outlines, and classifies Bluetooth-based threats. This classification system can help determine the severity of threats, provides precautionary methods, and presents reactionary strategies. Some threats may display characteristics of several classifications; however, they are classified based on their predominant characteristic.

9.9 Common Bluetooth Attacks

The pairing process is a main contributor to security issues found in Bluetooth. Attacks can be performed during different stages of the pairing process

Table 9.4 Bluetooth threat taxonomy

Classification	Method	Threats
Obfuscation	Techniques are used to hide the attack and prevent detection.	HCIConfig (Device Name) HCIConfig/BTClass (Class of Device) Bdaddr (Device Address) SpoofTooph
Surveillance	Device monitoring is done to collect information.	HCITool (Device Discovery) Sdptool (Service Discovery) Redfang Blueprinting Bt Audit War-Nibbling Bluefish BNAP BNAP/BlueProPro BlueScanner
Range Extension	Range of connectivity is extended so attacks can be conducted at a distance.	BlueSniping/Bluetoone
Sniffing	Sniffer is used to intercept data by capturing network traffic.	Merlin/FT4USB (External Based) BlueSniff (Frequency Based) HCIDump (Host Based)
Man-In-The-Middle	Attackers trick devices into thinking they are paired, when in reality they are both connected to the attacker.	Bthidproxy
Unauthorized Direct Data Access	Data stored in cloud is directly accessed due to vulnerabilities.	Bluesnarf/Bl000ver BTCrack/Btpincrack Car Whisperer HeloMoto Bluebugger HID Attack Btaptap
Denial of Service	Services are disrupted, making a machine or network unavaiable to users.	BlueSmack/Tanya Blueper BlueJacking/BlueSpam/Smurf vCardBlaster Signal Jamming BlueSYN/Pingblender (Multi-Vector DoS) Battery Exhaustion

(Continued)

Table 9.4   Continued

Classification	Method	Threats
Malware	Intrusive or harmful software is put on a computer to disrupt operations, steal data, or extort a target for ransom.	BlueBag Caribe CommWarrior Skuller
Fuzzer	Injects data into a stack or program and has the ability to detect bugs.	Bluetooth Stack Smasher/BluePAss BlueStab HCIDump Crash L2CAP Header Overflow Nokia N70 L2CAP DoS Sonyericson Reset Display

including before the pairing process has completed and after devices are paired. For example, attackers may be able to carry out Man-in-the-Middle attacks based on information they collected after pairing. Some of the more common attacks on Bluetooth are described below:

A. MAC Spoofing Attack

The attack is performed before encryption is established and during the formation of the piconet when link keys are being generated. Devices are able to authenticate each other by generating link-keys. During the attack, attackers can imitate another user. They also have the ability to terminate connections or intercept/modify data with the use of special tools. Figure 9.22 illustrates a MAC spoofing attack.

B. PIN Cracking Attack

The attack occurs during the device pairing and authentication process. An attacker uses a frequency sniffer tool to collect the RAND and the BD_ADDR of the targeted device. A brute-force algorithm (E22 algorithm) is then used to test all possible permutations of the PIN with the data previously collected until the correct PIN is found. Figure 9.23 illustrates the PIN cracking attack structure.

C. Man-in-the-Middle Attack

Man-in-the-Middle Attacks such as the one illustrated in Figure 9.24 below occur when devices are attempting to pair. During the attack, messages

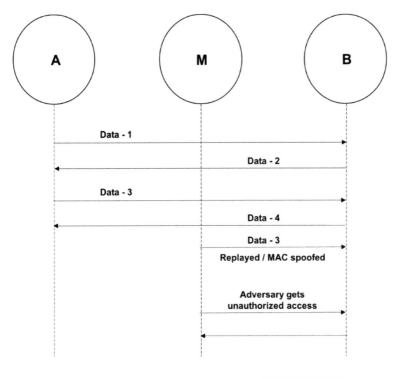

Figure 9.22 MAC spoofing attack.

are relayed unknowingly between the devices. This enables authentication without the shared secret keys. In a successful attack, the user believes the pairing was successful; however, this is not the case, as the two devices are paired to the attacker.

D. BlueJacking Attack

During a BlueJacking attack, the attacker sends unsolicited messages to a device to trick the user into using an access code. This enables the adversary to access files on the targeted device. The devices involved in the attack and the exact source of the message received need to be within a specific range, 10 m, for the attack to be successful. This attack is commonly used in crowded areas (e.g., airports, shopping malls, and train stations). While it does not usually involve the alteration of data, it could make devices susceptible to other attacks.

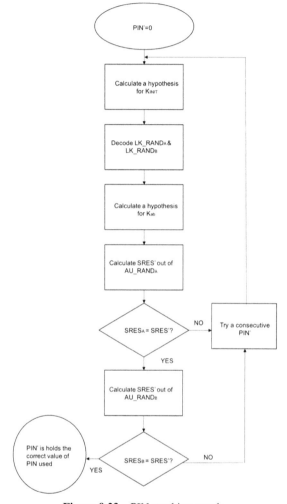

Figure 9.23   PIN cracking attack.

E. BlueSnarfing Attack

The attack involves hacking into a mobile phone and stealing any of the data stored in the phone's memory, such as contacts, calendar entries, images, etc. During the attack, the attacker connects by exploiting the OBEX File Transfer Protocol, a file transfer program used in Bluetooth. This enables the attacker to pair with the user's device. Figure 9.25 below illustrates a BlueSnarfing attack.

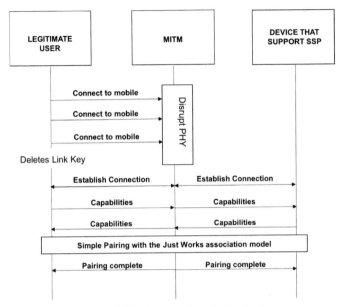

Figure 9.24 Man-in-the-middle attack.

E. BlueBugging Attack

The attack occurs in the RFCOMM protocol. Physical connections are made via L2CAP + base band and it emulates serial RS-232 connections. During the attack, the attacker connects to the target device without the knowledge of the owner. The attacker then takes over the device by gaining access to the device's set of "AT" commands, which are attention commands that send instructions to the module. This enables the attacker to execute commands as if he was the device owner. The attacker can also steal information and access the phone's services and settings.

F. BlueBump Attack

The attack occurs when there is a weakness in the handling of link keys. During the attack a business card is sent between the attacker and user. By forcing the user to accept the card, a trusted and authenticated connection is made. After the pairing, the user then has the ability to delete the link key; however, the user unknowingly still has an active connection to the attacker. The attacker is then able to pair with the device later without authenticating by simply requesting link-key regenerations. The attacker can continue to pair with the target device if the key is not deleted.

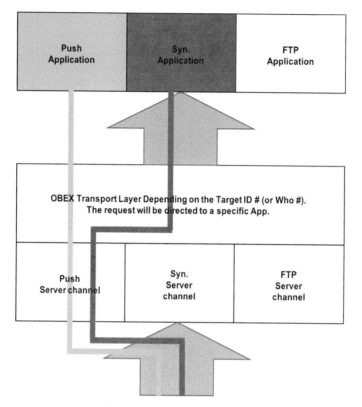

Figure 9.25   BlueSnarfing attack.

G. BlueDump Attack

During the attack, the attacker spoofs the BD_ADDR of one of the devices to connect with the other. During pairing, the target device sends an authentication request. The attacker responds with 'HCI_Link_Key_Request_Negative_Reply. This is a result of the attacker not having a link key. In some cases, the device targeted deletes its link key. It then goes into pairing mode.

F. BluePrinting Attack

The attack is carried out by combining the information that is revealed about a device to gain additional information, such as the manufacturer, device model, and firmware version. This attack can only be performed when the BD_ADDR of the device is known.

G. Blueover Attack

Blueover and its successor Blueover II are auditing tools that are used to determine if a Bluetooth device is vulnerable, but they can also be used to initiate a BlueBugging attack.

H. BlueBorne Attack

The attack is conducted by exploiting a stack buffer overflow flaw. By targeting the processing of pending client L2CAP configuration responses, the attacker is able to hijack Bluetooth connections. This enables them to control a targeted device's embedded content and functions. For the attack to be successful, only MAC and Bluetooth addresses are needed.

I. Fuzzing Attack

The attack occurs when the adversary attempts to cause a device to behave abnormally by sending malformed data packets and non-standard data to a device's Bluetooth radio. The attacker then watches how the device reacts to the data packets being sent. If the device's operations become sluggish or stop during these attacks, the attacker could infer that the protocol stack has vulnerabilities.

J. Off-Line PIN Recovery Attack

During the attack, the attacker tries to intercept the IN_RAND value, LK_RAND values, AU_RAND value, and SRES (signed response) value. The SRES value is a matching variable needed for authentication [9]. The attacker then uses brute-force to obtain a PIN that could be used to determine the correct SRES value, which is equal to the intercepted SRES value.

K. Brute-force BD_ADDR Attack

This attack is a scanning attack on the last three bytes of the BD_ADDR of a device. It is important to note that the first three bytes, which are known publicly, can be set as fixed.

L. Reflection/Relay Attack

The attack occurs when the adversary impersonates a device. The attacker is not looking for any undisclosed information during the attack. It simply authenticates the connection by reflecting/relaying device information.

M. Backdoor Attack

The attack occurs when establishing a trusted relationship during pairing. During the attack, the adversary does not appear in the register of paired devices on the target device. After a relationship is established, the attacker has access to the devices services and resources. This access is unbeknownst to the device owner. The BD_ADDR of the target device needs to be known for a backdoor attack to be successful. It also needs to be determined that the device targeted by the attacker is vulnerable to the attack.

N. Denial of Service Attacks

DDoS (Distributed Denial of Service) and ordinary DoS are two types of Denial of Service attacks. For ordinary DoS attacks, the attacker tries to crash the network or restart the system by sending packets to the targeted system. DDoS attacks can be done by a single attacker. These attacks can disable a network. They also can restrict network accessibility to a larger network. The attacks target the Physical Layer in the protocol stack or those above the Physical Layer. Some typical Denial of Service (DoS) attacks are BD_ADDR duplication, BlueSmack, BlueChop, L2CAP guaranteed service, battery exhaustion, and Big NAK (Negative Acknowledgment), which is an attack using a continuous retransmission loop.

O. Worm Attacks

The attacks occur when a malicious software or Trojan file sends itself to available Bluetooth devices. Examples of these attacks are:

1. Cabir Worm: A malicious software that targets Bluetooth technology. Mobile phones that use the Symbian series 60 interface platform are vulnerable to the attacks. For the attacks to be successful, the user must accept the worm. This causes the malware to install on the device. The worms are usually disguised in applications, which results in users unknowingly accepting them. Once installed, the software is the able to use the compromised device to search for and send itself to other available devices. The Mabir worm is a form of the Cabir worm. This worm replicates by using Multimedia Messaging Service messages and Bluetooth.
2. Skulls Worm: The Skulls Worm, a malicious SIS (Symbian Installation System) trojan file, targets Symbian mobile phones with the Series 60 platform. The worm poses as a Macromedia Flash player. The user must

open and install the SIS file for the worm to become active. It then searches for additional devices to infect and the process repeats itself.

3. Lasco Worm: The Lasco worm, is a combination of a Bluetooth worm and SIS file. It targets and infects Symbian mobile phones that support the Series 60 platform. The user must open and install the velasco.sis file. This prompts the activation of the worm. It can then begin searching for additional devices to infect and the process repeats itself.

P. Bluesmack Attack

The attack is a DoS attack on Bluetooth devices and is similar to the "Ping of Death" attacks that are carried out on IP-based devices, which are networked devices. It is done by sending pings that are approximately 600 bytes, as well as L2CAP echo requests to Bluetooth devices. This results in the input buffer to overflow and the targeted device to be knocked out.

Q. MultiBlue Attack

The attack occurs when an attacker has access to the device they wish to hack. The MultiBlue dongle, a bluetooth capable 4 GB thumb, is used to gain access to and take over the targeted device. The attacker then uses the MultiBlue application to see all discoverable devices within range and send pairing requests. The targeted device then presents a code, a pre-shared key, which needs to be entered into the MultiBlue application. This key is needed for authentication and encryption. The attacker then has control of the device.

R. HeloMoto Attack

The attack is a mix of the previously mentioned BlueSnarfing and Blue-Bugging attacks. A vulnerability caused by the erroneous implementation of "trusted devices" on Motorola devices is exploited during the attack. When exploited, the adversary's device is stored as a trusted device on the target device's trusted list. After connecting to the OBEX push profile, the attacker tries sending a vcard. The adversary is then able to elude authentication and connect to the target device's headset profile. BlueBugging is also used to take control of the device.

S. Bluecasing/War Nibbling Attack

The attack occurs when a phreaker, a telephone network hacker, uses laptops or PCs with high gain antennas and special software to discover and exploit vulnerabilities in Bluetooth phones to obtain access.

The ability to identify and describe these different attacks enables us to better understand the threats surrounding IoT Bluetooth devices. In addition, by classifying the threats, we can determine the severity of the various Bluetooth attacks. Next, we discuss ways to mitigate the risks and identify potential countermeasures used in response to these threats

9.10 Bluetooth Risk Mitigation and Countermeasures

The attacks that are described in the previous section discuss Bluetooth flaws that result in vulnerabilities, which can be exploited by an attacker to steal data, send messages, make phone calls, and connect to the Internet using the compromised device. We now discuss risk mitigation techniques and countermeasures.

9.10.1 Mitigation Techniques

Mitigating Bluetooth vulnerabilities differ significantly from mitigating vulnerabilities in a computer system. While application software patches are used to resolve vulnerabilities in computer systems, Bluetooth devices require upgrades in device firmware. These upgrades cannot be developed by the general public and/or user community. Therefore, Bluetooth devices will continue to be vulnerable to attacks even if mitigation solutions become available.

While all attacks cannot be prevented, and security is not guaranteed, there are countermeasures that can be used to secure Bluetooth communications. Some of those mitigation techniques are described below:

A. Enhancement of Bluetooth user awareness: it is necessary to educate Bluetooth users to ensure they have knowledge of the proper Bluetooth security practices. These security practices include:

(1) Default settings should be updated to achieve optimal standards.
(2) Ensuring devices are in and remain in a secure range. This is done by setting devices to the lowest power level.
(3) Using long and random PIN codes, which make the codes less susceptible to brute-force attacks.
(4) Changing the default PIN for devices and frequently updating this PIN (i.e., once every other month).
(5) Setting devices to undiscoverable mode by default, except as needed for pairing. Most active discovery tools require that devices be in discoverable mode to be identified. Devices set to

undiscoverable mode will not be visible to other Bluetooth devices. Devices previously configured, better known as trusted devices, will be able to connect and communicate while in this hidden mode.

(6) Turning off a device's Bluetooth when not needed or in use, especially while in certain public areas such as shopping malls, coffee shops, public transportation, clubs, bars, etc. This can prevent users from receiving advertisements from other Bluejackers.

(7) Refraining from entering passkeys or PINs when unexpectedly prompted to do so.

(8) Frequently updating software and drivers to have the most recent product improvements and security fixes.

(9) It is recommended that users refrain from using non-supported or not secure Bluetooth-enabled devices or modules. This includes Bluetooth versions 1.0 and 1.2.

(10) Pairing devices as needed. Users need to maintain that any pairing should take place in a secure non-public setting. This will help prevent attackers from intercepting pairing messages. As previously mentioned, a crucial part of Bluetooth security is pairing, so users should have knowledge regarding eavesdropping.

(11) Users should use SSP instead of legacy PIN authentication for the pairing exchange process when it is possible. This will help mitigate PIN cracking attacks.

(12) All lost or stolen Bluetooth devices should be unpaired from devices they had previously been paired with. Unpairing will prevent an attacker from accessing the users other devices through the Bluetooth pairing.

(13) Users should never accept transmissions from unknown or suspicious devices.
Content should only be accepted from trusted devices.

(14) All devices that are paired should be removed immediately after use.

(15) Devices should be monitored and kept at close range.

B. Instead of basing link keys on unit keys, they should be based on combination keys. This will prevent Man-in-the-Middle attacks.

C. Use link encryption for all data transmissions to prevent any eavesdropping, including passive eavesdropping. Use of the HID boot mode mechanism, a connectionless human interface device, should be avoided, as it sends traffic in plaintext.

D. Users should ensure all links are encryption-enabled when using multi-hop communication. Failure to do so could result in the entire communication chain being compromised.

E. Require mutual authentication for network connected devices. This will provide confirmation that the network connections are legitimate.

F. Lower the risk of broadcast interceptions by encrypting the broadcasts.

G. The maximum encryption key size should be used. In addition, a minimum key size should also be set–128 bits is recommended. The utilization of these minimum and maximum keys will protect devices from brute-force attacks.

H. To provide the highest level of security, Security Mode 3 is highly recommended. This mode of security, which is implemented at the link level, is one of the highest levels of Bluetooth security.

9.10.2 Applications for Protecting Bluetooth Devices

(1) Bluetooth firewall: The Firewall application protects devices, specifically Android devices, from all Bluetooth related attacks. Users are alerted upon any Bluetooth activity. The application also enables you to see Bluetooth capabilities on devices or specific apps.

(2) Bluetooth file transfer: This application only enables authorized devices to be connected.

References

[1] Bisikian, C. An overview of the Bluetooth wireless technology. *IEEE Commun. Mag.* **2001**, *39*, 86–94.

[2] Bluetooth Core Specification v5.0, Bluetooth Special Interest Group (SIG), Dec. 2016.

[3] Bluetooth Core Specification v4.1, Bluetooth Special Interest Group (SIG), Dec. 2013.

[4] S. (n.d.)., Bluetooth low energy (BLE) enabled devices market volume worldwide, from 2013 to 2020 (in million units), 2015. [Online]. Available: https://www.statista.com/statistics/750569/ worldwide-bluetooth-low-energy-device-market-volume/

[5] Harris, A., III; Khanna, V.; Tuncay, G.; Want, R.; Kravets, R. Bluetooth low energy in dense IoT environments. *IEEE Commun. Mag.* **2016**, *54*, 30–36.

[6] Darroudi, S.M.; Gomez, C. Bluetooth low energy mesh networks: A survey. *Sensors* **2017**, *17*, 1467.

[7] Hayajneh, T.; Almashaqbeh, G.; Ullah, S.; Vasilakos, A.V. A survey of wireless technologies coexistence in WBAN: Analysis and open research issues. *Wirel. Netw.* **2014**, *20*, 2165–2199.

[8] Sairam, K.; Gunasekaran, N.; Reddy, S.R. Bluetooth in wireless communication. *IEEE Commun. Mag.* **2002**, *40*, 90–96.

[9] Jordan, R.; Abdallah, C.T. Wireless communications and networking: An overview. *IEEE Antennas Propag. Mag.* **2002**, *44*, 185–193.

[10] Ferro, E.; Potorti, F. Bluetooth and Wi-Fi wireless protocols: A survey and a comparison. *IEEE Wirel. Commun.* **2005**, *12*, 12–26.

[11] Bluetooth Versions Walkthrough and Bluetooth 4.0 Low Energy Development Resources. Available online:

[12] https://www.cnx-software.com/2013/06/05/bluetooth-versions-walkthrough-and-bluetooth-4-0-low-e.

[13] Carettoni, L.; Merloni, C.; Zanero, S. Studying bluetooth malware propagation: The BlueBag project. *IEEE Secur. Priv.* **2007**, *5*, 17–25.

[14] Bluetooth SIG, "Specification of the Bluetooth system v1.1, Core", December 2000.

[15] Bluetooth SIG, "Specification of the Bluetooth system v1.1, Profiles", December 2000.

[16] Salutation Consortium, "Salutation Architecture Specification v2.1, Part 1", 1999.

[17] Sun Microsystems, "Jini Architecture Specification v1.2", December 2001.

[18] Microsoft Corporation, "Universal Plug and Play Device Architecture v1.0", June 2000.

[19] Jennifer Bray and Charles F Sturman, "Bluetooth, Connect without cables", Prentice Hall, 2001.

[20] Oliver Kasten and Marc Langheinrich, "First Experiences with Bluetooth in the Smart-Its Distributed Sensor Network", Swiss Federal Institute of Technology, October 2001.

[21] Eugene A Gryazin, "Service Discovery in Bluetooth", Helsinki University of Technology, 2000

[22] Choonhwa Lee and Sumi Helal, "Protocols for Service Discovery in dynamic and mobile networks", University of Florida, September 2000.

[23] Bluetooth SIG, "Mapping Salutation Architecture APIs to Bluetooth Service Discovery Layer", White paper, July 1999.

[24] Sun Microsystems, "Jini Architectural overview ", White paper, January 1999.

[25] John Rakesh, "UPnP,Jini and Salutation–A look at some popular coordination frameworks for future networked devices", California Software Laboratories, June 1999.

[26] Salutation Consortium, "Salutation Lite, Find-And-Bind Technologies for mobile devices", White paper, June 1999.

[27] Microsoft Corporation, "Understanding Universal Plug and Play", White paper, June 2000.

[28] Steven Czerwinski, Ben Zhau, Todd Hodes, Anthony Joseph and Randy Katz, "An Architecture for a Secure Service Discovery Service", University of California, Berkeley, 1999.

[29] E. Guttman, C. Perkins, J. Veizades and M. Day, "Service Location Protocol, Version 2" IETF, June 1999.

[30] M. Wahl, T. Howes, S. Kille, "Lightweight Directory Access Protocol (v3)" IETF, December 1997.

[31] Erik Guttman, "Service Location Protocol: Automatic Discovery of IP Network Services" IEEE Internet Computing, vol 3, no.4, pp 71–80, August 1999.

[32] Network Simulator web page, available 8/2/2002. http://www.isi.edu/ns nam/ns/

[33] BlueHoc web page, available 8/2/2002. http://oss.software.ibm.com/blu ehoc/

[34] Bahar, I., Frohm, E., Gaona, C., Hachtel, G., Macii, E., Pardo, A., Somenzi, F.: Algebraic decision diagrams and their applications. Formal Methods in System Design **10**(2/3), 171–206 (1997).

[35] Basagni, S., Bruno, R., Petrioli, C.: Device discovery in Bluetooth networks: A scatternet perspective. In: Proc. Networking 2002, *LNCS*, vol. 2345, pp. 1087–1092. Springer (2002).

[36] BlueHoc–An open-source Bluetooth simulator: www-124.ibm.com/de veloperworks/opensource/bluehoc

[37] Bluetooth specification, version 1.2, Bluetooth SIG, 2003, www.blueto oth.com

[38] Bryant, R.: Binary decision diagrams and beyond: Enabling technologies for formal verification. In: Proc. International Conference on Computer-Aided Design (ICCAD'95), pp. 236–243 (1995).

[39] Clarke, E., Fujita, M., McGeer, P., McMillan, K., Yang, J., Zhao, X.: Multi-terminal binary decision diagrams: An efficient data structure for

matrix representation. Formal Methods in System Design **10**((2/3), 149–169 (1997).

[40] Daws, C., Kwiatkowska, M., Norman, G.: Automatic verification of the IEEE 1394 root contention protocol with KRONOS and PRISM. International Journal on Software Tools for Technology Transfer (STTT) **5**(2–3), 221–236 (2004).

[41] Duflot, M., Fribourg, L., Herault, T., Lassaigne, R., Magniette, F., Messika, S., Peyronnet, S., Picaronny, C.: Probabilistic model checking of the CSMA/CD protocol using PRISM and APMC. In: Proc. 4th Workshop on Automated Verification of Critical Systems (AVoCS'04), *Electronic Notes in Theoretical Computer Science*, vol. 128(6), pp. 195–214. Elsevier Science (2004).

[42] Kasten, O., Langheinrich, M.: First experiences with Bluetooth in the Smart-Its distributed sensor network. In: Proc. PACT'01 (2001).

[43] Kwiatkowska, M., Norman, G.: Verifying randomized Byzantine agreement. In: D. Peled, M. Vardi (eds.) Proc. Formal Techniques for Networked and Distributed Systems (FORTE'02), *LNCS*, vol. 2529, pp. 194–209. Springer (2002).

[44] Kwiatkowska, M., Norman, G., Parker, D.: Controller dependability analysis by probabilistic model checking. In: Proc. 11th IFAC Symposium on Information Control Problems in Manufacturing (INCOM'04) (2004).

[45] Kwiatkowska, M., Norman, G., Parker, D.: PRISM 2.0: A tool for probabilistic model checking. In: Proc. 1st International Conference on Quantitative Evaluation of Systems (QEST'04), pp. 322–323. IEEE Computer Society Press (2004).

[46] Kwiatkowska, M., Norman, G., Parker, D., Sproston, J.: Performance analysis of probabilistic timed automata using digital clocks. In: K. Larsen, P. Niebert (eds.) Proc. Formal Modeling and Analysis of Timed Systems (FORMATS'03), *LNCS*, vol. 2791, pp. 105–120. Springer-Verlag (2003).

[47] Padgette, J. Bahr, M. Batra, M. Holtmann, R. Smithbey, L. Chen, and K. Scarfone, Guide to Bluetooth Security: Recommendations of the National Institute of Standards and Technology (Special Publication 800-121 Revision 2). USA: CreateSpace Independent Publishing Platform, 2017.

[48] Haataja, K. Hyppĺonen, S. Pasanen, and P. Toivanen, Bluetooth Security Attacks: Comparative Analysis, Attacks, and Countermeasures, ser. SpringerBriefs in Computer Science. Springer Berlin Heidelberg, 2013.

[Online]. Available: https://books.google.com/books?id=bFm6BAAA QBAJ

[49] D.-Z. Sun, Y. Mu, and W. Susilo, "Man-in-the-middle attacks on secure simple pairing in bluetooth standard v5.0 and its countermeasure," Personal Ubiquitous Comput., vol. 22, no. 1, pp. 55–67, Feb. 2018. [Online]. Available: https://doi.org/10.1007/s00779-017-1081

[50] National Institute of Standards and Technology. *Guide to Bluetooth Security: Recommendations of the National Institute of Standards and Technology*; Special Publication 800-121 Revision 1; National Institute of Standards and Technology: Maryland, MD, USA, 2008.

[51] NIST Special Publication 800-121 Revision 2: Guide to Bluetooth Security. Available online: https://nvlpubs.nist.gov/nistpubs/SpecialPublicati ons/NIST.SP.800-121r2.pdf.

[52] Gerez, S. Implementation of Digital Signal Processing: Some Background on GFSK Modulation.

[53] Available online: http://wwwhome.ewi.utwente.nl/~{}gerezsh/sendfile/ sendfile.php/gfsk-intro.pdf?sendfile=gfsk-intro.pdf.

[54] Bluetooth 5 FAQ: Everything You Need to Know. Available online: http s://www.macworld.com/article/3262664/hardware/bluetooth-5-faq-ev erything-you-need-to-know.html.

[55] Nateq Be-Nazir Ibn, M.; Tarique, M. Bluetooth security threats and solutions: A survey. *Int. J. Distrib. Parallel Syst.* **2012**, *3*, 127.

[56] Panse, T.; Panse, P. A Survey on Security Threats and Vulnerability attacks on Bluetooth Communication. *Int. J. Comput. Sci. Inf. Technol.* **2013**, *4*, 741–746.

[57] Hassan, S.S.; Bibon, S.D.; Hossain, M.S.; Atiquzzaman, M. Security Threats in Bluetooth Technology. *Comput. Secur.* **2017**, *74*, 308–322.

[58] Zou, Y.; Wang, X.; Hanzo, L. A survey on wireless security: Technical challenges, recent advances and future trends. *arXiv* **2015**, arxiv:1505.07919.

[59] Shrivastava, M. Analysis of security risks in Bluetooth. *Int. J. Comput. Acad. Res.* **2012**, *1*, 88–95.

[60] National Institute of Standards and Technology. *Guide to Bluetooth Security*; NIST 800-121-Rev 1; NIST: Gaithersburg, MD, USA, 2016.

[61] Mitchell, A. The Car Whisperer: Eavesdrop Onand Take Part in Nearby Bluetooth Conversations. 2005. Available online: https://www.theinter netpatrol.com/the-car-whisperer-eavesdrop-on-and-take-part-innearb y-bluetooth-conversations/.

[62] Dunning, J.P. Bluetooth Threat Taxonomy. Available online: https://vtec hworks.lib.vt.edu/bitstream/handle/10919/76883/etd-10242010-1630 02_Dunning_JP_T_2010.pdf?sequence=1&isAllowed=y)

[63] Bluetooth Security. Available online: https://cs.stanford.edu/people/ero berts/courses/soco/projects/2003-04/wireless-computing/sec_bluetooth .shtml

[64] Sen, J. Security and Privacy Challenges in Cognitive Wireless Sensor Networks. In *Cognitive Radio Technology Applications for Wireless and Mobile Ad hoc Networks*; Meghanathan, N., Reddy, Y.B., Eds.; IGI-Global: Hershey, PA, USA, 2013.

[65] Shaked, Y.; Wool, A. Cracking Bluetooth PIN. Available online: http: //www.eng.tau.ac.il/~{}yash/shakedwool-mobisys05/

[66] Saravanan, K.; Vijayanand, L.; Negesh, R.K. A Novel Bluetooth Man-In-The-Middle Attack Based on SSP using OOB Association model. *arXiv*, **2012**, arxiv:1203.4649.

[67] Moreno, A.; Okamoto, E. BlueSnarf Revisited: OBEX FTP Service Directory Traversal. Available online: https://link.springer.com/cont ent/pdf/10.1007%2F978-3-642-23041-7_16.pdf.

[68] Ahmed, M.; Musleh, A.; Marouf, A.; Mahmoud, A.; Abu-Amara, M. Bluetooth Security. Available online: https://www.slideshare.net/ram_a ri/bluetooth-security.

[69] Becker, A. *Bluetooth Security & Hacks*; Seminar ITS Ruhr-Universitat Bochum SS2007; Ruhr University of Bochum: Bochum, Germany, 2007; Available online: https://gsyc.urjc.es/~anto/ubicuos2/bluetoot h_security_and_hacks.pdf.

[70] Trifinite: BlueBump. Available online: https://trifinite.org/trifinite_stu ff_bluebump.html.

[71] BlueDump. Available online: https://trifinite.org/trifinite_stuff_blued ump.html.

[72] Trifinite: BluePrinting. Available online: https://trifinite.org/trifinite_stu ff_blueprinting.html.

[73] The Attack Vector "BlueBorne" Exposes Almost Every Connected Device. Available online: https://www.armis.com/blueborne/

[74] Tsira, V.; Nandi, G. Bluetooth technology: Security issues and its prevention. *Int. J. Comput. Appl. Technol.* **2014**, *5*, 1833–1837.

[75] Nawir, M.; Amir, A.; Yaakob, N.; Lynn, O. Internet of Things (IoT): Taxonomy of Security Attacks. In Proceedings of the 2016 3rd International Conference on Electronic Design (ICED), Phuket, Thailand, 11–12 August 2016.

[76] Haataja, K. Security Threats and Countermeasures in Bluetooth-Enabled Systems. Kuopio University Publications H. Business and Information Technology 13. 2009. Page 75. Available online: http://epublica tions.uef.fi/pub/urn_isbn_978-951-27-0111-7/urn_isbn_978-951-27-0111-7.pdf.

Sources

- Bluetooth Special Interest Group (SIG) (Oct. 21, 2019) http://www.blue tooth.com
- Bluetooth SIG. "Bluetooth Core Specification v5.1." Jan. 21, 2019. (Oct. 28, 2019). https://www.bluetooth.org/docman/handlers/downloaddoc.as hx?doc_id=457080
- Bluetooth SIG. "Understanding Bluetooth Range." (Oct. 27, 2019) https://www.bluetooth.com/bluetooth-technology/range/
- Fleishman, Glenn. "Inside Bluetooth 2.0." Macworld.com. Feb. 9, 2005. (Oct. 21, 2019) http://www.macworld.com/news/2005/02/09/bluetooth 2/index.php
- Mobile Resource Group. "Lions and tigers ... and Bluesnarfing." Credentialed Mobile Device Security Professional. 2019. (Oct. 28, 2019) https://cmdsp.org/2019/08/12/lions-and-tigers-and-bluesnarfing/
- PCMag.com. "Bluetooth." (Oct. 28, 2019) https://www.pcmag.com/en cyclopedia/term/38794/bluetoothPCMag.com. "Bluetooth versions." (Oct. 28, 2019) https://www.pcmag.com/encyclopedia/term/69982/blue tooth-versions
- PCMag.com. "Spread spectrum." (Oct. 28, 2019) https://www.pcmag.com/encyclopedia/term/51883/spread-spectrumm
- Shepter, John. "How Bluetooth cuts the cord." March 15, 2005. (Oct. 21, 2019) http://searchmobilecomputing.techtarget.com/generic/0,2955 82,sid40_gci1067872,00.html

10

Z-Wave Technology

10.1 Introduction: What is Z-Wave?

The Z-Wave protocol is an interoperable, wireless, RF-based communications technology designed specifically for control, monitoring and status reading applications in residential and light commercial environments.

Z-Wave is a mesh networking technology developed in 1999 by Danish company Zensys while looking to create a standard for wireless radio frequency (RF) communication for home devices, and which allows for wireless control of residential appliances and other devices, such as lighting control, security systems, thermostats, windows, locks, swimming pools and garage door openers. Like other protocols and systems aimed at the home and office automation market, a Z-Wave system can be controlled via the Internet from a smart phone, tablet or computer, and locally through a smart speaker, wireless keyfob, or wall-mounted panel with a Z-Wave gateway or central control device serving as both the hub controller and portal to the outside. Z-Wave provides the application layer interoperability between home control systems of different manufacturers that are a part of its alliance. There are a growing number of interoperable Z-Wave products; over 1,700 in 2017, and over 2,600 by 2019.

10.1.1 Z-Wave Operating Characteristics

Z-Wave devices do not use the same frequency as other home devices like wireless phones, which typically operate at 2.4 GHz. The frequency used by Z-Wave varies based on country; however, in the United States, Z-Wave operates at 908.42 MHz, and in Europe, it operates at 868.42 MHz frequencies. This means Z-Wave devices would not interfere with other household devices. List of used frequencies worldwide is given in Section 10.1.2.

With such frequency band, Z-Wave devices have a greater signal range compared with ZigBee that works in the higher band of 2.4 GHz. The range of a Z-Wave device is influenced by a number of factors which also affecting the range of ZigBee. The factors, as discussed while discussing ZigBee, start first by the presence of walls in the vicinity. Typical reported ranges are around 30 m (90 ft) indoors and 100 m (300 ft) in the open air.

Extending the normal range of these products is possible simply by adding more Z-Wave devices to the network. Because all Z-Wave devices are repeaters, the signal is passed along from one to the next and each time it is repeated, another 30 m (approximately) of range is gained. Up to three additional devices (hops) can be used to extend the signal before the protocol terminates the signal (called a *Hop Kill*).

10.1.2 Technical Characteristics: Z-Wave frequency bands

Radio frequencies

Z-Wave is designed to provide reliable, low-latency transmission of small data packets at data rates up to 100 kbps. The throughput is 40 kbps (9.6 kbps using old chips) and suitable for control and sensor applications, unlike Wi-Fi and other IEEE802.11-based wireless LAN (WLAN) systems that are designed primarily for high data rates. Communication distance between two nodes is about 30 m (40 m with 500 series chip), and with message ability to hop up to four times between nodes, it gives enough coverage for most residential houses. Modulation is frequency-shift keying (FSK) with Manchester encoding.

Z-Wave uses the Part 15 unlicensed industrial, scientific, and medical (ISM) band. It operates at 868.42 MHz in Europe, at 908.42 MHz in the North America and uses other frequencies in other countries depending on their regulations. This band competes with some cordless telephones and other consumer electronics devices, but avoids interference with Wi-Fi, Bluetooth and other systems that operate on the crowded 2.4 GHz band. The lower layers, MAC and PHY, are described by ITU-T G.9959 and fully backwards compatible. In 2012, the International Telecommunication Union (ITU) included the Z-Wave PHY and MAC layers as an option in its G.9959 standard for wireless devices under 1 GHz. Data rates include 9600 bps and 40 kbps, with output power at 1 mW or 0 dBm. The Z-Wave transceiver chips are supplied by Silicon Labs.

Table 10.1 gives frequency bands, data rate and channel bandwidth supported by Z-Wave technology throughout the world.

Table 10.1 Z-Wave frequency bands

Region	RF Center Frequency (G.9959/MHz)	Data Rate	Channel Width
Australia	f_{ANZ1}/919.80, f_{ANZ2}/921.40,	100/40/9.6 kbps	400/300/300 KHz
Brazil	Same as Australia		
Canada	Same as USA		
Chile	Same as USA		
China	f_{CN1}/868.40,	100/40/9.6 kbps	400/300/300 KHz
European Union	f_{EU1}/869.85 f_{EU2}/868.40	100/ 40/9.6 kbps	400/300/300 KHz
Hong Kong	f_{HK1}/919.80	100/ 40/9.6 kbps	400/300/300 KHz
India	f_{IN1}/865.20	100/ 40/9.6 kbps	400/300/300 KHz
Israel	f_{IL1}/916.00	100/ 40/9.6 kbps	400/300/300 KHz
Japan	f_{JP1}/922.50, f_{JP2}/923.90, f_{JP3}/926.30	100/100/100 kbps for all bands	400/400/400 KHz for all bands
Korea	f_{KR1}/920.90, f_{KR2}/921.70, f_{KR3}/923.10	100/100/100 kbps for all bands	400/400/400 KHz for all bands
Malaysia	f_{MY1}/868.10	100/40/9.6 kbps	400/300/300 KHz
Mexico	Same as USA		
New Zealand	Same as Australia		
Russia	f_{RU1}/869.00	100/40/9.6 kbps	400/300/300 KHz
Singapore	Same as EU		
South Africa	Same as EU		
Taiwan	Same as Japan		
UAE	Same as EU		
USA	f_{US1}/916.00, f_{US2}/908.40	100/40/9.6 kbps	400/300/300 KHz

10.1.3 Z-Wave Features

The main features of Z-Wave protocol are summarized in Table 10.2.

10.1.4 Z-Wave History

The Z-Wave protocol was developed by Zensys, a Danish company based in Copenhagen, in 1999. That year, Zensys introduced a consumer light-control system, which evolved into Z-Wave as a proprietary system on a chip (SoC) home automation protocol on an unlicensed frequency band in the 900 MHz

Table 10.2   Z-Wave features

Specification	Z-Wave Support
Standard	ITU-T G.9959 (PHY and MAC)
RF Frequency Range	868.42 MHz in Europe, 908.42 MHz in US
Data rate	9.6, 40, 100 kbps
Maximum Nodes	232
Architecture	Master and slave in mesh mode
MAC layer	CSMA/CA
RF PHY modulation	FSK (for 9.6 kbps and 40 kbps), GFSK with BT=0.6 (for 100 kbps)
Coding	Manchester (for 9.6 kbps), NRZ (for 40 and 100 kbps)
Distance	30 m in indoors, 100 m in outdoors

range. Its 100 series chip set was released in 2003, and its 200 series was released in May 2005, with the ZW0201 chip offering a high performance at a low cost. Its 500 series chip, also known as Z-Wave Plus, was released in March 2013, with four times the memory, improved wireless range, and improved battery life.

The technology began to catch on in North America around 2005, when five companies, including Danfoss, Ingersoll-Rand and Leviton Manufacturing, adopted Z-Wave. They formed the Z-Wave Alliance, whose objective is to promote the use of Z-Wave technology, with all products by companies in the alliance interoperable.

In 2005, Bessemer Venture Partners led a $16 million third seed round for Zensys.

In May 2006, Intel Capital announced that it was investing in Zensys, a few days after Intel joined the Z-Wave Alliance. In 2008, Zensys received investments from Panasonic, Cisco Systems, Palamon Capital Partners and Sunstone Capital.

Z-Wave was acquired by Sigma Designs in December 2008. Following the acquisition, Z-Wave's US headquarters in Fremont, California were merged with Sigma's headquarters in Milpitas, California. On January 23, 2018, Sigma announced it planned to sell the Z-Wave technology and business assets to Silicon Labs for $240 million, and the sale was completed on April 18, 2018.

In 2005, there were six products on the market that used Z-Wave technology. By 2012, as smart home technology was becoming increasingly popular, there were approximately 600 products using Z-Wave technology available in the US. As of January 2019, there are over 2,600 Z-Wave certified interoperable products.

10.2 Z-Wave Protocol

10.2.1 Overview

The Z-Wave protocol is a low bandwidth half duplex protocol designed for reliable wireless communication in a low cost control network. The protocols main purpose is to communicate short control messages in a reliable manner from a control unit to one or more nodes in the network.

The protocol is not designed to transfer large amounts of data or to transfer any kind of streaming or timing critical data.

Figure 10.1 shows the Z-Wave protocol stack. The stack consists of 5 layers (Figure 10.1(a)): PHY layer, MAC layer, Transport layer, network layer and application layer. The security layer is not defined in Z-Wave open protocol specifications and hence it is implementation specific. We note here that sometimes PHY and MAC layers combined in one layer letting the stack consists of four layers (Figure 10.1(b)).

The Z-Wave protocol layers main function is to communicate very short messages of few bytes long from a control unit to one or more Z-Wave nodes. It is a low bandwidth and half duplex protocol to establish reliable wireless communication. Z-Wave protocol stack need not have to take care of large amount of data as well as any kind of time critical or streaming data.

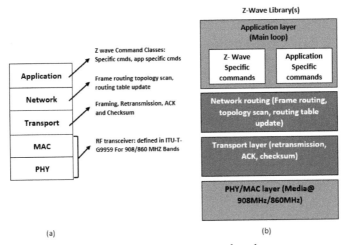

Figure 10.1 Z-Wave protocol stack.

The major functions of the protocol layers are (the details will be discussed next).

- Physical layer takes care of modulation and RF channel assignment as well preamble addition at the transmitter and synchronization at the receiver using preamble.
- MAC layer takes care of HomeID and NodeID, controls the medium between nodes based on collision avoidance algorithm and backoff algorithm.
- Transport layer takes care of transmission and reception of frames, takes care of retransmission, ACK frame transmission and insertion of checksum.
- Network layer takes care of frame routing, topology scan and routing table updates.
- Application layer takes care of control of payloads in the frames received or to be transmitted.

The details of the stack layers will be considered in detail in the chapter.

10.2.2 Z-Wave Basic Device Classes: Controller and Slave Nodes

Z-Wave network has two basic types of device (Figure 10.2):

- **Controllers**–devices that control other Z-Wave devices.
- **Slaves**–devices that are controlled by other Z-Wave devices.

Controller devices are the nodes in a Z-Wave network which initiates control commands. It also sends out the commands to other nodes. The slave devices are the nodes which replies based on command received and also execute the commands. Slave nodes also forward the commands to other nodes in the network. This makes it possible for controller to establish communication with the nodes who are not in radio frequency region. Controllers are factory programmed with a Home ID, this cannot be changed by the user. Slaves do not have a pre-programmed Home ID as they take the Home ID assigned to them by the network.

The simplest Z-Wave network is a single controllable device and a primary controller. Additional devices can be added at any time, as can secondary controllers, including traditional hand-held controllers, key-fob controllers, wall-switch controllers and PC applications designed for management and control of a Z-Wave network. A Z-Wave network can consist of up to 232 devices, with the option of bridging networks if more devices are required.

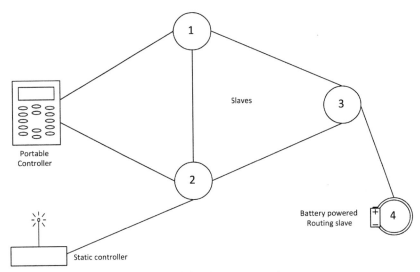

Figure 10.2 Z-Wave network.

The controllers can be classified to "portable controller" and "static controller, also classified as "primary" and "secondary." Slaves also can be classified into "slaves" and "routing slaves", that is, "slaves with routing capabilities."

The difference between the controller, slave and routing slave, is their knowledge of the network routing table and their ability to send messages to the network.

In the following, more details on the devices are given.

10.2.2.1 Controllers

The controller device will have full routing table for the mesh network and it will host it. Hence controller can communicate with all the nodes of Z-Wave network. There are two types of controllers: primary and secondary.

The controller which creates new Z-Wave network initially will become primary controller. This primary controller is master controller in the network and there will be only one in each Z-Wave network. Primary controller will have capability to include and exclude the nodes in the network. Hence primary controller always keeps latest topology of the network. Primary controller also takes care of managing allocation of node IDs.

The controllers which are added to the Z-Wave network using the primary controller are known as secondary controllers. They do not have capability to include or exclude any nodes. They will get copies of the routing tables from primary controller.

10.2.2.1.1 Portable Controller

A portable controller is a controller, which is designed to change position in the Z-Wave network. The portable controller uses a number of mechanisms to estimate the current location and hereby calculating the fastest route through the network. An example of a portable controller could be a remote control.

10.2.2.1.2 Static controller

A static controller is a fixed controller that mustn't change position in the network and has to be powered up all the time. This controller has the advantage that Routing slaves can report unsolicited status messages to it, and it also has the advantage of always knowing where it is located in the network. A static controller will typically be a secondary controller in a Z-Wave network. An example of a static controller could be an Internet gateway that monitors a Z-Wave system.

10.2.2.1.3 Static update controller

A Z-Wave network can optionally have a static controller with enabled Static Update Controller (SUC) functionality to distribute network topology updates. A SUC is a static controller that will receive notifications from the primary controller regarding all changes made to the network topology. In addition the SUC is capable of sending network topology updates to other controllers and routing slaves upon request. It is the application in a primary controller that requests a static controller to become a SUC. There can only be one SUC in a Z-Wave network.

10.2.2.1.4 SUC ID server

A Z-Wave network can optionally have a SUC with enabled node ID server functionality (SIS). The SIS enables other controllers to include/exclude nodes in the network on its behalf. The SIS is the primary controller in the network because it has the latest update of the network topology and capability to include/exclude nodes in the network. When including additional controllers to the network they become inclusion controllers because they have the capability to include/exclude nodes in the network on behalf of the SIS. The inclusion controllers network topology is dated from last time a node was included or it requested a network update from the SIS and therefore it can't be classified as a primary controller.

10.2.2.1.5 Installer controller

An Installer controller is a portable controller that has additional functionality, which enables it to do more sophisticated network management and network quality testing than other controllers. An example of an installer controller could be an installation tool used by an installer to install a Z-Wave network at a customer site.

10.2.2.1.6 Bridge controller

A Z-Wave network can optionally have a bridge controller. A bridge controller is an extended static controller, which incorporates extra functionality that can be used to implement controllers, targeted for bridging between the Z-Wave network and other networks. The bridge controller device stores the information concerning the nodes in the Z-Wave network and in addition it can control up to 128 virtual slave nodes. A virtual slave node is a slave node that corresponds to a node, which resides on a different network type. An example of a bridge controller could be a bridge between an UPnP network and a Z-Wave network to link broadband and narrowband devices together in a home entertainment application.

10.2.2.2 Slaves

The slave devices/nodes in Z-Wave network receive the commands and performs action based on the commands. These slave nodes are unable to transmit information directly to the other slave nodes or controllers unless they are instructed to do so in the commands. The slave nodes do not compute routing tables. They can store routing tables. They will act as a repeater.

10.2.2.2.1 Slave

Slave nodes are nodes in a Z-Wave network that receives commands and performs an action based on the command. Slave nodes are unable to send information directly to other slaves or controllers unless they are requested to do so in a command. An example of a slave node could be a light dimmer.

10.2.2.2.2 Routing slave

Routing slaves has the same overall functionality as a slave. The major difference is that a routing slave can send unsolicited messages to other nodes in the network. They store a number of static routes for use when sending unsolicited messages to a limited number of nodes. An example of a routing slave node could be a thermostat or a Passive Infrared (PIR) movement sensor.

Table 10.3   Properties of the Z-Wave device models

	Neighbors	Route	Possible Functions
Controller	Knows all neighbors	Has access to complete routing table	Can communicate with every device in the network, if route exists
Slave	Knows all neighbors	Has no information about routing table	Can only reply to the node which it has received the message from. Hence, it cannot send unsolicited messages.
Routing Slave	Knows all neighbors	Has partial knowledge of routing table	Can reply to the node which he has received the message from and can send unsolicited messages to a number of predefined nodes he has a route too

Table 10.4   Typical applications for slaves

Slave	Fixed installed mains powered devices like wall switches, wall dimmers or Venetian blind controllers
Routing Slave	Battery-operated devices and mobile applicable devices as for example sensors with battery operation, wall plugs for Schuko and plug types, thermostats and heaters with battery operation and all other slave applications

10.2.2.2.3 Enhanced slave

Enhanced slaves have the same functionality as routing slaves and they are handled in the same way in the network. The difference between routing slaves and enhanced slaves is that enhanced slaves have a real time clock and an EEPROM for storing application data. An example of an enhanced slave node could be a weather station

Differences between controller, slave, and routing slave are given in Table 10.3. Table 10.4 gives the typical applications of slaves

The primary controller includes other nodes into the network by assigning them its own Home ID. If a node accepts the Home ID of the primary controller this node becomes part of the network. The primary controller also assigns an individual Node ID to each new device that is added to the network. This process is known as **Inclusion**.

Different classes provide the device with a certain role in the Z-Wave network. Inside a Basic Class, Generic and Specific device classes are used to achieve the wanted functionality in the control network. In the Z-Wave protocol, the unique identification of the devices is used through a 32-bit ID. This ID value cannot be changed as it is written in the device chipset by the device manufacturer. A Z-Wave network has only one primary controller device at a time. Each of the 232 nodes of this network can also be a repeater for forwarding data to its neighbors, mediating a connection. Battery-powered nodes do not enjoy this facility. In an environment with a certain level of device drift or even when a device is removed from the network for some reason, the network topology may change. Changing network topology can lead to problems in packet forwarding and packet routing in the network. To minimize this effect, routing tables should be kept up-to-date, optimized and any new topology detected; Z-Wave supports the discovery and suitability of the new network topology. This is possible by keeping the routing table up to date on each device and showing all neighboring devices. When a node changes its position or is removed from the network, a topology failure can start an automatic topology and healing procedure to detect the new topology and define the best routes to update the routing tables. This mechanism is subjected to unauthorized modification of routing table attacks by rouge nodes.

10.2.2.3 Home ID and Node ID

The Z-Wave protocol uses a unique identifier called the Home ID to separate networks from each other. The Home ID is a 32 bit unique identifier that is preprogrammed in all controller devices. All slave nodes in the network will initially have a home ID that is zero, and they will therefore need to have a home ID assigned to them by a controller in order to communicate with the network. Controllers in a network can exchange home ID's so more than one controller can control slave nodes in a network.

Node ID's are used to address individual nodes in a network, they are only unique within a network defined by a unique home ID. A node ID is an 8 bit value and like home ID's they are assigned to slave nodes by a controller.

Nodes with different Network IDs cannot communicate with each other. The Node ID is the address of a single node in the network.

Table 10.5 gives comparison between home ID and node ID

10.2.2.4 Z-Wave chip

Managing the routing of messages on a network is complicated, especially if some devices in the network are mobile. Z-Wave uses source routing, so any device capable of initiating communication must know which routes are currently possible, choose the best route and then embed the routing

Table 10.5 Home ID and node ID comparison

	Definition	In the Controller	In the Slave
Home ID	The Home ID is the common identification of a Z-Wave network	The Home ID is already set as factory default	No Home ID at factory default
Node ID	The Node ID is the individual identification (address) of a node within a common network	Controller has its own Node ID predefined (typically 0x01)	Assigned by the primary controller

information into the messages that it sends. Such routing algorithms have been well developed for a variety of networks, but they require a lot of code. Lots of code means lots of memory on a chip and therefore higher build costs for devices that use the chip. Z-Wave chips have 32K or 64K bytes of flash memory, with provision for adding more memory externally. To reserve as much memory as possible for the application code in a Z-Wave device, the Z-Wave communications stack must be kept as small as possible. This is one of the main reason that Z-Wave defines, as mentioned before, a number of different device types that have varying capabilities and stack sizes: Controllers, Routing Slaves and Slaves.

The Z-Wave chip is optimized for battery-powered devices, and most of the time remains in a power saving mode to consume less energy, waking up only to perform its function. With Z-Wave mesh networks, each device in the house bounces wireless signals around the house, which results in low power consumption, allowing devices work for years without needing to replace batteries. For Z-Wave units to be able to route unsolicited messages, they cannot be in sleep mode. Therefore, battery-operated devices are not designed as repeater units. Mobile devices, such as remote controls, are also excluded since Z-Wave assumes that all repeater capable devices in the network remain in their original detected position.

10.2.2.5 Z-Wave frame structure

Figure 10.3 shows the Z-Wave frames of the different layers. The figure shows PHY/MAC frame, 4 types of frames at transport layer and application layer frame. As shown in the figure, **Z-Wave frame** consists of a preamble part, SOF (Start of Frame), Frame data and EOF (End of Frame) symbol. The data part is Manchester codes or NRZ coded based on data rate. MAC layer controls the RF spectrum. Data part comes from the upper layers and Z-Wave frame as mentioned in formed at the MAC/PHY layers. After this is done the

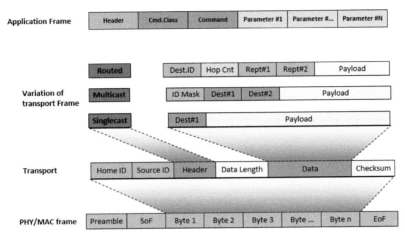

Figure 10.3 Z-Wave frames.

Z-Wave frame as depicted is transmitted by the RF antenna after necessary radio frequency conversion as desired using RF Transceiver.

10.2.2.5.1 Z-Wave messaging

The minimum length of a properly formatted Z-Wave message is 9 bytes, but a routed message requires 12 bytes plus repeater data plus the payload. The message protocol includes routing, frame acknowledgment, collision avoidance with random back off and a frame checksum with retransmission if necessary. The Z-Wave network is self-organizing and self-healing. To achieve self-organization, Z-Wave nodes have software that discovers the node's neighbors and informs the network's Static Update Controller (SUC) about them. A Source Routing Algorithm (SRA) in devices capable of initiating communication finds message pathways and generates routes based on a network topology database. Self-healing requires software to dynamically generate new routes around temporarily unavailable nodes. Moving nodes have software routines that can request new neighbor searches automatically. This software, which is part of the Z-Wave stack, resides in on-chip memory.

10.2.3 Z-Wave Physical Layer

The ITU-T recommendation G.9959 contains physical and MAC layer specifications for sub GHz radio communication including the Z-Wave protocol. It also outlines some aspects of the Z-Wave transport layer such as frame

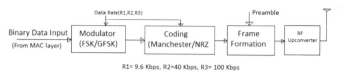

R1= 9.6 Kbps, R2=40 Kbps, R3= 100 Kbps

Figure 10.4   Z-Wave layer transmitter.

Table 10.6   Z-Wave data rate and accuracy

Data Rate Designation	Bit Rate	Symbol Rate	Accuracy
R1	9.6 kbps	19.2 Kbaud	±27 ppm
R2	40 kbps	40 Kbaud	±27 ppm
R3	100 kbps	100 Kbaud	±27 ppm

formats and Beam control which is necessary to communicate with Z-Wave door locks.

The following are the functions performed at Z-Wave physical layer (Z-Wave PHY):

- to assign RF profile to the Z-Wave physical channel
- to activate and deactivate RF transceiver
- Transmission and reception of data frames, that is, payload
- clear channel assessment
- radio frequency selection
- link quality check based on received frames

There are three different data rates supported in Z-Wave Physical layer. They are 9.6 kbps (designated as 'R1'), 40 kbps (designated as 'R2') and 100 kbps (designated as 'R3'). Based on this data rate different physical layer configuration is made.

Figure 10.4 depicts a Z-Wave layer transmitter.

As shown in Figure 10.4, Z-Wave physical layer consists of modulation and coding blocks. Based on the data rates R1, R2 or R3 different configurations of the Z-Wave PHY is made as mentioned in Table 10.6.

After the modulation and coding of the input MAC layer data, frame is formed as per format outlined in Figure 10.4. In order to do this at physical layer preamble is inserted before 'SOF' field.

As mentioned in the Table 10.7, FSK modulation is employed for R1 and R2 rates while GFSK is used for R3 data rate. For R1 Manchester coding is used while for R2/R3 NRZ coding types is used.

Table 10.7   Modulation and coding format in Z-Wave physical layer

Data Rate Designation	Modulation	Coding	Separation	Symbols
R1	FSK	Manchester	40 KHz ± 10%	Binary
R2	FSK	NRZ	40 KHz ± 10%	Binary
R3	GFSK (BT = 0.6)	NRZ	58 KHz ± 10%	Binary

Figure 10.5   MAC frame.

All the data are transmitted in little endian format. They are transmitted in blocks of 8 bit with MSB is sent first after Manchester coding. Manchester coding will help in obtaining DC free signal.

At the Z-Wave physical layer receiver, preamble is used for synchronization before decoding and demodulation is carried out to retrieve the MAC layer data. After the decoding of the data, they will be passed to the upper layers for further processing.

10.2.4 Z-Wave MAC Layer

The Z-Wave MAC layer controls the radio frequency medium. The functions performed at Z-Wave MAC layer are:

- unique network ID number (HomeID)
- up to 232 nodes in one network
- collision avoidance algorithm
- backoff algorithm
- automatic retransmission for reliable data transfer
- support for low-power operation via dedicated wakeup patterns.

10.2.4.1 Z-Wave MAC frame types

The data stream is Manchester coded and consists of a preamble, start of frame (SOF), frame data and an end of frame (EOF) symbol. The frame data are the part of the frame that is obtained from transport layer at the transmitter and given to the transport layer at the receiver.

All data are sent in little endian format.

The MAC layer is independent of the RF media, frequency and modulation method but the MAC layer requires either access to the frame data when

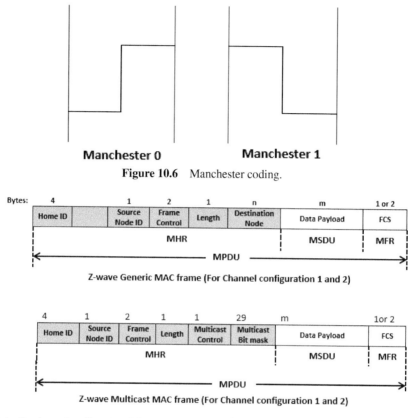

Figure 10.6   Manchester coding.

Z-wave Generic MAC frame (For Channel configuration 1 and 2)

Z-wave Multicast MAC frame (For Channel configuration 1 and 2)

Note: For channel configuration 3, 'sequence number' (1 byte) is added after 'Length' field above

Figure 10.7   MAC layer frame types.

received or to the whole signal in binary form either as an decoded bit stream or to the Manchester coded bit stream.

Data are transmitted in blocks of 8bit, most significant bit first, and the data are Manchester coded in order to have a DC free signal.

10.2.4.2 Collision avoidance

The MAC layer has a collision avoidance mechanism that prevents nodes from starting to transmit while other nodes are transmitting. The collision avoidance is achieved by letting nodes be in receive mode when they are not transmitting, and then delay a transmit if the MAC layer is currently in the data phase in the receiver. The collision avoidance is active on all types of nodes when they have the radio activated.

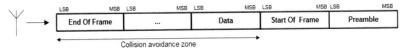

Figure 10.8 Collision avoidance zone.

MAC frame fields

The MAC layer is independent of the RF media, frequency and modulation method but the MAC layer requires either access to the frame data when received or to the whole signal in binary form either as an decoded bit stream or to the Manchester coded bit stream.

The transmission of the frame is delayed a random number of milliseconds.

Let us understand the Z-Wave MAC frame in some more details. The Z-Wave MAC layer uses frame formats based on channel configuration, that is, 1/2 and 3. Figure 1 depicts generic MAC frame format and multicast frame format for channel configuration 1 and 2. For channel configuration 3, 'sequence number' (1 byte) field is added after the 'length' field.

In Figure 10.7, channel configuration-1 supports 1 channel (Ch.B), channel configuration-2 supports 2 channels (Ch.A and Ch.B) and channel configuration-3 supports 3 channels (Ch.A, Ch.B, Ch.C).

Each MAC frame (MPDU) consists of MHR, MAC payload and MFR.

- MHR consists of addresses, frame control and length information
- MAC payload contains data as per frame type. Ack frames do not have payload field.
- MFR contains a FCS(Frame Check Sequence).

HomeID

This field is 4 bytes in length. Z-Wave HomeID specifies unique network identifier. All nodes in a Z-Wave network have the same HomeID. It is assigned by a primary node during inclusion.

Source Node ID

This field is an 8 bit unique identifier of a node. Along with HomeID, NodeID identifies the node of the originated frame.

Frame Control

This field is 16 bits in length. This frame control field contains information defining the frame type, addressing fields and other control flags.

Header type subfield

This field defines different frame type, that is, single cast, multicast, ACK, routed frame etc.

Length

It is 1 byte size and indicates length of the whole MPDU in bytes. A receiving node does not accept more bytes than the maximum length allowed for the actual data rate.

Sequence Number:

This field is a number provided by higher layers when transmitting. The valid range is 0x00 to 0xFF.

Destination Node ID:

The destination NodeID is used to address individual nodes.

0x00–(Uninitialized NodeID)
0x01–0xE8 (NodeID)
0xE9–0xFE (Reserved)
0xFF–(Broadcast NodeID)

Data Payload

This field is variable size in length. It contains information specific to individual frames. An acknowledgment frame do not have this field.

FCS

An 8-bit frame checksum is used for checking frame correctness for R1 and R2 data rates at the receiver. This error detection technique will help in finding the erroneous frame and hence will initiate retransmission in the Z-Wave network.

Multicast frames:

Multicast frames carry a destination bit map

10.2.5 Z-Wave Transport Layer

The Z-Wave transfer layer controls the transfer of data between two nodes including retransmission, checksum check and acknowledgments.

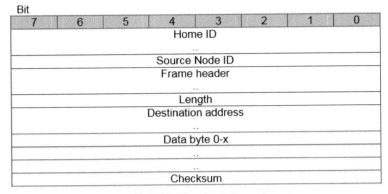

Figure 10.9 Z-Wave basic frame format.

Frame Layout

The Z-Wave transfer layer contains 4 basic frame formats used for transferring commands in the network. All 4 frames uses the following frame layout:

Z-Wave transport layer is mainly responsible for retransmission, packet acknowledgment, waking up low power network nodes and packet origin authentication. The Z-Wave transport layer (or transfer layer) consists of four basic frame types, see Figure 10.7. These are used for transferring commands in the network. All the frames use the same format (Figure 10.9):

Transport Frame = {HomeID, Source NodeID, Header, length, Data byte (0 to X), Checksum}

As shown, each of the transport Z-Wave frame contains the 32 bits Home ID that identifies the associated Z-Wave network, 8 bits source node ID, frame header that defines frame type (single-cast, multi-cast, routed) and control flag such as low power transmission, 8 bits payload length followed by the payload and the 8 bits frame checksum value.

Transport layer relies on a frame checksum value to detect and discard erroneous frames. The Z-Wave protocol uses the following checksum algorithm from the ITU-T G.9959 standard:

```
BYTE GenCheckSum(BYTE *Data, BYTE Length){
    BYTE CheckSum = 0xFF;
    for (; Length > 0; Length--){
    CheckSum ^= *Data++;}
return CheckSum;}
```

An overview of Z-Wave transport frame format and fields is shown in Figure 10.3. Frame retransmission occurs when an acknowledgment frame was not received from the destination node before the frame expiration time. Beam frames that are used to wake up battery powered Z-Wave nodes are controlled by the transport layer. Some battery powered Z-Wave devices including door locks needs to listen for the incoming commands from the network controller, but keeping their radio on will drain the battery quickly. In order to preserve the battery power, the device enters sleep mode and periodically turns on its radio, looking for beam frames. The transmitting node sends several back to back beam frames in 100 ms intervals to ensure that the sleeping device will notice one of those frames when waking up and therefore it will keep its radio on to receive subsequent transmissions.

When in secure transmission mode, an 8-byte frame authentication header is added to the end of the frame just before the frame checksum filed. No public specifications for the Z-Wave origin authentication header were available before our research. By using Z-Force tool and performing binary code analysis of a Z-Wave controller appliance firmware we found the cryptographic algorithm and parameters to calculate this header value.

As mentioned above, the transport layer has 4 frame types. The 4 frame types of transport layer is explained below:

10.2.5.1 Singlecast frame type

These type of frames are transmitted to one specific Z-Wave node. The frame is acknowledged so that transmitter will know whether the frame is received or not.

A singlecast transmission has the following frame flow.

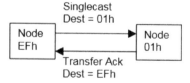

Figure 10.10   Singlecast.

If this frame or its ACK is lost or damaged then the singlecast frame is retransmitted. In order to avoid potential collisions with parallel systems the retransmissions are delayed with a random delay. The random delay must be in steps of the time it takes to send a frame of the maximum frame size and receive the Transfer Ack. The singlecast frame can optionally be used without acknowledgment in a system where reliable communication is not required.

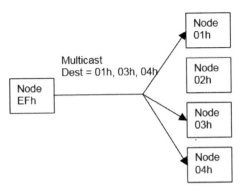

Figure 10.11 Case of multicast.

10.2.5.2 ACK frame type

The transfer acknowledge is a Z-Wave singlecast frame where the size of the data section is zero.

10.2.5.2.1 Multicast frame type

These frames are transmitted to more than one node, that is, ranging from 1 to 232 nodes. This type of frame does not support acknowledgment concept. Hence this type is not used for reliable communication.

The multicast destination address is used to address selected nodes without having to send a separate frame to each node. Note that a multicast frame doesn't get acknowledged so this type of frame can't be used for reliable communication. If reliable communication is needed a multicast must be followed by a singlecast frame to each destination node.

10.2.5.2.2 Broadcast frame type

Broadcast frames are received by all nodes in a network, and the frame is not acknowledged by any nodes.

Note that a broadcast frame doesn't get acknowledged so this type of frame cannot be used for reliable transfer. If reliable communication is needed a broadcast must be followed by a singlecast frame to each destination node.

10.2.6 Z-Wave Network Layer (Routing Layer)

Z-Wave protocol forms a mesh network with one primary controller device and up to 232 nodes each of which can act as a packet repeater–with the exception of battery powered nodes–to route Z-Wave data even when the two communicating parties cannot establish a direct radio link between

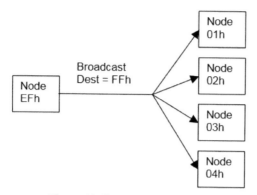

Figure 10.12   Case of broadcast.

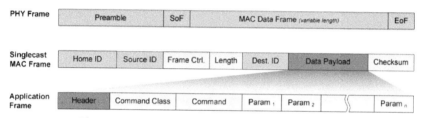

Figure 10.13   Z-Wave frame format in different layers.

each other. In order to determine the best route to a destination node, each device in the Z-Wave network maintains a network topology that indicates all other devices in proximity. When device locations at home changes or they are removed from the network, this topology can become wrong and cause routing issues in the network. The Z-Wave protocol supports automatic topology discovery and healing to detect new network location and routes and optimize the routing tables. Although, Z-Wave routing mechanism and topology discovery might be subjected to attacks such as unauthorized modification of routing tables by rouge nodes, we did not perform security tests in network layer, as our research was focused on the encryption and origin authentication that are handled by transport and application layers

The Z-Wave routing layer controls the routing of frames from one node to another. Both controllers and slaves can participate in routing of frames in case they are always listening and have a static position. The layer is responsible for both sending a frame with a correct repeater list, and also to ensure that the frame is repeated from node to node. The routing layer is also responsible for scanning the network topology and maintaining a routing table in the controller.

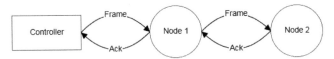

Figure 10.14 Routed singlecast.

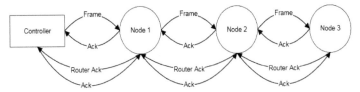

Figure 10.15 Routed acknowledgment.

10.2.6.1 Frame Layout

The Z-Wave routing layer has 2 kinds of frames that are used when repeating of frames is necessary.

10.2.6.1.1 Routed singlecast frame type

The Z-Wave routed singlecast is a one-node destination frame with acknowledge that contains repeater information. The frame is repeated from one repeater to another until it reaches its destination.

10.2.6.1.2 Routed acknowledge frame type

The Z-Wave route acknowledge is a routed singlecast frame without payload that is used to tell the controller that the routed singlecast has reached its destination.

10.2.6.2 Routing table

The routing table is where a controller keeps the information from the nodes about the network topology. The table is a bit field table where all information about what nodes that can see each other is kept. This routing table is built by primary controller based on information received from all the nodes in the Z-Wave network. Figure 10.16 illustrates a network topology and the resulting routing table.

10.2.6.3 Route to node

Finding the route to a node is a difficult task because a portable controller is defined as a device that will be moved around a lot (e.g., a remote control) Therefore a portable controller will always try to reach a node without routing

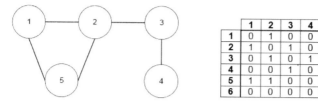

Figure 10.16   Network topology and routing table.

and if that fails the portable controller will use several techniques to find the best route to the node.

Example 10.1: Routing table

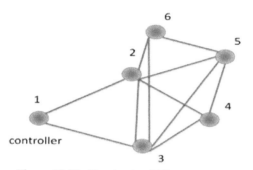

Figure 10.17   Routing in a Z-Wave network.

Figure 10.2 shows a Z-Wave meshed network, with one controller and five nodes. The controller can communicate directly with node 2 and 3. There is no direct connection to node 4, 5 and 6. If the controller needs to communicate with node 4, the communication to node 4 works either via node 2 or via node 3.

The routing for this network is shown in Table 10.8–the rows contain the source nodes and the columns contain the destination nodes. A cell with "1" indicates that the nodes are neighbors and a "0" shows there is no direct communication path. The table also shows the connection between Source Node 1 and destination Node 4. The cell between Node 1 and 4 is marked "0". Therefore the network routes the signal via Node 3 which is in direct range of both Node 1 and Node 4. This option is shown in Figure 10.18a. It is also possible for the controller to reach node 4 through node 2 as shown in Figure 10.18b.

Table 10.8   Routing table for the Z-Wave network

Source node	To 1	To 2	To 3	To 4	To 5	To 6
1	X	1	1	0	0	0
2	1	X	1	1	1	1
3	1	1	X	1	1	1
4	0	1	1	X	1	0
5	0	1	1	1	X	1
6	0	1	1	0	1	X

Source node	To 1	To 2	To 3	To 4	To 5	To 6
1	X	1	1	0	0	0
2	1	X	1	1	1	1
3	1	1	X	1	1	1
4	0	1	1	X	1	0
5	0	1	1	1	X	1
6	0	1	1	0	1	X

Source node	To 1	To 2	To 3	To 4	To 5	To 6
1	X	1	1	0	0	0
2	1	X	1	1	1	1
3	1	1	X	1	1	1
4	0	1	1	X	1	0
5	0	1	1	1	X	1
6	0	1	1	0	1	X

(a)   (b)

Figure 10.18   Route between controller and node 4 (a) through node 3, (b) through node 2.

Example 10.2

Another example is the Z-Wave network given in Figure 10.19.a. In this figure, Node 6 can only communicate with the rest of the network using Node 5 as a repeater. Since the controller does not have a direct connection to Node 5, the controller needs to use one of the following routes: "**1 -> 3 -> 4 -> 5 -> 6**," shown in Figure 10.4b, or "**1 -> 2 -> 5 ->6**" which is shown in Figure 10.19c.

10.2.7 Application Layer

The Z-Wave application layer is responsible for decoding and executing commands in a Z-Wave network. The only part of the application layer that is described in this overview is the assignment of Home ID's and Node ID's and the replication of controllers. The rest of the application layer is implementation specific, and can be different from one implementation to another.

The frame format used in **application layer** consists of following fields (Figure 10.20):

Frame Format = {Single/Multi, broadcast frame header, Application command class,Application command, Command parameter1-to-X}

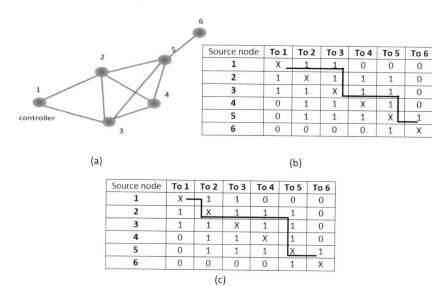

(a)   (b)

(c)

Figure 10.19   Alternative Z-Wave network routing.

| Bit |
7	6	5	4	3	2	1	0
Single/Multi/Broadcast frame header
Application command class
Application command
Command parameter 1
Command parameter 2
...
Command parameter x

Figure 10.20   Z-Wave application frame format.

The application command class defines class of commands the command belong to:

00h-1Fh (This command class reserved for Z-Wave protocol)

20h-FFh (This command class reserved for Z-Wave application)

All the Z-Wave frame types except the acknowledgment frame contain an application command.

10.2.7.1 Frame layout

The frame format used in the Z-Wave application layer is described in this section.

Command class

Z-Wave networks are composed of controllers, sensors, and actuator devices. To ensure device compatibility, the application layer protocol is well-defined (OpenZwave, 2016); however, not all devices require the ability to participate in every application layer transaction. For example, a light switch does not need to know how to respond to a request for a temperature reading. Consequently, the application layer is partitioned by functionality into a series of *command classes*. As a fact, Command Classes are the heart of the operability layer, they define groups of functionality that is implemented in a standard way. Within each command class is a subset of the application layer commands pertaining to the class. A given Z-Wave device belongs to one or more command classes and the associated application layer protocol functionality is included during compilation of its firmware image. A device announces this set of supported classes to the controller during the pairing operation. Each command class is a unique byte value (OpenZwave, 2016). For example, a light switch may announce that it uses the *Binary Switch* command class, which provides commands to get, set, and report the state of the switch (OpenZwave, 2016). Given the announcement, the controller is now aware of a subset of commands that the light switch obeys. Commands sent to the light switch from unsupported command classes are ignored.

Z-Wave certification ensures that these functions are implemented correctly, and therefore ensures that different devices are compatible. Currently approximately 120 command classes are defined.

Application Command class: The application command class specifies which class of commands the command belongs to.

Currently defined command classes are as follows:

Command Class	Description
00h-1Fh	Reserved for the Z-Wave Protocol
20h-FFh	Reserved for the Z-Wave Application

The application command specifies the specific command or action within the command class.

Command parameter 1-x:

The command parameters contain any parameters associated with the specified command. The number of parameters depends on the command. All frame types except acknowledge can contain an application command.

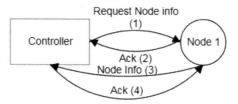

Figure 10.21 Get node info frame flow.

10.2.7.2 Node information

Because a controller in a Z-Wave network should be able to control many different kinds of nodes, it is necessary to have a frame that describes the capabilities of a node. Some of the capabilities will be protocol related and some will be application specific. All nodes will automatically send out their node information when the action button on the node is pressed. A controller can also get the node information from a node by requesting it with a "get node information" frame.

10.2.7.2.1 Node information frame flow

The node information frame is send out by a node each time its action button is pressed. The frame is sent out as a broadcast to any controller/node that might be interested in the information. A controller can also request the node information from a node by sending a get node information frame to it.

10.3 Z-Wave Network Operation

10.3.1 Z-Wave Network Setup

As discussed before, Z-Wave allows Only One Master Controller (Primary controller). Because Z-Wave is a routed network, new Z-Wave devices must be enrolled into the network before they can be used. One master controller is allowed because the master controller maintains the database that represents the network's topology. Whichever device you choose as your master controller, to add secondary controllers to Z-Wave network, it is a must go through a multistep controller replication process. The result is that only the master controller can add or remove Z-Wave devices from the Z-Wave network. If any time is required to modify an existing network it will be needed to repeat the controller replication procedure in order to copy the new master controller network topology database to all of the secondary controllers.

10.3.1.1 Including and removing device in the network: 'pairing' operation

A distinguishing aspect of Z-Wave is the manner in which devices enter and leave the network. A device must be "included" to the Z-Wave network before it can be controlled via Z-Wave. This *inclusion* or *pairing* process is similar to Bluetooth device pairing (See Chapter 9). A user wishing to add a device to the network first puts the Z-Wave controller and the new device into a pairing mode. While placing a device in pairing mode is device specific, this is usually achieved by pressing a sequence of buttons on the controller and on the device being added to the network or physically resetting the device. This sequence only needs to be performed once, after which the device is always recognized by the controller. While in the pairing mode, the controller adds any device found to also be in pairing mode. Depending on the controller implementation, it may present a list of discovered devices and allow the user to select ones to add. Alternatively, the controller may add all discovered devices within physical proximity without discrimination. Examples of each type of controller include the Mi Casa Verde Vera (Vera, 2016) and Aeon Labs Z-Stick (Z-Stick, 2016), respectively.

Removing a device from a Z-Wave network is accomplished in a similar manner. The controller learns the signal strength between the devices during the inclusion process, thus the architecture expects the devices to be in their intended final location before they are added to the system. Typically, the controller has a small internal battery backup, allowing it to be unplugged temporarily and taken to the location of a new device for pairing. The controller is then returned to its normal location and reconnected.

10.3.1.2 Z-wave network enrollment and button associations

After adding a new Z-Wave device to the network by enrolling it with the network master controller, there still has to be a choice of which button or buttons on the master controller will operate the device. In other words, there are two issues to resolve when adding a new Z-Wave device to an existing Z-Wave network–network enrollment and button association. The method for accomplishing these two tasks depends on the controller. For example, the Intermatic controller requires two distinct steps–first you enroll a new device into the Z-Wave network, then you associate buttons to control the device. It is needed to use the controller instruction manual to find out how to perform these steps, because the controller's buttons do not have obvious labels for performing this setup and the display cannot show arbitrary text prompts. In contrast, the HomePro handheld controller merges the two setup steps into one by using the concept of HomePro groups and HomePro scenes. There are six numbered buttons on the HomePro handset, with each button able to control a HomePro group or scene. The difference between a group and a scene is that scenes can remember brightness settings. To add a new Z-Wave device to the HomePro master controller, you traverse a text menu tree and follow text prompts on the display. The controller's software can recognize whether a device is new to the network or if it is already enrolled in the network and just being associated with more than one HomePro group button. The software automatically enrolls devices that are new to the network without involving the user, so users believe they are simply performing button associations.

Z-Wave Association Process

To support the creation of associations between buttons on a controller and actions of a controlled device, Z-Wave specifies that:

- The network must provide an Association Wizard,
- There must be a sanity check of requested associations, and
- All nodes must be able to present their supported capabilities.

Software routines to support these features, to the extent that product developers choose to develop and include them, further enlarge memory requirements.

10.3.1.3 Button pairs versus on/off toggles

The Intermatic controller uses on/off button pairs to control devices, while the HomePro handset uses single buttons that alternate between on/bright and off/dim. Neither controller is "guest friendly," meaning that a person who had never seen one before would not be able to figure out how to use it without some trial and error.

10.4 Z-Wave Security

Z-Wave is based on a proprietary design, supported by Sigma Designs as its primary chip vendor, but the Z-Wave business unit was acquired by Silicon Labs in 2018. In 2014, Mitsumi became a licensed second source for Z-Wave 500 series chips. Although there have been a number of academic and practical security researches on home automation systems based on ZigBee and X10 protocols, research is still in its infancy to analyze the Z-Wave protocol stack layers, requiring the design of a radio packet capture device and related software to intercept Z-Wave communications. An early vulnerability was uncovered in AES-encrypted Z-Wave door locks that could be remotely exploited to unlock doors without the knowledge of the encryption keys, and due to the changed keys, subsequent network messages, as in "door is open", would be ignored by the established controller of the network. The vulnerability was not due to a flaw in the Z-Wave protocol specification but was an implementation error by the door-lock manufacturer.

On November 17, 2016, the Z-Wave Alliance announced stronger security standards for devices receiving Z-Wave Certification as of April 2, 2017. Known as Security 2 (or S2), it provides advanced security for smart home devices, gateways and hubs. It shores up encryption standards for transmissions between nodes, and mandates new pairing procedures for each device, with unique PIN or QR codes on each device. The new layer of authentication is intended to prevent hackers from taking control of unsecured or poorly-secured devices. According to the Z-Wave Alliance, the new security standard is the most advanced security available on the market for smart home devices and controllers, gateways and hubs.

In general, Z-Wave protocol provides packet encryption, integrity protection and device authentication services and is gaining momentum against ZigBee protocol with regards to home automation. This is partly due to interoperability of devices and shorter time to market on the vendor side. Another benefit is that it is less subjected to signal interference compared to the ZigBee protocol, which operates on the widely populated 2.4 GHz band shared by both Bluetooth and Wi-Fi devices.

The following are features used in Z-Wave security:

- Message Freshness: 64 bit Nonce
- 128 bit random network key: K_n
- Encryption: AES-OFB
- Data Authentication: AES-CBCMAC
- 128 bit Cipher and MAC Keys: Derived from K_n
- Custom key Establishment Protocol, as shown below.

Z-Wave data origin authentication is based on the cipher block chaining message authentication code, that is, CBC-MAC technique. This technique calculates a message authentication code, that is, MAC from a block cipher algorithm such as AES. Figure 10.22 describes K-exchange protocol used in Z-Wave security.

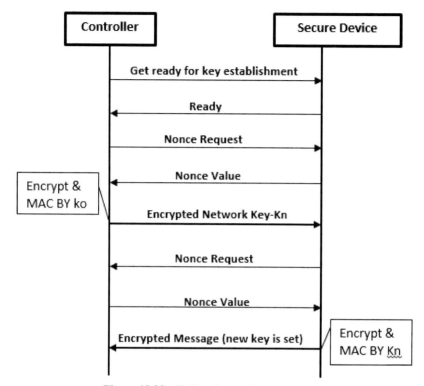

Figure 10.22 Z-Wave key exchange protocol.

10.4.1 Vulnerabilities

The Z-Wave protocol suffers from vulnerabilities some of them are considered next:

10.4.1.1 Impersonation

Impersonation attacks violate the source integrity of the protocol. With the exception of the controller, Z-Wave devices implicitly trust the source and destination fields of the MPDU frame. This makes it trivial to impersonate frames originating from the controller or another device.

Devices using the Z-Wave security layer have some protection against outsider impersonation. Devices specify, at a command class granularity, which command messages must use the secure frame. Secure frames are signed and encrypted using keys exchanged during network inclusion. An outsider who is not in possession of the authentication and encryption keys is unable to transmit a valid secure frame. Regardless of the chosen source

ID, the outsider is unable to impersonate the origin of a command message if the destination requires that it is sent in a secure frame. However, the outsider may still perform impersonation attacks on the device using commands from a supported command class that is not required by the device to use the security layer.

Summary

- Low Powered RF communications technology that supports full mesh networks without the need for a coordinator node.
- Operates in the sub-1 GHz band; impervious to interference from Wi-Fi and other wireless technologies in the 2.4-GHz range (Bluetooth, ZigBee, etc.):
    - Details related to the frequency coverage are given in Table 10.1.
    - The Z-Wave PHY and MAC layers are defined by ITU-T Recommendation G.9959.
    - Details regarding the frequencies used by Z-Wave can be found in Z-Wave Alliance Recommendation ZAD12837, "Z-Wave transceivers–Specification of spectrum-related components."
- Designed specifically for control and status applications, supports data rates of up to 100 kbps, with AES128 encryption, IPV6, and multi-channel operation.
- Full interoperability through layer 6 with backwards compatibility to all versions.
- Successfully bridged and trialed with OpenADR, SEP 1, SEP 1.1 and other Smart Energy protocols.
- Shares the same position in the NIST / SGIP Catalog of Standards as the IEEE 802.11 and 802.15 and 802.16 families.

Z-Wave Market Facts

- Approximately 600 companies and over 2400 interoperable products (certified devices) available, 100 million Z-Wave products worldwide.
- Extensively used in residential systems throughout numerous business spectrums, including ADT, Alarm.com, AT&T, DSC, GE/Interlogics, Honeywell, Lowes, Verizon, Vivint, and other prominent service providers worldwide.
- Found in thousands of hotels, cruise ships, and vacation rentals; including 65,000 devices in the flagship Wynn Hotel in Las Vegas, NV.

- Actively supported by over 700 manufacturers and service providers throughout the world.
- Designed specifically for control, monitoring and status operations; no interference from Wi-Fi or other 2.4 GHz wireless technologies in similar band.

Z-Wave PLUS

- Introduced in 2015 and backward compatible with the original standard.
- Mixture of software and hardware changes.
- Improved devices with lower battery consumption, coupled with improved protocol functionality:
  - Increased data rate from 40 kbps to 100 kbps
  - Lower power consumption for better battery life
  - Higher output power for better range
  - Better routing options with addition of Explorer frames
  - Additional command classes to support new device management

References

[1] Zensys, 90310020x, Z-Wave Device Class Specification
[2] WHITEPAPER: Compared, INSTEON
[3] C. W. Badenhop, S. R. Graham, B. W. Ramsey, B. E. Mullins, L. O. Mailloux, "The Z-Wave routing protocol and its security implications" https://doi.org/10.1016/j.cose.2017.04.004
[4] M. B. Yassein, W. Mardini and A. Khalil, "Smart homes automation using Z-wave protocol," 2016 International Conference on Engineering & MIS (ICEMIS), Agadir, 2016, pp. 1–6. doi: 10.1109/ICEMIS.2016.7745306
[5] Sigma Designs. (n.d.). Z-Wave development Kit. Retrieved June 2013, from Sigma Designs public web site: http://www.sigmadesigns.com/upl oads/documents/zwave_dev_kit_br.pdf
[6] "Introduction to the Z-Wave Security Ecosystem", available from Sigma Design, Inc., on the company website: http://z-wave.sigmadesigns.com/ wp-content/uploads/2016/08/Z-Wave-Security-White-Paper.pdf
[7] M. B. Yassein, W. Mardini and A. Khalil, "Smart homes automation using Z-wave protocol," 2016 International Conference on Engineering & MIS (ICEMIS), Agadir, 2016, pp. 1–6. doi: 10.1109/ICEMIS.2016.7745306

11

Smart Home Protocols: Comparison

11.1 Introduction

For smart home wireless networking, there are numerous technologies competing to become the standard of choice. Some of these technologies are popular as: Wi-Fi, Wi-Fi HaLow, Bluetooth, Bluetooth 5, Insteon, Thread, ZigBee, and Z-Wave. Each of these technologies has its advantages, disadvantages and limitations. Wi-Fi consumes a lot of power, and Bluetooth is limited in signal range and number of devices. Z-Wave is the technology that starts to become the one for home automation. The standards that are competing with Z-Wave include Wi-Fi HaLow, Bluetooth 5, Insteon, Thread and ZigBee.

Z-Wave has a long open-air operating range at 90 meter (outdoor) and 24+ meter (indoor). Insteon has a large number of maximum devices capability at 17.7 million (to ZigBee's 65,000 and Z-Wave's 232). Thread has a fast data transmission rate at 250 kbps. Z-Wave has better interoperability than ZigBee, but ZigBee has a faster data transmission rate. Thread operates on the busy Wi-Fi standard frequency of 2.4 GHz, while Z-Wave operates at 908 MHz in the US, which has reduced noise and a greater coverage area. ZigBee operates on both 915 MHz and 2.4 GHz frequencies. Thread, Insteon, ZigBee and Z-Wave are mesh networks. The Z-Wave MAC/PHY is globally standardized by the International Telecommunications Union as ITU 9959 radio, and the Z-Wave Interoperability, Security (S2), Middleware and Z-Wave over IP specifications were all released into the public domain in 2016, making Z-Wave highly accessible to Internet of Things (IoT) developers. Because of the many similarities between Z-Wave and ZigBee, some more detailed comparison is given next. Since Thread and Z-Wave belong to IoT, the two will be compared after giving some enough knowledge on Thread

In this book, we introduced three of the smart home protocols: Bluetooth, ZigBee and Z-Wave. In our book "Serial Communication protocols and standards" we introduced Wi-Fi and WiMax. Because of the popularity of the other protocols, mainly, Insteon, Thread, Intellon, X10 and UPB, we are going to introduce them to the reader in brief but with enough knowledge. This is needed to be able to compare the different smart home and IoT protocols.

The aim of this Chapter is to compare between the different technologies mainly: Z-Wave, ZigBee, Bluetooth, Insteon, Intellon, X10, Thread and Universal Powerline Bus (UPB). The comparison does not target defining which technology is the best, but to guide the reader while he is selecting his technology.

11.2 ZigBee versus Z-Wave

The comparison starts by comparing ZigBee and Z-Wave since both of them are detailed in Chapters 7, 8 and 10. The material of the two chapters are taking as base for the comparison.

11.2.1 ZigBee versus Z-Wave: Specifications and Capabilities

The ZigBee and Z-Wave are short-range wireless technologies used for remote monitoring and control. However, their specifications and applications are different. Both technologies are ideal for home-area networks (HANs), which started to be widely spread worldwide. We start the comparison between the two technologies by the next table (Table 11.1) that gives some of the specifications and capabilities of the two technologies followed by brief discussion of the table contents.

a. ZigBee Alliance, Z-Wave Alliance, and Sigma Designs

ZigBee: ZigBee was established in the late 1990s as an alternative to Wi-Fi and Bluetooth for some applications. The IEEE 802.15.4 standard was completed in 2003 and updated in 2006. The ZigBee Alliance was established in 2004 to maintain the standard, continue its development, and provide interoperability testing.

Table 11.1   ZigBee and Z-Wave specifications and capabilities

Technology	Frequency	Modulation	Data Rate	Range	Application
ZigBee	2.4 to 2.483 GHz	OQPSK	250 kbps	10 m	Home automation, smart grid, remote control
Z-Wave	908.42 MHz	GFSK	9.6/40 kbps	30 m	Home automation, security

The ZigBee Alliance is a consortium of companies and other organizations supporting the development of the standard and promoting its use. There are more than 220 members. The Alliance performs testing to certify that all products comply with the standard.

Z-Wave: Before it was acquired by Sigma Designs in 2008, Zensys developed Z-Wave as a proprietary wireless standard. Sigma Designs makes ICs and other products for power-line communications (PLC) as well as wireless. The standard is not open like many wireless standards, but it is available to Zensys/Sigma Design customers. Recently, the International Telecommunications Union (ITU) included the Z-Wave PHY and MAC layers as an option in its new G.9959 standard, which defines a set of guidelines for sub-1-GHz narrowband wireless devices.

The Z-Wave Alliance, consortium of more than 160 companies that design and sell wireless home control products based on the Z-Wave standard, plays a major role in the Z-Wave system. Currently, more than 575 interoperable products are available in 22 countries.

b. Topology
ZigBee: As a wireless mesh networking technology, ZigBee can be used in direct communications, but most applications are based on a star or tree topology mesh network. A master coordinator node controls other connectEed nodes. If a node cannot communicate with another node, the two may communicate by way of links to other nodes within range acting as repeaters. ZigBee can support up to 65k nodes.

Z-Wave: Concerning Z-Wave, it is also a wireless mesh networking technology enables any node to talk to other adjacent nodes directly or indirectly through available relays. A master controller node controls any additional nodes. The nodes communicate directly with one another if they are within range. If two nodes that want to communicate are not within range, they can link with another node that both can access and exchange information. A Z-Wave network can have up to 232 nodes. Multiple controllers can be set up to partition a network as required for different functions.

c. Frequency and Modulation
ZigBee: ZigBee devices operate in the unlicensed industrial, scientific, and medical (ISM) bands. The most popular configuration is in the 2.4-GHz band, where the standard defines sixteen 5-MHz channels of operation. Maximum data rate is 250 kbps using offset quadrature phase shift keying (OQPSK).

Optional specifications provide for operation at 915 MHz (in the U.S.) with 40 kbps using binary phase shift keying (BPSK) modulation. A European version uses 868 MHz with 20 kbps.

Z-Wave: Z-Wave uses the Part 15 unlicensed ISM band. It operates at 908.42 MHz in the U.S. and Canada but uses other frequencies in other countries depending on their regulations. The modulation is Gaussian frequency shift keying (FSK). Available data rates include 9600 bits/s and 40 kbps.

d. Power and Range

ZigBee: ZigBee's typical power of 1 mW or 0 dBm limits the free space range to about 10 m, although longer ranges are possible depending on conditions. A big issue is co-existence with Wi-Fi and Bluetooth, which use the same band. Most transceivers have some co-existence mechanism to minimize interference. The available 16 channels typically allow the radio to find a frequency with minimum interference.

Z-Wave: Output power is 1 mW or 0 dBm. As with any wireless technology, the range of transmission depends on the environment. In free space conditions, a range of up to 30 m is possible. The through-wall range is considerably less, of course.

e. Security

Z-Wave protocol provides packet encryption, integrity protection and device authentication services. It is less subjected to signal interference compared to the ZigBee protocol, which operates on the widely populated 2.4 GHz band shared by both Bluetooth and Wi-Fi devices.

11.2.2 ZigBee versus Z-Wave: The Differences

Some of the differences between ZigBee and Z-Wave are given next.

a. ZigBee can connect more devices.

ZigBee can support way more devices at a given time–65,000+ devices, or nodes on the mesh network, while Z-Wave, on the other hand, caps out at 232 devices. As a fact, 232 devices are still more than enough for most households.

b. ZigBee is open.

ZigBee is an open standard (open-source software) run by the ZigBee Alliance, whereas Z-Wave is run by Silicon Labs, which has stricter controls to make sure every Z-Wave device works with every Z-Wave controller. Actually Z-Wave is proprietary software supported and certified by the Z-Wave Alliance. The main reason to this strict control is reliability, though the downside is a lack of compatibility.

Late 2019 there was some news concerning this point coming from both Z-Wave and ZigBee. The news from Z-Wave was that Z-Wave is set to open up the standard in 2020; "as a ratified, multi-source wireless standard available to all silicon and stack vendors for development."

Concerning the news of ZigBee–Apple, Google, and Amazon announced, alongside the ZigBee Alliance, the creation of the Connected Home over IP project; an initiative to simplify development for manufacturers and increase compatibility for consumers in the smart home world.

c. ZigBee allows for more hopping.

ZigBee and Z-Wave are both mesh networks–meaning the signals can hop from gadget to gadget around the home and each device or sensor does not need to connect to Wi-Fi–but they usually have a central hub which connects to the internet. Z-Wave allows up to four 'hops' between the controller and the device, whereas ZigBee does not have a limit.

d. Z-Wave has better range.

For a given power level of 0 dBm, Z-Wave's range is greater than ZigBee simply because the lower operating frequency supports it with pure physics (Friis formula). Z-Wave has a larger range, at 30 m (100 ft), than ZigBee's 10–20 m (33–66 ft) so it might work better for larger houses and buildings. Z-Wave's latest platform can connect even further, at up to 100 m from point to point.

e. ZigBee and Z-Wave are equal on security.

Both ZigBee and Z-Wave use the same AES-128 symmetric encryption. The operation showed that using AES-128 for encryption, secure the systems from hacking.

f. Z-Wave is more unified.

ZigBee operates on the 915 MHz frequency in the US and the 2.4 GHz frequency in the UK, which might look familiar, as that is a major frequency for Wi-Fi, too.

One concern on ZigBee is that it is broken up into several protocols. There is Home Automation, Smart Energy Profile, SEP energy management, Light Link, digital health, home hospital care, and more. Unfortunately, ZigBee devices of different protocols cannot exactly talk to each other well.

ZigBee 3.0, however, unifies all these protocols so that they can work together. Lux, for example, offered the first thermostat that got ZigBee 3.0 approval, allowing it to talk to any device that has a ZigBee badge on it. This is how Z-Wave actually works–it is a unified standard.

g. Z-Wave works with minimal interference.

Z-Wave operates at the low frequency 918/960 MHz band, meaning interference is minimal for Z-Wave and possible on ZigBee which is working in the same band of W-Fi.

Concerning speed, ZigBee is faster with data rates at 40–250 kbps versus Z-Wave's 9.6–100 kbps.

h. ZigBee is more flexible.

One of the great benefits of ZigBee is its flexibility. It was designed so specific application software known as profiles could be developed and deployed. Profiles connect to the ZigBee stack and make it faster and easier for manufacturers to create wireless products for very specific applications. Available profiles include home automation, smart energy, telecommunications, health care, remote control (RF4CE, or radio frequency for consumer electronics), building automation, and retail services.

ZigBee and Z-Wave target the same general applications. Of the two, ZigBee is by far the more versatile since it can be configured for virtually any short-range wireless task. Profiles are readily available to minimize development time for common applications. On the other hand the protocol is far more complex, resulting in longer development times. Z-Wave uses a far simpler protocol, so development can be faster and simpler.

Z-Wave chips are available from only one source, Sigma Designs. They sell only to OEMs, ODM, and other major clients. More than 500 consumer home control products are available in stores like Home Depot and Lowes, but many don't state that Z-Wave is used.

ZigBee chips are available from Ember, Freescale, Microchip Technology, and Texas Instruments. Complete, ready to use ZigBee modules are also available from multiple sources like Atmel, CEL, Digi, Jennic, Lemos, and RFM.

11.2.3 Z-Wave versus ZigBee: The Devices

With ZigBee, there around 2,500 products from 400 members of the Zig-Bee Alliance. Z-Wave has more than 3,200 supported devices and more companies signed up–700 members.

When considering the popular, premium smart home brands and products, we would say ZigBee slightly has the edge–but which one is best for you depends on what smart home kit you already own and what you're looking to add.

Devices that work with ZigBee

Some well-known smart home brands and devices that support ZigBee (otherwise known as ZigBee certified products) include the following:

- Philips Hue
- Samsung SmartThings
- Amazon Echo Plus (with Alexa for voice controls)
- Hive Active Heating and accessories
- Honeywell thermostats
- Ikea Tradfri
- Belkin WeMo Link
- Yale smart locks
- Sengled smart lights
- ADT Security Hub
- Wink hub
- Somfy blinds and drapery motors
- GE Appliances
- LG SmartThinq
- Lux Konoz

Devices that work with Z-Wave

Many big name brands also support the Z-Wave standard–and the reader can see all the certified products on the official website. Here are some of the highlights:

- Samsung SmartThings
- Wink hub
- Honeywell thermostats
- Hogar Milo (with Google Assistant for voice controls)
- ADT Security Hub
- August smart locks

- Yale smart locks
- Logitech Home Harmony Hub Extender
- Somfy
- GE Appliances
- LG SmartThinq
- Kwikset smart locks

11.3 Z-Wave and ZigBee versus INSTEON

It is important to start this section by giving the reader the main characteristics of Insteon technology. This will help comparing Z-Wave and ZigBee with Insteon .

11.3.1 INSTEON Overview (www.insteon.com)

INSTEON enables simple, low-cost devices to be networked together using the powerline, radio or both. All INSTEON devices are peers, meaning that any device can transmit, receive or repeat other messages without requiring a master controller or complex routing software. Adding more devices makes an INSTEON network more robust, because INSTEON devices repeat each other's messages by simulcasting them at precisely the same time, so with more devices the INSTEON signal gets stronger. On the powerline, INSTEON devices can also be compatible with legacy X10 devices (See X10 latter). First deployed in 2005, there are now millions of INSTEON nodes in use around the world in many different products, including dimmers and switches, handheld and tabletop remotes, thermostats, sprinkler controllers, energy monitoring devices, sensors and low-voltage input/output interfaces. INSTEON networks easily connect to other, larger networks, such as LANs and Wi-Fi. Many different user interfaces, including smartphones, PCs and third-party controllers, can manage an INSTEON network. Applications running on such devices not only can talk to INSTEON products, but can interoperate with cameras, sensors and other devices using different networking protocols, including TCP/IP, ZigBee and Z-Wave.

11.3.2 INSTEON Main Characteristics compared with other Technologies

a. INSTEON Uses Both Powerline and Radio Communications

Radio communication is far from 100% reliable, as cell phone and Wi-Fi users are frequently reminded from firsthand experience. Z-Wave's single media design is clearly sub-optimal to INSTEON's dual-mesh design. Z-Wave radios, because they use narrowband FSK (frequency shift keying) signaling, are considerably less sophisticated than cell phones or Wi-Fi. Furthermore, because they share the unlicensed 900 MHz ISM (industrial, scientific and medical) band, they are required to transmit at low power. Metal in the home can block or reflect radio waves–installing radio wall switches in metal junction boxes can cause particular difficulties. That is why INSTEON is dual-mesh–if radio fails, powerline provides a backup and vice-versa. In combination, two independently simple, low-cost signaling methods can be much more reliable than sophisticated, high-cost methods employed on single media with no backup. Because it is radio-only, Z-Wave had to resort to complex network self-organization, self-healing and routing procedures in an attempt to maximize reliability. INSTEON, on the other hand, even though it employs the same sort of narrowband FSK radios in the 900 MHz range as Z-Wave, achieves greater overall reliability with far less complication, because it relies on dual-mesh powerline backup and repeated message simulcasting.

b. INSTEON Uses Simulcasting Instead of Routing

Simulcasting is much simpler than routing and more robust because multiple devices simulcasting the same message add to the signal power. For a complete discussion of how simulcasting works see INSTEON, the Details, available at www.insteon.com. As explained above, routing entails complexity. Devices that initiate communication using a source routing algorithm, which is what Z-Wave uses, must know the topology of the network. Maintaining and distributing a network topology database is not trivial, especially with mobile devices. To keep costs down, Z-Wave has defined different classes of devices, some of which cannot participate in routing.

c. INSTEON is a Peer-to-Peer Network

INSTEON devices are two-way simulcasting repeaters, which means they all handle INSTEON messages in exactly the same way, with no need for network controllers or routers. A Z-Wave network, on the other hand, must contain a Static Update Controller (SUC) along with other kinds of devices, including slaves or routing slaves. If you want more than one controller in your Z-Wave network and you don't have a SIS (SUC I.D. Server), you must choose a single controller as a master and use that controller alone for network maintenance. To set up your other controllers as secondaries you must go through an involved controller replication procedure. Thereafter, any time you add or delete network devices using your single master controller, you must repeat the entire controller replication process for all of your other controllers. In contrast, INSTEON lets you add or remove INSTEON devices of any kind at any time, because no matter how a device appears to the user, to the INSTEON network all devices are peers. Therefore, setting up INSTEON devices requires far less customer knowledge and involvement.

d. INSTEON Requires No Network Enrollment

As a consequence of routing using network controllers, Z-Wave requires that new devices be enrolled in the Z-Wave network before controller buttons can be associated with device functions. Some Z-Wave vendors have found ways to hide the network enrollment step from users by combining it with the control association procedure, but others have elected to keep the two procedures separate. This situation is confusing if consumers have Z-Wave products from different vendors who have implemented different setup methods. With INSTEON this problem doesn't arise because there is simply no need for network enrollment. There is no routing and there is no network controller. At the factory, INSTEON devices are each given a unique I.D. number that serves as a permanent network address. All INSTEON devices automatically become part of an INSTEON network and start simulcasting repeated messages as soon as a customer powers them up.

INSTEON Benefits

The biggest benefit is the redundancy the system provides. If wireless is down or there is some sort of interference, the system easily changes to powerline communications and vice-versa. Unfortunately, if the power goes out then you lose both forms of communication, but you'll also lose power to your accessories so nothing works anyway.

Another benefit is that you don't have to run new cables since the INSTEON protocol includes powerline connectivity. It uses your existing electrical wires to pass messages. There is no need to open walls or snake wires through existing construction.

INSTEON Limitations

While INSTEON makes hundreds of different accessories for their network, one big drawback is the lack of third-party options. Some do exist, but for the most part you are locked into the INSTEON ecosystem. If the company goes out of business then you won't have much opportunity to grow or upgrade.

11.3.3 INSTEON versus Z-Wave and ZigBee: Comparison

Considering the INSTEON Characteristics given above and the characteristics of Z-Wave and ZigBee given in Chapters 7, 8 and 10 we can conclude the following.

INSTEON Compared with Z-Wave

The main differences between INSTEON and Z-Wave are that

- INSTEON is dual-mesh (it uses Both Powerline and Radio Communications), Z-Wave is radio only;
- INSTEON propagates messages by simulcasting, Z-Wave routes messages;
- INSTEON uses backwards and forwards compatible Statelink commands, Z-Wave uses commands that must be understood by the responder;
- INSTEON devices are peers, Z-Wave has network controllers and slaves; and
- INSTEON requires no separate network enrollment step.

INSTEON Compared with ZigBee

There are several noteworthy differences between INSTEON and ZigBee that have consequences in the home control marketplace. The main differences are that:

- INSTEON is dual-mesh, ZigBee is radio only;
- INSTEON propagates messages by simulcasting, ZigBee routes messages;
- INSTEON messages are backwards and forwards compatible, ZigBee messages are specific to revision level and profile;
- INSTEON devices are peers, ZigBee has full-function devices and reduced-function devices;
- INSTEON requires no separate network enrollment step;
- INSTEON uses FSK radios in one band per market, ZigBee uses DSSS in two bands per market.

Taken together, these distinctions entail higher cost and complexity for ZigBee radio. Granting that the elevated cost buys ZigBee radio devices with somewhat higher performance than INSTEON radios, INSTEON's dual-mesh architecture provides a powerline backup that ZigBee can't rely on when its radio-only messaging does fail. On balance, ZigBee's cost and complexity make it better suited for large industrial and commercial networks, whereas INSTEON is optimized for the home.

11.3.4 Z-Wave, ZigBee, and INSTEON Compared: Summary

Table 11.2 summarizes the comparison between the three technologies: Z-Wave, ZigBee and INSTEON

11.4 Z-Wave versus THREAD Protocol: Comparison

Today there are a multitude of IoT (Internet of Things) protocols available in the marketplace. In this section, two protocols from different time periods are selected to be compared. The two protocols selected are: Z-Wave which is one of the oldest and the most commercially successful IoT protocol, and Thread which is the latest protocol released for commercialization.

Z-Wave is based on propriety standards, most of which is not publicly available, although some have been reverse engineered by researchers. Thread on other hand is based completely on open standards. All Z-Wave modules are made by a single company, while Thread modules are expected to be available from multiple vendors. Z-Wave has a large installed base and has proven to be a commercial success. Thread is new and has an open protocol, but Thread based devices are not yet readily available in the market for users.

For the reader to follow the comparison between Thread and Z-Wave, he must have good knowledge about:

- Meaning of IoT
- Z-Wave, and
- Thread

Chapter 10 covers Z-Wave. This section gives the reader enough knowledge about the other two topics: IoT and Thread.

11.4.1 Internet of Things (IoT)

What is an IoT system? An IoT system is a concept of connecting sensors and actuators into a coherent network with Cloud computing services that can provide services to the user.

The Internet of Things (IoT) aims to transform people's lives through smart homes and businesses. In the home, the goal is a network of connected

Table 11.2   Z-Wave versus ZigBee versus INSTEON comparison table

Property	Z-Wave	ZigBee	INSTEON
Media	Radio only	Radio only	Powerline and Radio
Module types	Controller Static controller (SUC) Slave Routing slave Enhanced slave Installer Bridge	Network coordinator (1 per network) Full-function device (FFD) Reduced-function device (RFD)	All are peers
Message propagation	Routing tables	Routing tables	Simulcasting repeaters
Network Topology	Routed mesh	Partial routed mesh–only FFDs are repeaters Partial star–RFDs are star networked each to a single FFD	Full mesh
RF interoperability	Yes, 908.42 MHz	No, 915 MHz and 2.4 GHz	Yes, 908.42 MHz
Command interoperability	No, application and revision specific commands	No, 8 profiles and revisions thereof	Yes, Statelink
RF data rate, bps	9600 instantaneous	20 K instantaneous at 868 MHz 40 K instantaneous at 915 MHz 250 K instantaneous at 2.4 GHz	38400 instantaneous
RF PHY	FSK in ISM Band	DSSS in two ISM bands	FSK in ISM Band
Acknowledged	Yes	yes	Yes
Addressing	32-bit home ID 8-bit node ID (232 devices per network)	64-bit IEEE address 16-bit short address (65,536 devices)	24-bit pre-assigned module ID (16,777,216 devices)
Network enrollment	Part of installation procedure	Part of installation procedure	Automatic
Groups	Controller database	?	Databases in modules
Commands	Command classes	Application profiles	2-byte (65,536)

(Continued)

Table 11.2 Continued

Device types	Device classes	Application profiles	3-byte (16,777,216)
Global clock	No	Optional beaconing	No
Retries	Yes	yes	Yes
Message length	Variable	Variable, 128 byte maximum	15, 33 byte
Collision avoidance	Yes	Yes	Yes
Triac control	Yes	?	Yes
SRAM	2048 byte	?	256 bytes
Flash	32K bytes	128K bytes typical	4K x 14 bits (7K bytes)
EEPROM external	Up to 24K bytes	?	4K bytes
EEPROM internal	0	?	256 bytes
Watchdog	Yes	?	Yes
Powerline data rate, bps	None	None	13,165 instantaneous PL 2,880 sustained PL 1,440 standard msg payload 1,698 extended msg payload 1,034 user data
Powerline PHY	None	None	131.65 KHz carrier–BPSK
Powerline phase bridging	N/A	N/A	RF or hardware
X10 compatible	N/A	N/A	yes

appliances, lights, climate controls, security, and entertainment systems, all of which work together to make life more convenient and rewarding for consumers. In commercial buildings, IoT aims to improve the efficiency, functionality, automation, and safety of buildings such as offices, healthcare facilities, hotels, and schools.

An IoT system is depicted in Figure 11.1. In an IoT system, the IoT devices can be localized in a home or spread across the city. The IoT devices

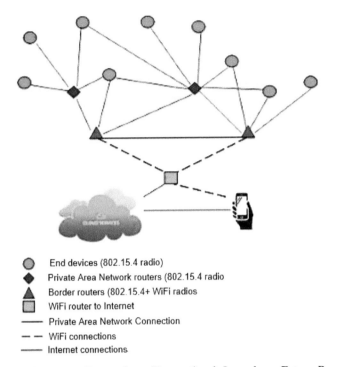

End devices (802.15.4 radio)
Private Area Network routers (802.15.4 radio
Border routers (802.15.4+ WiFi radios
WiFi router to Internet
—— Private Area Network Connection
— — WiFi connections
—— Internet connections

Figure 11.1   IoT system (Image from "International Journal on Future Revolution in Computer Science & Communication Engineering ISSN: 2454–4248 Volume: 3I").

are connected in a Private Area Network (PAN) using wireless communication protocols. The PAN is then connected through the router(s) to Internet allowing it to accessing the Cloud computing services. The data processed in the Cloud can result in necessity for physical action. The action may be as simple as sending a text message or as complex as activating a set of actuators to accomplish some task. The action is communicated from the Cloud computer back to the IoT actuator device(s) through the router and the IoT PAN. An IoT system (Figure 11.1) has following features:

- **Device Identification**. An individual device in an IoT PAN has a unique address.
- **Sensing**. The IoT PAN has a number of sensing devices that collect data from a monitored event or environment. Examples of sensors include detectors for motion, light, vibration, pressure, temperature, acceleration, magnetic field, infrared, proximity, distance, biometric parameters and many others.

- **Response**. The IoT PAN may also have a number of devices that respond, that is, perform actions. Examples include relays, actuators and switches.
- **Communication**. The IoT devices in the PAN communicate wirelessly with each other. The sensor data are communicated, by the router, to a central location, usually using Cloud computing, for processing and evaluation. Results of data evaluation are then communicated back to the IoT PAN for any required physical action.
- **Computation**. Cloud computing services are usually used to process the IoT sensor data. Cloud computing formulates a response and sends to the IoT router, if needed. The IoT router then uses the PAN wireless communication links, to deliver the message.
- **Services**. The IoT PAN is ubiquitous and provides a number of services: item identification, collecting and communicating sensor data, event monitoring, performing physical action(s), and device collaboration.
- **Semantics**. Cloud processing software analyzes the data and provides a context-aware response for the IoT system.

All these capabilities in the IoT PAN are provided under practical constraints. Some of the major constraints on the IoT PAN are:

- **Low power**. Many of IoT devices are battery powered; replacing batteries is expensive, inconvenient, and unreliable. Even if the IoT devices are attached to the power grid, low power is required for a couple of reasons. First, the IoT devices are continuously operating and therefore even small amount of power consumption can quickly add up. Second, a typical IoT installation consists of hundreds of devices. The aggregate power consumption of these devices operating continuously can be quite significant.
- **Low cost**. Cost is always a major factor in the consumer adoption of any product. Low cost promotes a wider use of the smart technologies.
- **Security**. Beside physical security of the IoT devices, authenticity, privacy, confidentiality, and integrity of data are important consideration in an IoT system.
- **Communication**. The IoT faces communication constraints on multiple fronts. The IoT PAN radio operates on low power, short range transmission, in a restricted public frequency spectrum. The operating environment is noisy with other equipment generating radio interference, and there are many physical obstacles. The IoT communication also faces the challenge of frequent network route changes as IoT

devices are moved, some of the IoT devices are mobile, plus the users may move obstacles in its communication paths. Additionally, the IoT data packets are small, making it harder to include redundancy for fault tolerance.

- **User interface**. Most IoT devices either have a primitive or no user interface. The lack of graphical user interface (GUI) constraints the communication between the user and the device. GUI-based components, like smart-phone, smart-tablet, a computer or even a voice controlled device is normally required to set up and communicate with the IoT device.

IoT Private Area Network

There are a number of IoT protocols currently available, Z-Wave, ZigBee, Thread, IoTivity, AllJoyn, and among others. Although particulars for each of these protocols are different, they share many similarities. All these protocols establish a wireless Private Area Network (PAN). PAN network topology is built using connectivity through wireless radios. Topologies commonly used for PAN are point-to-point, star and mesh. The mesh topology is the most popular due to its robustness, scalability and fault resilience. A typical PAN with mesh topology can be seen in Figure 11.1. Generally, an IoT network have three types of nodes: End devices, PAN routers, Border routers. Data from IoT devices can be transmitted across multiple hops, as determined by a routing protocol, before arriving at gateway Border router(s). The End devices normally consist of sensors or actuators. The PAN routers serve multiple purposes, besides being sensors or actuators, PAN routers also act as communication router for the End devices. The PAN routers have additional capability to connect with other PAN routers and/or Border routers. The Border router(s) have an additional capability to act as gateway to Internet/Cloud by connecting to Internet router. The End devices can communicate with either the PAN network router or directly with Border routers. An example of a Border router is Nest thermostat, which has connectivity to both PAN and Wi-Fi router, plus it is also an IoT sensor and actuator.

11.4.2 Thread Protocol

Thread was announced by Thread group Inc. in 2015, so it is a relatively new IoT protocol. Hardware boards with wireless radios, plus software libraries for Thread are now available for research and development. Consumer devices are not yet widely available, although Nest thermostat and some ZigBee devices are Thread compatible. Thread is an open standard. Thread specification is available from the Thread group. Thread was founded by seven companies (ARM (acquired by SoftBank Group), Big Ass Fans, Freescale (acquired by NXP Semiconductors), Nest (acquired by Alphabet/Google), Samsung, Silicon Labs, and Yale). Today, Thread group has more than 200 member companies. Thread intends to consolidate IoT protocols by working with other IoT alliances. Thread is specially designed for Home Automation and supports a wide variety of home use applications: appliances, access control, climate control, energy management, lighting, safety, and security.

Thread is designed to address the unique interoperability, security, power, and architecture challenges of the IoT.

- **Thread is a low-power wireless mesh networking protocol,** based on the universally-supported Internet Protocol (IP), and built using open and proven standards.
- **Thread enables device-to-device and device-to-cloud communications** and reliably connects hundreds (or thousands) of products and includes mandatory security features.
- **Thread networks have no single point of failure,** can self-heal and reconfigure when a device is added or removed, and are simple to setup and use.
- **Thread is based on the broadly supported IEEE 802.15.4 radio standard,** which is designed from the ground up for extremely low power consumption and low latency.

A key enabler for the IoT is interoperability. Thread addresses this challenge by providing a certification program that validates a device's conformance to the specification as well as its interoperability against a blended network comprised of multiple certified stacks.

The latest Thread specification is v1.1.1 2017.Thread is based on open standards such as IEEE 802.15.4 (2450 MHz), IPv6, 6LoWPAN, wireless PAN MESH network topology with no single point of failure. The wireless PAN operates at 2.4 GHz with a data rate up to 250 kbps and using O-QPSK modulation. Thread has a maximum range of 30 m per hop, with a default hop limit of 36 hops. Thread has the ability to connect 250+ devices in a single PAN. Thread can have multiple Border routers, which connect to Internet, this eliminates Border router failure as a single point of failure. Each IoT device in Thread has an IPv6 address. The IPv6 address is compressed using 6LoWPAN, which is a standard for low power devices. Readers might notice

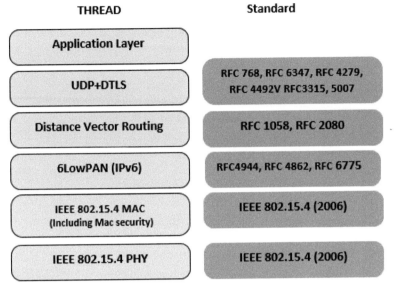

Figure 11.2 Thread protocol stack.

similarities between Thread and the ZigBee protocol. As a matter of fact, ZigBee libraries can run as an application on Thread.

11.4.3 Thread General Characteristics

The Thread stack (Figure 11.2) supports IPv6 addresses and provides low-cost bridging to other IP networks and is optimized for low-power / battery-backed operation, and wireless device-to-device communication. The Thread stack is designed specifically for Connected Home and commercial applications where IP-based networking is desired and a variety of application layers can be used on the stack. These are the general characteristics of the Thread stack:

- **Simple network installation, start-up, and operation:** The Thread stack supports several network topologies. Installation is simple using a smartphone, tablet, or computer. Product installation codes are used to ensure only authorized devices can join the network. The simple protocols for forming and joining networks allow systems to self-configure and fix routing problems as they occur.
- **Secure:** Devices do not join the network unless authorized and all communications are encrypted and secure. Security is provided at the

network layer and can be at the application layer. All Thread networks are encrypted using a smartphone-era authentication scheme and Advanced Encryption Standard (AES) encryption. The security used in Thread networks is stronger than other wireless standards the Thread Group has evaluated.

- **Small and large home networks:** Home networks vary from several to hundreds of devices. The networking layer is designed to optimize the network operation based on the expected use.
- **Large commercial networks:** For larger commercial installations, a single Thread network is not sufficient to cover all the application, system and network requirements. The Thread Domain model allows scalability up to 10,000s of Thread devices in a single deployment, using a combination of different connectivity technologies (Thread, Ethernet, Wi-Fi, and so on).
- **Range:** Typical devices provide sufficient range to cover a normal home. Readily available designs with power amplifiers extend the range substantially. A distributed spread spectrum is used at the Physical Layer (PHY) to be more immune to interference. For commercial installations, the Thread Domain model allows multiple Thread networks to communicate with each other over a backbone, thus extending the range to cover many mesh subnets.
- **No single point of failure:** The Thread stack is designed to provide secure and reliable operations even with the failure or loss of individual devices.
- **Low power:** Devices efficiently communicate to deliver an enhanced user experience with years of expected life under normal battery conditions. Devices can typically operate for several years on AA type batteries using suitable duty cycles.
- **Cost-effective:** Compatible chipsets and software stacks from multiple vendors are priced for mass deployment, and designed from the ground up to have extremely low-power consumption.

11.4.3.1 Network stack: Thread versus z-wave stack

The IoT protocol layers are shown in Figure 11.2. Figure 11.3 compares the Thread stack with that of Z-Wave. Figure 11.4 is a network security comparison. The main security for IoT protocols are in Perception Layer, Data Link Layer and Transport Layer. Although data in Cloud is an important part of the IoT system, Cloud is not part of IoT protocols.

Network Layer Comparison

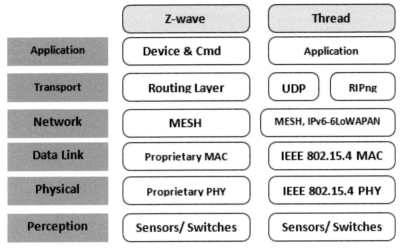

Figure 11.3 Network layer comparison.

Network Security Comparison

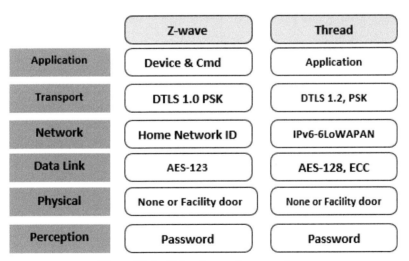

Figure 11.4 Network security comparison.

Note: We refer the reader to the references given at the end of the chapter to get more detailed information. The following discussion is summary of some of the references.

A. Perception Layer

Security Perception layer includes WSN (Wireless Sensor Network), IMD (Implantable Medical Devices), RFID (Radio Frequency Identification), NFC (Near Field Communication), temperature sensors, pressure sensors, luminance sensors, vibration sensors, accelerometer magnetometer, gyroscope, etc. Z-Wave and Thread protocol have devices with similar functionality at perception layer, but are incompatible with each other. The communication standard for each protocol is very different. Perception layer requires physical security of devices. Physically manipulated faulty nodes may have unauthorized attachments, or reprogrammed boot to change their behavior. This can be secured by running fault detection algorithms. Encryption related data must protected by using encryption before storing it in memory. To add new devices in PAN, strong passwords, biometric recognition (fingerprints, iris, voice, face, etc.) or a synchronized key generator must be used. Currently, none of the major IoT protocols address the issue of protecting data stored in the device registers and memory.

B. Physical and Data Link Layer (PHY and MAC) Security

In Z-Wave PHY and MAC layers are proprietary. In Thread PHY and MAC layers are defined by IEEE 802.15.4. Although IEEE 802.15.4 allows other frequencies, Thread only uses the 2450 MHz related portion of the standard. This is the lowest layer for which there are protocol-based security specifications. Support for this encryption is built into the radio hardware. Almost all IoT protocols offer AES (Advanced Encryption Standard) 128-bit. However, AES does not scale well with hundreds of devices in the network and requires very large keys. Therefore, IoT protocols implement, ECC (Elliptic Curve Cryptography, RFC 4492). ECC is an asymmetric, public-key method that scales well and provides a higher level of security for the same number of bits. Other low power public key encryption algorithm Rabin's Scheme and NtruEncrypt are also good candidates for IoT. Z-Wave security has an interesting history when it comes to data encryption. Initial Z-Wave, series 100, had TDES (Triple Data Encryption Standard). Later in series 200 silicon, encryption was dropped according to Knight [see the references]. Encryption was again added back and now it uses AES (Advanced Encryption Standard). Since then Z-Wave has taken steps to improve security features and recently has been evaluated by UL for security applications according to press release by Sigma Design, "Z-Wave modules models ZM5101, ZM5202, and ZM5304 with protocol SDK version 6.60 have been evaluated to ULĎs standards for home security". This paper considers Z-Wave protocol, with AES-128, as it exists today. Z-Wave Smart Hub to Internet connection is encrypted with AES-based pre-shared key (TLS 1.1, PSK). Z-Wave accomplishes secure key exchange using Elliptic Curve Diffie-Hellman (ECDH).Thread uses AES-128 and ECC for data link encryption. AES does not scale well with hundreds of devices, as expected in the Thread network and requires very large keys. Therefore, Thread uses ECC (Elliptic Curve Cryptography, RFC 4492). ECC is an asymmetric, public-key method that scales well and provides a higher level of security for the same number of bits.

C. Transport Layer

Security Transport layer mainly consists of routing protocol and data transport protocol. Z-Wave routing protocol is propriety; however, researcher have reverse engineered most of it. Z-Wave routing has 4 possible hops plus one final hop to the destination node. Thread uses standard routing RIPng [RFC 2080]. RIPng is a distance vector routing protocol [RFC 1058] for IPv6. RIPng has maximum of 15 hops. There are two main data protocols currently being used at transport layer in IoT PANs, CoAP (Constrained Application Protocol), and MQTT (Message Queuing Telemetry Transport). Both of these protocols were specifically designed for low power IoT type devices. Actual transport protocol used by CoAP is a UDP (User Datagram Protocol), which enforces use of DTLS (Datagram Transport Layer Security, IETF RFC 6347). CoAP has a set of security modes and mandatory-to-implement ciphers. CoAP borrows some concepts from REST (Representational State Transfer) protocol. The transport protocol used by MQTT is TCP, which enforces TLS (Transport Layer Security, IETF RFC 5246). MQTT is a publisher/subscriber style protocol, requiring a server-broker. A typical MQTT session consists of establishing a connection, authentication, communication, termination. MQTT provides authentication and authorization scheme, but does not have any security implementation requirements. MQTT is designed to operate in a secure network, and thus has no defined security mechanism. MQTT sends username and password in clear text and thus relies on Transport layer encryption, like TLS, to provide security. MQTT should not be used for global network as it does not scale well. TLS is expensive protocol and not well suited for very low power devices. MQTT has TCP port 1883 reserved for non-encrypted and TCP port 8883 reserved for encrypted communication using TLS. Although the DTLS and TLS have similar concepts, CoAP with DTLS is preferred in IoT protocol due to low power IoT devices having limited memory and processing capabilities. Both CoAP and MQTT can be operated in "no security" mode, "pre-shared key" mode and "certificate" mode. CoAPs (CoAP secured) can also be configured in "raw public key" [IETF RFC 7250], but MQTT implementation is not yet available. In pre-shared key mode CoAPs hashes pre-shared keys with a list of corresponding communication nodes. Although use of certificate mode is well established, its use in IoT is discouraged due to resource constraints. However, there is a one big advantage in use of certificates; the certificate can be revoked if the IoT device is compromised. In raw public key mode the IoT device holds an asymmetric key pair but without a certificate. Normally, the asymmetric key pair is generated and embedded in the device by the manufacturer. This asymmetric key pair needs to be validated in out-of-band (OOB) mechanism using public key. A device can have multiple raw public keys. Z-Wave and Thread both utilize DTLS encryption for the transport layer. If the radio hardware does not support data encryption than Transport layer encryption must be used. Using "no security" mode is acceptable if the network is running on a VPN, or if Data layer encryption is being used. Encrypting the data twice, at Data layer and Transport layer can decrease the data compression ratio, which leads to more package transmissions and higher power consumption. Thread Mesh is self-configuring and each link is individually encrypted. Every device in Thread holds the credentials which allow it to be part of the network. However, before the device can receive the credentials it has to be authenticated by the Commissioner (an authentication server). Commissioner itself must be authenticated. Commissioner authentication requires a onetime Commissioning Session, a secured client/server socket connection, between Commissioner and the Border Router via DTLS (RFC 6347) or TLS (RFC 5246). Using the advertised UDP port, during discovery, Commissioner provides PSKc (Pre-Shared Key for Commissioner) credential. The Border router with human interaction than authenticates the Commissioner. A new device wishing to join the Thread network must transmit an unsecured Discovery Request message. A router responds with a Discovery Response message including the joining UDP port. The device will perform DTLS handshake to establish a secure session with router. The router will relay UDP messages to the Border router, which in turn will relay them to the Commissioner. The device and Commissioner then exchange token to establish trust. Commissioner inspects the device IID (Interface Identifier) and credentials.

If the Commissioner is satisfied with the responses from the device, it will be provisioned with the appropriate data and services, and also provided with KEK (Key Encryption Key). Once it is authenticated by the Commissioner, router will provide the device network credentials. Each Thread node also receives a master key when joining. Two different 16-bit keys, one for MAC and other for DTLS, are generated using Hashed Message Authentication Mode with SHA-256 algorithm (HMAC-SHA256) produces 32-bit output [RFC 6234] and the master key. The key set are rotated based on key index changes, or the key rotation timer expiry, or incoming messages matches the next key.

Concluding Remarks on Z-Wave versus Thread

Both the protocols, Z-Wave and Thread, fulfill the requirements, features, and capabilities of an IoT protocol. Both protocols have a reliable MESH configuration for PAN. Z-Wave has tried to protect the PAN from intruders by hiding the information in the propriety standards. These same propriety standards also make it hard to judge the quality of Z-Wave standards, but researchers have reverse engineered some aspects of the PAN. Thread is based on open standards, which allows everyone to study and detect any security flaws. Figure 11.3 shows the network layer comparison, which clearly highlights the differences in PHY, MAC, Routing between Z-Wave and Thread. Network security comparison is shown in Figure 11.4, where the differences are less discernible. The reason for similarity in security is that Z-Wave seems to have been upgrading its devices. Z-Wave has been commercially successful with its Z-Alliance. Thread, with its Thread Group Inc., still has to prove itself in the marketplace.

11.5 Wi-Fi versus INSTEON and Z-Wave

11.5.1 Wi-Fi Overview

In recent years, Wi-Fi (IEEE 802.11) has become the de facto standard for broadband networking of wireless LANs (local area networks) in the home, in offices and at an increasing number of commercial 'hotspots' around the world.

Since Wi-Fi is so common in homes today, it makes sense that home automation accessories take advantage of its availability. While it is important that your hub devices include Wi-Fi, it isn't the best protocol for all smart home accessories.

Wi-Fi is a wireless communication standard developed and maintained by the Wi-Fi Alliance. The standard allows interoperability and communications among different devices.

Wi-Fi: Technology

Wi-Fi communications travel across radio waves in the 2.4 GHz or 5 GHz spectrum. Compatible devices include a receiver and transmitter plus connected antennae.

The theoretical range is relatively far, but for practical home usage, the maximum range is around 60-feet. Of course, range can be affected by many things including physical barriers, interference, transmitter power, and antenna quality.

Communications throughput is quite high; speeds range from 10 Mbps to 100 Mbps. Wi-Fi is capable of streaming high-definition audio and video.

Wi-Fi has a theoretical limit of 256 connected devices. Network congestion may become a problem before 256 devices connect, especially if some of those devices demand large amounts of bandwidth. You may have experienced this if multiple people in your home try to stream high-definition video at the same time.

Wi-Fi: Benefits

The main benefits of Wi-Fi are transmission speed, range, and the overall availability of the technology. As mentioned earlier, most people already have a Wi-Fi router in their home and adding a router is way less expensive than running Ethernet cables around the house. Wi-Fi provides easy and fairly reliable access to the internet anywhere in your home.

Wi-Fi: Limitations

From a home automation standpoint, there are a few drawbacks. First, Wi-Fi is susceptible to interference since there are so many devices competing for bandwidth–phones, tablets, watches, scales, streaming media devices, thermostats, and many others all want to communicate over Wi-Fi. The bands become even more congested as we add more devices to our homes.

Another problem is the power consumption of Wi-Fi devices. Wide range and high speeds require a lot of power to operate. While the power usage doesn't affect our utility bills very much, the consumption is a problem for small sensors and other home automation accessories that need to run on batteries. Plug-in items can easily use Wi-Fi, but we don't want to plug in every smart home accessory, especially very small items.

Let us discuss range as a limitation. This may seem odd, since we also talked about range as a benefit. However, Wi-Fi range can be limiting in larger homes or those made from dense materials like concrete block. If this is the case for you, you have probably already installed additional Wi-Fi routers to create a mesh network.

Finally, security is something to consider. While Wi-Fi has plenty of security features to help keep your wireless network safe, they have to be set up and maintained over time. Wi-Fi itself isn't a security risk, but when people disable (or fail to enable) proper security measures then the network can be compromised. Always make sure your network is secure and routinely check for router software updates.

The Wi-Fi Alliance understands the limitations with their technology and they are developing new Wi-Fi standards that will make it more beneficial for use with home automation. Wi-Fi 6 includes new features that improve range and power efficiency, while still maintaining data throughput. The new Wi-Fi Ha-Low is directed at the smart home market and is being developed to lower power consumption even further while also increasing the communication range. Wi-Fi will become more reliable for smart homes when these new versions are released. However, you will need to purchase a new router to take advantage of the technologies.

Wi-Fi: Compatible Hubs
Most of the hubs available in the market include Wi-Fi for easy wireless connectivity.

11.5.2 Z-Wave Compared with Wi-Fi

Z-Wave also beats Wi-Fi in terms of network interference. Like Bluetooth, Wi-Fi devices compete with one another, so signal strength and network speeds suffer when there are many devices connected. However, Wi-Fi can carry more information. Z-Wave technology sends and receives plenty of data for smart devices like light bulbs, motion detectors and other small appliances, but devices that send lots of information require the larger capacity of a Wi-Fi network. For example, an HD video surveillance camera sends more data than a low-power Z-Wave or Bluetooth network can handle.

11.5.3 INSTEON Compared with Wi-Fi

Using Wi-Fi to network together low-cost devices like light switches is theoretically possible, but Wi-Fi is overkill by a wide margin. The slowest

Wi-Fi specification, 802.15b, delivers data at 11 Mbps, far in excess of what a light switch needs to dim a light or a thermostat needs to control a heater. TCP/IP transport, arguably the gold standard for packet routing, requires a minimum of 30 bytes of overhead per packet, with 2K of RAM buffers and at least 14K of code space just for the software stack. Proposed reduced-overhead and lower-power variations of IP signaling, such as 6LoWPAN and OSIAN, are still over-complex for dense command and control and they have yet to be proven by adoption in the market place. Wi-Fi radios work very well, but they are complicated and so power-hungry that the only way to implement battery-operated devices is with rechargeable batteries and frequent recharging. Advanced as it is, Wi-Fi nevertheless has no powerline backup. As with other single-media networks, if communication fails for any reason, the only recourse with Wi-Fi is to retry. In contrast, INSTEON's data rate is optimized for home control. It is fast enough that users don't experience a noticeable delay when controlling devices, but not so fast that simulcast messages might jam each other. And simulcasting is how INSTEON avoids routing altogether. All INSTEON devices within range repeat1 identical messages at precisely the same time, with each device adding to the strength of the signal, so simulcasting is not only radically simpler than routing, it is more robust. Simplicity translates into low-cost–the INSTEON engine for communicating both by powerline and radio occupies a mere 2600 bytes, with an entire application like a lamp dimmer residing in only 7K bytes, all running in a low-end microcontroller such as the PIC16f638.Perhaps the strongest argument against increasing the cost of devices by building Wi-Fi into them is that you can get the same benefit but without incurring the cost in every device. With a single gateway between a Wi-Fi network and INSTEON, all of the devices on an INSTEON network can still appear as part of the Wi-Fi network. Wi-Fi is great at networking together computers, broadband access points,

11.6 Bluetooth Compared with Z-Wave, ZigBee, Wi-Fi, and INSTEON

a. Bluetooth Compared with Z-Wave

The biggest improvement that Z-Wave makes over Bluetooth is signal strength. Bluetooth signal is prone to interference and interruption because all Bluetooth devices send and receive information on the same 2.4 GHz band. They compete with one another for bandwidth. And unless you run a 5 GHz

Wi-Fi network in your home, they're competing with devices on your Wi-Fi network too.

With Z-Wave, every Z-Wave signal repeater works together to make the network stronger. Every device actually strengthens the signal. The more devices you have, the easier it is to create a strong network that's capable of bypassing obstacles and passing through walls, ceilings and floors.

b. Bluetooth Compared with ZigBee

The next table lists some characteristics of each protocol. The table shows the similarities between them. The two technologies are similar in that:

- They operate in the 2.4 GHz Industrial, Scientific and Medical (ISM) unlicensed frequency band and are based on the IEEE standard 802.15 for Wireless Personal Area Networks (WPANs). Bluetooth is based on standard 802.15.1 and ZigBee is based on standard 802.15.4 for low rate devices (LR-WPAN).
- They can automatically form ad hoc networks as devices within range are detected.
- They only address devices in their own networks and do not currently connect to the internet.
- They both modulate signals for radio transmission and packetize data.

Characteristics of ZigBee, Bluetooth Classic, Bluetooth Smart

Feature	ZigBee	Bluetooth Classic (BT)	Bluetooth Smart
Design Focus	Wireless networking among sensors	Wireless keyboards, mouse, headsets	Wireless sensor and fitness devices
IEEE Standard	802.15.4	802.15.1	802.15.1
Network Type	Mesh, ZigBee PRO	Piconet, Master/Slave; Scatternet	Scatternet
Distance	75–100 m line of sight	10 m (33 ft) min	>10 m >33 ft
Nodes Connected, max	65000	8	N/A
Operating Band	2.400–2.4835 GHz ISM band 16 channels, 5 MHz apart 2 MHz used Direct Spread Spectrum	2.400–2.4835 GHz ISM band 79.1 MHz channels Frequency Spread Spectrum	2.400–2.4835 GHz ISM band 40.2 MHz channels Frequency Spread Spectrum

Feature	ZigBee	Bluetooth Classic (BT)	Bluetooth Smart
Throughput	0.03 Mbps	1–3 Mbps	0.27 Mbps
Latency with Connect	15 ms	100 ms–3 s	3–6 ms
Type of Data	Operational instructions Low data rate	Continuous streaming All types of data; text, multimedia Relatively high speeds	Burst
Voice	No	Yes	No
Security	EAP (Extensible Authentication Protocol)	56/128 bit and application layer user defined	128-bit AES (Advanced Encryption Standard) with Counter Mode CBC-MAC and application layer user defined
Power Consumed (dependent on application)	30 mW	100 mW	0.01–0.5 W
Modulation	Direct Sequence Spread Spectrum	Frequency Hopping Spread Spectrum	Gaussian Frequency Shift Keying

c. Bluetooth Compared with INSTEON

Bluetooth, because it was designed for personal connectivity, is much faster and more complex than needed for home control applications. Considering Bluetooth's cost, limited range, point-to-point topology (not a mesh) and lack of powerline backup, INSTEON is a superior choice for networking together mass-market devices like light switches. Bluetooth coexists with INSTEON without any issues, because Bluetooth occupies the 2.4 GHz band and INSTEON radio uses 915 MHz.

d. Bluetooth versus Wi-Fi

Bluetooth and **Wi-Fi** are different standards for wireless communication.

Bluetooth technology is useful when transferring information between two or more devices that are near each other when speed is not an issue, such as telephones, printers, modems and headsets. It is best suited to low-bandwidth applications like transferring sound data with telephones (i.e., with a Bluetooth headset) or byte data with hand-held computers (transferring files) or keyboard and mice.

Wi-Fi is better suited for operating full-scale networks because it enables a faster connection, better range from the base station, and better wireless security (if configured properly) than Bluetooth.

Comparison table: Bluetooth versus Wi-Fi.

Bluetooth versus Wi-Fi Comparison Table		
	Bluetooth	Wi-Fi
Frequency	2.4 GHz	2.4, 3.6, 5 GHz
Cost	Low	High
Bandwidth	Low (800 kbps)	High (11 Mbps)
Specifications authority	Bluetooth SIG	IEEE, WECA
Security	It is less secure	Security issues are already being debated.
Year of development	1994	1991
Primary Devices	Mobile phones, mouse, keyboards, office and industrial automation devices. Activity trackers, such as Fitbit and Jawbone.	Notebook computers, desktop computers, servers, TV, Latest mobiles.
Hardware requirement	Bluetooth adaptor on all the devices connecting with each other	Wireless adaptors on all the devices of the network, a wireless router and/or wireless access points
Range	5–30 m	With 802.11b/g the typical range is 32 m indoors and 95 m (300 ft) outdoors. 802.11n has greater range. 2.5 GHz Wi-Fi communication has greater range than 5 GHz. Antennas can also increase range.
Power Consumption	Low	High
Ease of Use	Fairly simple to use. Can be used to connect up to seven devices at a time. It is easy to switch between devices or find and connect to any device.	It is more complex and requires configuration of hardware and software.
Latency	200 ms	150 ms
Bit rate	2.1 Mbps	600 Mbps

11.7 X10 Technology

X10 is a wired home automation protocol that uses your existing electrical wires to send and receive commands–it is a powerline technology. The protocol was developed in 1975 by Pico Electronics and remains in use today, although the technology has not advanced to keep up with newer systems. It is too unreliable and inflexible to be useful today as an infrastructure home-control network. Although the X-10 Ltd. company and factory are no longer in business, there are still numerous legacy devices still in use.

11.7.1 X10: Overview

X10: Technology

Commands are sent across existing AC electrical wiring in bursts, which are small packets of data that provide identifiers and commands. Early X10 devices were one-way communicators–they were only capable of receiving commands. More advanced (and more expensive) accessories can receive commands and also respond with a status message.

You can only install 16 accessories per house code, but there are 16 house codes available so the theoretical system limit is 256 accessories. You will have to install a filter to keep commands from other homes on the power grid from controlling your devices.

Furthermore, X10 defines only 16 different Command Codes, but not all devices can respond to all X10 commands. The six most common X10 commands are On, Off, Dim, Bright, All Units Off and All Lights On. X10 transmits one bit of information at each powerline zero crossing (every 8.33 milliseconds). A simple command-plus-address message contains 100 bits and takes 833 milliseconds to send (8/10 of a second). A one-millisecond burst of 120 KHz carrier signifies a one bit and the absence of a carrier signifies a zero bit. An X10 message consists of a 4-bit Start Code followed by an 8-bit House Code followed by a 10-bit Key Code. Each message is sent twice, followed by 6 zero-crossings of silence before starting another message. (The silence interval can be omitted if certain X10 Commands, such as Bright or Dim, are being repeated.)

X10: Benefits

X10 is easily added to existing homes without the need for additional communications wiring. In addition, it is more reliable than wireless technology and can communicate to any location that has an electrical outlet, socket, or switch. The technology is also relatively inexpensive and easy to install so if you just want a few simple light controls then X10 may be an option.

Radio transmitters have been developed for X10 so you can have remote control from within your home. In addition, there are some X10 controllers that connect to the internet to give you control from outside your home.

X10: Limitations

The X10 protocol has not kept up with other technological advances so there are some drawbacks. First, internet-based control requires more advanced setups and is not as easy to use as newer technology. In addition, X10 occasionally experiences interference from noise on the AC wave. While X10 can work across multiple electrical phases, extra components are required for this.

Developed as it was in the 1970s, X10 is too limited to function as a command and control infrastructure network in the mass market. Not only is X10 powerline signaling dreadfully slow, but X10 communication has no built-in mechanism to verify that X10 messages got through–X10 is open loop. Although X10 did eventually define Status Request and Status Response Commands, very few X10 devices actually employ them and those that do take even more time to send and receive the extra messages. Installing more than a few X10 devices in a network is not for the faint of heart. It usually requires blocking couplers, repeaters and filters–and then is highly likely to function properly for just a short while. X10 is fundamentally subject to false positives, both because of its signaling design but also of its limited addressing schema. Many X10 customers have lived through a neighbor inadvertently controlling products in their homes.

INSTEON protocol avoided the limitation of X10. This is giving in the comparison of INSTEON and X10.

X10: Compatible Hubs

We recommend that you only purchase an X10 controller if you already have X10 accessories and need to replace your current controller. If you are starting a home automation system from scratch, then we suggest looking at the other protocols above because they offer more advanced features.

Simple local controllers that are relatively inexpensive can be purchased on Amazon or Smarthome.com.

If you already have a local X10 system and you want to add remote/internet control, check out the Universal Devices ISY994ix (Amazon or Smarthome.com), but keep in mind that this requires some advanced setup.

11.7.2 INSTEON Compared with X10

We mentioned above the limitation of X10. INTEON also has some limitations given as follows:

- INSTEON uses both powerline and radio communications.

Because INSTEON powerline messages are repeated using radio and radio messages are repeated on the powerline, INSTEON messaging is highly reliable while still using simple signaling methods on both the powerline and the radio. Dual-band products also naturally couple the electrical phases without the need for additional products.

- INSTEON devices are simulcasting repeaters.

Because INSTEON devices simultaneously repeat1 each other's messages, adding more devices to an INSTEON network adds more energy to the INSTEON signal. Thus, as an INSTEON network grows it becomes more reliable, while adding X10 devices weakens an X10 network.

- INSTEON is closed loop.

All INSTEON devices can both listen and talk and the INSTEON protocol requires that all INSTEON messages that are not broadcast be acknowledged. INSTEON is faster than X10. At each powerline zero crossing (every 8.33 milliseconds), INSTEON sends 24-bits of information, while X10 sends only ¡ bit. Thus INSTEON's raw signaling rate is faster than X10's by a factor of 48.

- INSTEON has a large address and command space.

With room for more than 16 million addresses and 65 thousand basic commands, INSTEON is clearly in a different league compared to X10 with only 256 addresses and 16 commands.

11.8 Universal Powerline Bus (UPB)

UPB is an acronym for Universal Powerline Bus, which is a home automation protocol that sends communications over existing AC electrical cables in your home. It is a low-speed powerline-only network. UPB is developed by

California-based Powerline Control Systems. The company sells commercial and residential lighting control systems.

11.8.1 UPB an Overview

UPB: Technology

The wired system communicates via pulses within the alternating current on standard electrical wiring. UPB devices send messages over the powerline by coding data in the timing of high amplitude pulses. It is similar to the X10 technology, but is considered to be more efficient and more reliable. The intent was for UPB to replace X10, but the technology hasn't caught on and there aren't many devices on the market.

UPB: Benefits

The main benefit of UPB is that it can easily be added to a home without the need for new wiring. In addition, it is very reliable (up to 99%) and reaches anywhere in the home that has a power outlet, light, or switch. Devices communicate peer-to-peer so a central controller isn't necessary. The technology can also be set up to work on both phases of an electrical system.

UPB: Limitations

Unfortunately, there are a number of significant disadvantages. First, it is very complicated to set up a wireless controller that connects to UPB so remote or phone-based control is not available. In addition, there aren't a lot of product options available.

Because UPB only communicates via the powerline, if for any reason powerline messaging fails, UPB devices can only retry and hope for the best. Unfortunately, electrical noise from many types of lamp dimmers and other electrical equipment can easily masquerade as a UPB signal pulse, causing interference that can last indefinitely. While the powerline reliability is positive, we simply cannot recommend UPB in an era where more advanced protocols are available. INSTEON protocol solved this problem as given in the comparison.

UPB: Compatible Hubs

All of the lighting control devices communicate peer-to-peer so a central hub isn't required. If you are interested in UPB for lighting control in your home, check out Pulseworx by Powerline Control Systems.

11.8.2 INSTEON Compared with UPB

The most notable differences between INSTEON and UPB are that:

- INSTEON is dual band, and UPB is powerline only;
- INSTEON repeats1 messages by simulcasting, UPB does not repeat messages making it point-to-point;
- INSTEON cannot be jammed by triac transients; and
- INSTEON device setup does not require special tools.

Being dual mesh network, INSTEON uses both radio and powerline to back each other up. Moreover, INSTEON devices repeat each other's messages by simulcasting them in precise timeslots synchronized to the powerline zero crossings. With each new device adding to the signal strength, an INSTEON network grows in reliability as more devices are added. There is no integrated UPB chip available. Instead, the UPB protocol is implemented on general-purpose microcontrollers with discrete external components. Product development entails writing application code that runs on the microcontroller.

11.9 Intellon

11.9.1 Intellon: Overview

Acquired by Atheros (www.atheros.com) in September 2009, Intellon, of Ocala, Florida, got its start in 1989 by developing the winning designs for adoption as the CEBus powerline and RF signaling standards. Although CEBus did not find acceptance in the marketplace, Intellon went on to improve its spread spectrum technology to the point where over a hundred megabits per second could be transported reliably over the powerline. About Intellon Technology Intellon's broadband technology, called PowerPacket, is at the heart of the HomePlug 1.0 broadband powerline networking specification. As a founding sponsor and member of the board of directors of the HomePlug Powerline Alliance, Intellon is a primary driver of the even-more-advanced HomePlug AV technology. As one might expect, PowerPacket is very sophisticated and complex to be able to transport so much data reliably in such a hostile environment. At the physical (PHY) level, PowerPacket uses multiple carrier OFDM (orthogonal frequency division multiplexing, also used by Wi-Fi and DSL), with interleaving and both Viterbi and Reed-Solomon forward error correction (FEC). The control software dynamically negotiates an optimum payload data rate by choosing carriers, switching between DQPSK and DBPSK modulation and varying the FEC rate. The signal is spread from 4.5 MHz to 21 MHz on the powerline, with a digitally filtered PSD (power spectral density) to match HomePlug's requirements. The media access control (MAC) layer uses CSMA (carrier sense multiple access) with a randomized contention window and exponential back off, like Ethernet. Other MAC services include priority resolution, ACK/NAK (acknowledgements), ARQ (automatic repeat requests), segmentation of slow frames for QoS (quality of service) and 56-bit DES (digital encryption standard) encryption using cipher block chaining. This technology is clearly a tour de force. Highly advanced, it is well suited for its intended broadband applications, but it is far too expensive to find its way into device s like light switches.

11.9.2 Intellon Compared with INSTEON Technology

Intellon's broadband technology does not compete with INSTEON as explained above. Intellon does still offer legacy CEBus-inspired SSC (spread spectrum carrier) chips, but the marketplace has not adopted CEBus for reasons explained in the following section, so the chips are mainly used in proprietary building automation and vehicle brake control systems. Consequently, INSTEON does not overlap Intellon's technology in the marketplace, although both signals can coexist on the powerline without interference.

References

[1] Aston, Kevin. "That 'Internet of Things' Thing", Available: http://www.rfidjournal.com/articles/view?4986, accessed on June 2, 2017.

[2] Wright, J. (2011). Practical ZigBee Exploitation Framework. toorcon 2011.

[3] Sigma Design press release, "Sigma Designs Announces Break-Through Z-Wave(R) UL Component Recognition", http://www.sigmadesigns.com/news/sigma-designs-announces-break-through-z-waver-ul-component-recognition, accessed on June 2, 2017.

[4] M. Knight, "Wireless security–How safe is Z-wave?," in Computing & Control Engineering Journal, vol. 17, no. 6, pp. 18–23, Dec.-Jan. 2006. doi: 10.1049/cce:20060601

[5] M. Al-Zyoud, T. Atkison and J. Carver, "An Overview of Emerging Privacy Issues in the Internet of Things," 2016 International Conference on Computational Science and Computational Intelligence (CSCI), Las Vegas, NV, 2016, pp. 212–217. doi: 10.1109/CSCI.2016.0047

[6] C. W. Badenhop, S. R. Graham, B. W. Ramsey, B. E. Mullins, L. O. Mailloux, "The Z-Wave routing protocol and its security implications" https://doi.org/10.1016/j.cose.2017.04.004

[7] Threadgroup Inc. website: http://Threadgroup.org/What-is-Thread/Connected-Home

[8] H. Sharma and S. Sharma, "A review of sensor networks: Technologies and applications," 2014 Recent Advances in Engineering and Computational Sciences (RAECS), Chandigarh, 2014, pp. 1–4. doi: 10.1109/RAECS.2014.6799579

[9] M. B. Yassein, W. Mardini and A. Khalil, "Smart homes automation using Z-wave protocol," 2016 International Conference on Engineering & MIS (ICEMIS), Agadir, 2016, pp. 1–6. doi: 10.1109/ICEMIS.2016.7745306

[10] Sigma Designs. (n.d.). Z-Wave development Kit. Retrieved June 2013, from Sigma Designs public web site: http://www.sigmadesigns.com/uploads/documents/zwave_dev_kit_br.pdf

[11] M. Zareei, A. Zarei, R. Budiarto and M. A. Omar, "A comparative study of short range wireless sensor network on high density networks," The 17th Asia Pacific Conference on Communications, Sabah, 2011, pp. 247–252. doi: 10.1109/APCC.2011.6152813

[12] "Introduction to the Z-Wave Security Ecosystem", available from Sigma Design, Inc., on the company website: http://z-wave.sigmadesigns.com/wp-content/uploads/2016/08/Z-Wave-Security-White-Paper.pdf

[13] OpenZwave. (n.d.). Retrieved June 2013, from open-zwave Google code site: https://code.google.com/p/open-zwave/

[14] International Telecommunication Union (ITU). (2012). G.9959 : Short range narrow-band digital radio communication transceivers–PHY and MAC layer specifications.

[15] Texas Instruments. (2009, December). CC1110 Mini Development Kit. Retrieved from http://www.ti.com/tool/cc1110dk-mini-868

[16] I. Chatzigiannakis and A. Strikos, "A decentralized intrusion detection system for increasing security of wireless sensor networks," 2007 IEEE Conference on Emerging Technologies and Factory Automation (EFTA 2007), Patras, 2007, pp. 1408-1411. doi: 10.1109/EFTA.2007.4416949

[17] Texas Instruments. (2013, May). SmartRF Studio. Retrieved from http://www.ti.com/tool/smartrftm-studio

[18] Morais, A., & Cavalli, , A. (2011). Route Manipulation Attack in Wireless Mesh Networks. Advanced Information Networking and Applications (AINA).

[19] G. Gaubatz, J. P. Kaps, E. Ozturk and B. Sunar, "State of the art in ultra-low power public key cryptography for wireless sensor networks," Third IEEE International Conference on Pervasive Computing and Communications Workshops, Kauai Island, HI, 2005, pp. 146–150. doi: 10.1109/PERCOMW.2005.76

[20] S. Zamfir, T. Balan, I. Iliescu and F. Sandu, "A security analysis on standard IoT protocols," 2016 International Conference on Applied and Theoretical Electricity (ICATE), Craiova, 2016, pp. 1–6. doi: 10.1109/ICATE.2016.7754665

[21] M. B. Yassein, W. Mardini and A. Khalil, "Smart homes automation using Z-wave protocol," 2016 International Conference on Engineering & MIS (ICEMIS), Agadir, 2016, pp. 1–6. doi: 10.1109/ICEMIS.2016.7745306

[22] "The Thread Group Expands Influence through Partnership with ZigBee Alliance, TCLA and Innovation Enabler Award", (Accessed June 7, 2017) http://mysmahome.com/news/5905/the-Thread-group-expands-influence-through-partnership-with-zigbee-alliance-tcla-and-innovation-enabler-award-2/

[23] HackersOnBoard. "Black Hat 2013–Honey, I"m Home!!–Hacking Z-Wave Home Automation Systems." Online video clip. YouTube. YouTube, 19 Nov 2013. Web 1 Nov. 2016. URL: https://www.youtube.com/watch?v=KYaEQhvodc8

[24] M. B. Tamboli and D. Dambawade, "Secure and efficient CoAP based authentication and access control for Internet of Things (IoT)," 2016 IEEE International Conference on Recent Trends in Electronics, Information & Communication Technology (RTEICT), Bangalore, 2016, pp. 1245–1250. doi: 10.1109/RTEICT.2016.7808031

[25] J. Singh, T. Pasquier, J. Bacon, H. Ko and D. Eyers, "Twenty Security Considerations for Cloud-Supported Internet of Things," in IEEE Internet of Things Journal, vol. 3, no. 3, pp. 269–284, June 2016. doi: 10.1109/JIOT.2015.2460333

[26] L. Nastase, "Security in the Internet of Things: A Survey on Application Layer Protocols," 2017 21st International Conference on Control Systems and Computer Science (CSCS), Bucharest, Romania, 2017, pp. 659–666. doi: 10.1109/CSCS.2017.101

[27] M. B. Yassein, M. Q. Shatnawi and D. Al-zoubi, "Application layer protocols for the Internet of Things: A survey," 2016 International Conference on Engineering & MIS (ICEMIS), Agadir, 2016, pp. 1–4. doi: 10.1109/ICEMIS.2016.7745303

Index

About the Authors

Prof. Dawoud Shenouda Dawoud has a BSc (1965) and MSc (1969) from Cairo University in Communication Engineering. He completed his PhD in Russia in 1973 in the field of Computer hardware where he succeeded to own 3 Patents in the field of designing new types of memory which was the beginning of the FPGAs. In 1984, he was promoted to full Professor at the Egyptian Academy of Science and Technology, National Electronic Research Institute. During the period from 1973 to 1990, he supervised more than 5 PhDs and 15 MSc degrees all of them focused in the fields of computer and embedded system designs. During the period from 1990-1999 he established the Faculty of Engineering at the University of Botswana. During this period he supervised 3 PhDs and 7 MSc degrees. In the year 2000 he became Professor of Computer Engineering and Head of the Computer Engineering Department at the University of KwaZulu Natal, Durban, South Africa. For 10 years he was supervising research in the field of Security of Mobile Ad hoc Networks. He supervised 2 PhDs and many MSc degrees in this field and published more than 30 papers. At the same period, he was visiting the National University of Rwanda to run an MSc program in Communication. He supervised about 15 MSc students during these 4 years before moving to the National University of Rwanda in 2010 to become the Dean of the Faculty of Engineering.

In 2011 he moved to Uganda where he became the Dean of the Faculty of Engineering at the International University of East Africa (IUEA), where he currently remains. During this time, he also served as the Vice Chancellor of IUEA, for a period of 3 years.

Across his career, he has published over 200 Journal and Conference papers, as well as books in the fields of computer engineering, microcontroller system design, embedded system design and Security of Mobil Ad hoc Networks.

Peter Dawoud has been working in the area of biometrics and computer vision for the last six years, leading research and products within Microsoft in the space. After completing his Bachelor's degree in computer engineering, Peter spent time researching applied cryptographic and steganographic systems for his Master's Degrees. Upon completion of these degrees, Peter has been focused on the productization of biometric systems for consumer and enterprise scenarios and later focused on the principles and use of centralized cloud biometrics systems in the enterprise and commercial scenarios.